The Berlin Group and the Philosophy of Logical Empiricism

BOSTON STUDIES IN THE PHILOSOPHY AND HISTORY OF SCIENCE

VOLUME 273

For further volumes:
http://www.springer.com/series/5710

Nikolay Milkov • Volker Peckhaus
Editors

The Berlin Group and the Philosophy of Logical Empiricism

Springer

Editors
Nikolay Milkov
Department of Philosophy
University of Paderborn
33098 Paderborn
Germany

Volker Peckhaus
Department of Philosophy
University of Paderborn
33098 Paderborn
Germany

ISSN 0068-0346
ISBN 978-94-007-5484-3 ISBN 978-94-007-5485-0 (eBook)
DOI 10.1007/978-94-007-5485-0
Springer Dordrecht Heidelberg New York London

Library of Congress Control Number: 2013933306

Printed on acid-free paper

Springer is part of Springer Science+Business Media (www.springer.com)

Preface

The Berlin Group for scientific philosophy was active between 1926 and 1933 with Hans Reichenbach, Walter Dubislav and Kurt Grelling as its leading members. It organized the Society for Empirical/Scientific Philosophy as a forum for communicating with the preeminent scientists and the educated public of the time. In 1930 Hans Reichenbach together with Rudolf Carnap launched the legendary journal for scientific philosophy *Erkenntnis*.

Interest in the Berlin Group has grown appreciably in recent years, something clear from the ever-increasing number of articles on the Group.[1] This book is designed to help meet this growing interest in ways that significantly contribute to a better understanding of the seminal role that the Berlin Group played in the emergence of the philosophy of science as a discipline.

To date, only a single book has appeared on the Berlin Group, a volume in German edited by Lutz Danneberg, Andreas Kamlah, and Lothar Schäfer: *Hans Reichenbach und die Berliner Gruppe* (Braunschweig: Vieweg 1994). By contrast with that text, the chapters of the present collection do not concentrate only on Reichenbach's scientific philosophy. This volume is the first to assess the scientific philosophy of Walter Dubislav, to which it devotes three chapters. The work of Kurt Grelling is explored in two further chapters. Grelling, like Dubislav, was an accomplished philosopher of mathematics and science who is scarcely known today, particularly in Anglophone philosophical circles. Two other chapters probe the relation of Kurt Lewin and Carl Hempel to the Berlin Group. Also included is original essay on the thought of Paul Oppenheim, who went on to become a prominent figure in the philosophy of science in the USA from the 1940s through the 1960s. Enriching the historical and theoretical range of this collection are essays that shed light on the intellectual debt that the Berlin Group owed to the precursor of the German tradition of scientific philosophy, Jacob Friedrich Fries, and to Ernst Cassirer.

---

[1] Cf. Rescher (1997, 2006), Stadler (2011).

The idea for this book originated with a conference on the Berlin Group held at the University of Paderborn on September 3–5, 2009. The text, however, is no mere record of the proceedings. Rather, the editors selected from among the many conference papers those that in their judgment are of outstanding scholarly merit and likely to be of enduring historical and philosophical value. In addition, the collection includes studies solicited expressly for this volume from the some of the most distinguished authorities in the field.

Acknowledgements

The Paderborn conference that made this book possible was generously supported by the Fritz Thyssen Foundation (Cologne). Special thanks go to Anja Westermann and Nadine Sand (University of Paderborn) for their help by editing the text. Matthew Grellette (McMaster University) helped us to improve the English of six chapters.

The "family tree" of the Berlin Group and its successors in the USA was already presented in Rescher (2006, 282) which explains why we did not produced here. What we added to it in this volume are the pictures of members of its "Founding Generation", Hans Reichenbach, Walter Dubislav, Kurt Grelling, and Paul Oppenheim, and of the two representatives of the "Middle Generation": Carl Hempel and Olaf Helmer. Excerpts form letters of and to Carnap and Reichenbach are printed in the book with permission of the Special Collections Department, University Library System, University of Pittsburgh. All right preserved.

References

Rescher, Nicholas. 1997. H_2O: Hempel–Helmer–Oppenheim, an episode in the history of scientific philosophy in the 20th century. *Philosophy of Science* 64: 234–260.

Rescher, Nicholas. 2006. The Berlin School of Logical Empiricism and its legacy. *Erkenntnis* 64: 281–304.

Stadler, Friedrich. 2011. The road to *Experience and Prediction* from within: Hans Reichenbach's scientific correspondence from Berlin to Istanbul. *Synthese* 181: 137–155.

Contents

Contributors

Arkadiusz Chrudzimski Department of Philosophy, University of Szczecin, Szczecin, Poland

Jeremy Heis Department of Logic and Philosophy of Science, University of California in Irvine, Irvine, CA, USA

Andreas Kamlah Institute of Philosophy, University of Osnabruck, Osnabruck, Germany

Anita Kasabova Department of Anthropology, New Bulgarian University, Sofia, Bulgaria

Nikolay Milkov Department of Philosophy, University of Paderborn, Paderborn, Germany

Thomas Müller Department of Philosophy, University of Utrecht, Utrecht, The Netherlands

Flavia Padovani English and Philosophy Department, Drexel University, PA, USA

Volker Peckhaus Department of Philosophy, University of Paderborn, Paderborn, Germany

Helmut Pulte Institute of Philosophy, University of Bochum, Bochum, Germany

Erich H. Reck Department of Philosophy, University of California, Riverside, CA, USA

Nicholas Rescher Department of Philosophy, University of Pittsburgh, Pittsburgh, PA, USA

Michael Stöltzner Department of Philosophy, University of South Carolina in Columbus, Columbus, SC, USA

Christian Thiel Institute of Philosophy, University of Erlangen–Nürnberg, Erlangen, Germany

Temilo van Zantwijk Institute of Philosophy, University of Jena, Jena, Germany

Paul Ziche Department of Philosophy, University of Utrecht, Utrecht, The Netherlands

Part I
Introductory Chapters

Chapter 1
The Berlin Group and the Vienna Circle: Affinities and Divergences

Nikolay Milkov

> My collaboration with the Vienna Circle does not mean an agreement with the number of naiveties which it conveyed to us from Vienna (and to which I also count Schlick's *Ethics*), but that this union is a result of the compulsion of the isolation in which the school philosophy put the exact philosophers.[1]

1.1 Asymmetry in the History of the Vienna and Berlin Scientific Philosophy

The Vienna Circle and the Berlin Group were allied schools of scientific philosophy that together strove against what they understood to be a philosophical traditionalism that lost touch with the real world. The term "logical empiricism," as this scientific philosophy came to be called in the last years,[2] can be seen as the philosophy of the two Germanic capitals, Berlin and Vienna. Both cities were at the forefront of the modernity and, prior to the Second World War, leading centers of science and research. The cultural milieu in which the new scientifically oriented philosophy was nurtured departed perceptibly from what had long been the traditional seedbed of Germanophone philosophy, namely the small university town such as a Marburg or a Heidelberg, a Graz or a Jena.

---

[1] Reichenbach's letter to Heinrich Scholz from 13.10.1931 [HR 013-31-06].

[2] For criticism of this term see the last paragraphs of Sect. 1.8.

N. Milkov (✉)
Department of Philosophy, University of Paderborn, Paderborn, Germany
e-mail: nikolay.milkov@upb.de

N. Milkov and V. Peckhaus (eds.), *The Berlin Group and the Philosophy of Logical Empiricism*, Boston Studies in the Philosophy and History of Science 273, DOI 10.1007/978-94-007-5485-0_1,

The present chapter's objective is to correct the historical record, which till now has failed to present the rise and evolution of logical empiricism with due regard to its full complexity. The usual understanding is that the Vienna Circle dominated the scientific philosophy of the twentieth century's third and fourth decades. As we shall see, however, this view reflects what, at best, is only a superficial historical reading of scientific philosophy during that period. More exacting analysis yields a far different picture—one in which the Berlin Group, following a research program all its own, figures as an equal partner with the Vienna Circle in promulgating, around 1930, the scientific philosophy in the German-speaking world. Hans Reichenbach, for one, frequently underscored the Berlin Group's autonomy as a principal player in the emergence of logical empiricism. What's more, he emphasized the lead role that the Group in fact took in originally formulating the doctrine, citing as his personal contribution to this effort his introduction of the method of "logical analysis of science" in his book *The Theory of Relativity and Apriori Knowledge* (Reichenbach 1920, 74 ff.). Reichenbach also called attention to the circumstance that "*die Erkenntnis* ... was founded in Berlin [not in Vienna] and edited from there."[3]

The asymmetry that marks the currently accepted view of scientific philosophy in the German speaking countries around 1930 is in large part owing to the absence of any monograph on the history of the Berlin Group. That of the Vienna Circle, on the other hand, is preserved in a number of widely read texts, some authored by well-established scholars like Viktor Kraft, Oswald Hanfling, Rudolf Haller, Friedrich Stadler and Thomas Uebel. It is true that by the late 1980s publications began to appear that covered the history of the Berlin scientific philosophy as well as the particular projects of Reichenbach and his Group (Leitko 1987). And the wave of enthusiasm that German reunification aroused had an impact on scholarship which included newly issued collections of papers on the history of the Berlin Group,[4] including letters and documents—most notably among them a document describing the sessions of the Society for Empiric/Scientific Philosophy.[5] As a whole, however, these publications amount to little more than preliminary work on the history of the Berlin Group. Even collectively, they do not give us a comprehensive picture of the ideas it introduced, debated, and developed. Little wonder, then, that they have failed to correct the mainstream view, in which for many years now Reichenbach has often been cast as a "logical positivist and Vienna Circle insider" (Moran 2008, 180).

[3]Reichenbach's letter to Ernst von Aster, June 3, 1935 [HR-013-39-34]. Indeed, the manuscripts submitted to *Erkenntnis* were to be sent to Berlin, not to Vienna. This is reflected in the fact that on the cover of the first four volumes of the journal, Reichenbach's name was printed in bigger characters than the name of the official co-editor Carnap.

[4]We mean here above all Danneberg et al. (1994), Haller and Stadler (1993), Hentschel (1991), and Poser und Dirks (1998).

[5]Cf. Danneberg and Schernus (1994). We shall speak about this double naming of the Society a little bit later.

As we shall see, however, not only did Reichenbach differentiate himself from the logical positivists, he regarded himself as their "friendly opponent."[6]

1.2 Why the Asymmetry?

To be sure, no conspiracy was responsible for the disparity in prominence between the Vienna Circle and the Berlin Group, and the conspicuous absence of the latter from the received historical record. This situation devolved largely from one theoretical and three external factors that actually have little to do with strengths or weaknesses of the Berlin Group's philosophical program.[7]

The theoretical factor that made the Vienna Circle's activities the more visible was Ludwig Wittgenstein's philosophy of language, a doctrine that owed a great deal to Gottlob Frege's work on the topic. Wittgenstein's influence proved catalytic in the Circle's effort to articulate a set of related topics and problems which made possible its "planned co-operative project [*die planmässige Kollektivarbeit*]" (Neurath 1932, 208). This initiative came to life in a long series of discussions on themes such as the nature of truth, protocol sentences, and physicalism.

The Berlin Group's project was perceptibly different from that of the Vienna Circle. The Berliners' plan was to explore philosophical problems with scientists and mathematicians in their specific disciplines, its modus operandi being, as Reichenbach declared, "to gather together a group of men working with empiricist methods and fully conscious of their intellectual responsibility" (Reichenbach 1936a, 159). The objective of the Vienna Circle, by contrast, was to advance specific theories: for example, to reach consensus on the question of "protocol sentences." Invariably the discussions in Vienna were passionate, the participants being committed to the imperative of struggling to hammer out a common theory. It is true that they never reached such a common theory. Their energetic debates, however, part of which went to press, called attention to themselves in ways not seen in the Berlin Group.

Reichenbach had in mind just these key differences with his Vienna colleagues when he remarked that

> in the line of their more concrete working-program, which demanded analysis of specific problems in science, [the Berliners] avoided all theoretical maxims like those set up by the Vienna school and embarked upon detailed work in logistics, physics, biology, and psychology. (Reichenbach 1936a, 144)

---

[6]In order to understand what the predicate "friendly" meant in German philosophy around 1930, we must recall how hostile the relation between, what later were called, continental and analytic philosophers (for example, between Carnap and Heidegger) was.

[7]On this point, we agree with Peter Simons that "the way philosophical disputes get decided and the way subsequent history is written depend little on the dialectical strength, adequacy or sophistication of the position posed" (Simons 1997, 442).

We further consider the theoretical differences between the Vienna Circle and the Berlin Group (in Sect. 1.7, below). As for the principal external factors that explain the preponderant public interest in the Vienna Circle, three stand out as of particular importance:

(i) The Vienna Circle became prominent the moment its manifesto appeared in August 1929, which riveted attention as a *succès de scandale* in the philosophical community and beyond. In brief, the manifesto had a radical, clearly spelled out thesis that shocked the general educated public, namely that traditional philosophy is not false—it is *senseless*. The Circle's objective was no less than to eliminate metaphysics. Little wonder, then, that within two years after publishing its program the Circle had secured itself a prominent place in the philosophical literature.[8] A couple of years later, the young Alfred Ayer's book *Language, Truth, and Logic* (1936) made the ideas of the Vienna Circle attractive to the Anglophone students of philosophy.

It deserves note that the Vienna Circle's prominence was due in no small measure to the rhetorical skills of a charismatic person experienced at courting public opinion: such was Otto Neurath, lead author of the manifesto. Predictably enough, critics emerged to rebuke the Vienna Circle for its resort to rhetoric. No one did that better Wittgenstein, who issued the following challenge in a 1929 letter to Friedrich Waismann: "the Vienna School should not prostitute itself like all Vienna institutions want to do on all occasions... The Vienna school must not *say* what it achieves, but *show* it!" (Mulder 1968, 389). In contrast, the Berlin Group exercised a kind of intellectual modesty.

(ii) Among the accidental factors impacting the progress of a new philosophical movement is the strength of character and individual temperaments of its exponents. Here again, the Berliners were at a disadvantage. Only one of its members, Reichenbach, fully developed his philosophical program. The most distinguished example of those who failed to do so was Kurt Grelling (1886–1942). In 1908 he discovered what we know today as the "Grelling Paradox." Two years later he wrote a brilliant dissertation under Hilbert and Zermelo on the axioms of arithmetic, as well as an influential treatise on probability. Instead of persisting in his efforts to secure a university position, however, he went to Munich to study economics in 1910.[9] Back in Göttingen in 1913, Grelling further pursued his studies under Leonard Nelson.[10] After a break with Nelson in 1922, he removed to Berlin where from the autumn of 1926 he worked under his old acquaintance, Reichenbach.[11]

---

[8]Cf. Kaila (1930), Petzäll (1931), and Bloomberg and Feigl (1931).

[9]Hempel's impression, too, was that Grelling "didn't *want* to enter a university career. I don't quite understand why" (Hempel 2000, 6; my italics—N. M.).

[10]On Grellings's work with Leonard Nelson, see Sect. 1.4, below.

[11]On the contacts between Grelling and Reichenbach before 1926, see Sect. 1.5.

In his Berlin period, Grelling became what was later known as a "mainstream analytic philosopher," trenchantly critiquing the wide range of major new books and papers in scientific philosophy—most notably Carnap's *Aufbau,* Reichenbach's *The Philosophy of Space and Time,* and Dubislav's *Die Definition* (cf. Grelling 1929, 1930, 1933). He also served as a managing editor of *Erkenntnis.* But Grelling achieved maturity as an original philosopher only in the last years of his life (after 1938). Stimulated by discussions with Paul Oppenheim during that period,[12] Grelling produced original ideas in formal ontology (cf. Grelling 1939; Grelling and Oppenheim 1937/1938, 1939).[13] Tragically, the late flowering of his philosophical talent ended in Auschwitz's ovens on September 18, 1942.

By contrast with Grelling, Walter Dubislav (1895–1937) quickly advanced as an independent author. By 1931 he was Extraordinary Professor at the Technical University in Berlin. The political changes in Germany in 1933, however, marked a break in his career—and in his life: after Hitler came to power, Dubislav published scarcely anything. Apparently, the reason was that Dubislav, who unlike Reichenbach and Grelling was not Jewish but "Aryan," "believed that his connection with it [the journal *Erkenntnis;* but also with the Society of which he took the helm upon Reichenbach's departure] would be harmful for his career."[14] Sadly, Dubislav was all too prescient on this score.

Decades later, Olaf Helmer remembered Dubislav as "a brilliant logician and teacher" who "began to exhibit what were then considered to be paranoid tendencies, abetted no doubt by the political circumstances of the time" (quoted in Luchins 2000, 238). Consequently, when Reichenbach departed Berlin in the summer of 1933, the young members of the Berlin Group, Hempel and Helmer, did not ask Dubislav but other academics (Wolfgang Köhler and Georg Feigl, respectively) to supervise their dissertations. In 1937, Dubislav committed suicide under tragic circumstances.

(iii) A third "external" factor accounting for the dissimilar fates of the Vienna Circle and the Berlin Group was political. Whereas Hitler came to power in Berlin in January 1933, he did not force Austria into the German Reich (*schließ es an*) for more than 5 years (in March 1938). This afforded the members of the Vienna Circle more of an opportunity than the Berliners in the face of fascist tyranny to regroup and maneuver for a more or less organized exodus.

---

[12]On Paul Oppenheim cf. Sect. 1.5 (iii), n. 38, and Chaps. 12 and 13.

[13]Cf. Chaps. 10 and 11. Typically, this turn was preceded by an argument with Reichenbach. Cf. Sect. 1.6, (b).

[14]A letter of Felix Meiner to Reichenbach from 5.12.33 [HR 013 21 33].

1.3 The Berlin Group and the Society for Empirical/Scientific Philosophy

Some scholars (cf. Hoffmann 1994, 2007) represent the Berlin Group and the Society for Empirical/Scientific Philosophy as one and the same entity. Others, whose position is shared here, insist that these were two clearly different communities (cf. Danneberg and Schernus 1994, 394; Gerner 1997, 85 f.). That the Group was not identical with the Society explains why some Group members, Kurt Grelling, for one, did not lecture at the Society. On the other hand, leading members of the Board of the Society, like Richard von Mises, were clearly not a part of the Berlin Group.

The difference between the Berlin Group and the Society was somewhat analogous to that between the Vienna Circle and the Ernst Mach Association. Each school evidently regarded its respective society or association as a forum for communicating with the educated public at large (Neurath 1929, 305).

One clear difference with the Vienna Circle was that the Berlin Group was not formally organized. Whereas the Berlin Group was an informal gathering of thinkers which originated with a seminar that Reichenbach had led at the University of Berlin starting the autumn of 1926 (a couple of years later Reichenbach was conducting joint seminars with Dubislav[15]), the Vienna Circle convened regularly for meetings chronicled in detailed minutes.

First documents recording the birth of the Berlin Group date from the beginning of 1928.[16] That it soon achieved a high degree of organizational integrity is inferable from the fact that when in September 1929 Reichenbach declined an offer to become an (Extraordinary) Professor at the German University of Prague, one of the reasons he gave was his membership in a Berlin discussion group.[17]

It is largely thanks to Reichenbach that we have a published record of Berlin Group activities. Worth noting in this connection is that all his work that supplies data on the Group appeared after he left Germany in 1933. Two years later in Paris it was "in the name of the Berlin Group" that Reichenbach welcomed the *Congrès international de philosophie scientifique* (Reichenbach 1936b, 16). A programmatic paper of his that appeared in 1936, titled "Logistic Empiricism in Germany and the Present State of its Problems," provides the most detailed description extant of the Berlin Group (and it serves as a primary source in much of what follows below). As late as 1953, Reichenbach did not forget to mention the Group in the preface to the German edition of *The Rise of Scientific Philosophy* (Reichenbach 1953, 9).

Carl Hempel remembered the Berlin Group as "a small closed discussion group of scholars [that] imposed no membership restrictions. Reichenbach, Dubislav, and Grelling were the leading figures" (Hempel 1991, 6). Besides Reichenbach,

---

[15]Cf. Danneberg and Schernus (1994, 396, n. 26).

[16]Cf. Reichenbach's letter to Heinrich Scholtz from 05.01.1928 [015-41-15].

[17]Cf. Gerner (1997, 106).

Grelling, Dubislav and Alexander Herzberg,[18] at different periods the Group included Fritz London, Wolfgang Köhler and Kurt Lewin. Among the younger members were Carl Hempel, Olaf Helmer, Valentin Bargmann and Martin Strauss.

Important for understanding the character of the Berlin Group is the tradition to organize and take part in formal discussion groups that had its roots for the Berliners in Leonard Nelson's Göttingen-based Neo-Frisian Group to which most of the older generation of Berliners belonged.[19] Such discussion groups generated a remarkable collaborative spirit, as the following account attests:

> Reichenbach gave one the sense that one was a member of a team. His seminar was an open forum; he didn't sit there and have the answer, but he said, 'What can we do about this?' He had an idea, but he was open to counterproposals and also to criticism. So it was exhilarating. One had a sense of participating in an attack on an important problem. (Hempel 2000, 6 f.)

Typically, with his arrival in Istanbul, Reichenbach launched "a colloquium held by a small circle of scholars speaking German" (Hempel 1991, 10). But he deplored the circumstance that the group was "only a weak substitute for the circle in Berlin" (ibid.). A fact most material to the present discussion is that Reichenbach *needed* such a circle.

If anything, the same is still more true of Kurt Grelling. In conjunction with two of his seminars and a colloquium that he privately conducted, Grelling organized a new "Berlin Circle" in 1936, which included Franz Graf Hoensbroech, Leopold Löwenheim and Jürgen von Kempski on its rolls (Peckhaus 1994, 63). More than this: in the very teeth of Nazi persecution Grelling organized a colloquium in 1941in Gurs internment camp in South (Vichy) France (on the French–Spanish border). He led the colloquium until 1942, when, even as Oppenheim and Hempel and others were attempting through diplomatic channels to rescue him with an appointment at the New School for Social Research in New York, he was transported to Auschwitz and murdered (ibid., 66 ff).

Among other things, the foregoing facts make it clear that the Berlin Group was limited neither geographically to Berlin, nor temporally to the period of 1926–33.

Unlike the Berlin Group, the Society for Empirical/Scientific Philosophy, with which members of the Group developed formative ties, was formally organized if not officially registered in court records of the time. It had a president, board, manager and a scrupulously compiled list of members. It defined its activity through lectures and discussions, hosting from 10 to 20 talks per year. The Society usually met on Tuesdays at the famous Charité hospital. Contrary to common belief, the "Empirical Philosophy" in the Society's name did not refer to the variety of scientific philosophy to which the Berlin Group subscribed. The Ernst-Machian Josef Petzoldt, who founded the Society in February 1927, set it up as the Berlin chapter (the "Berlin Local Group") of the "International Society for Empirical Philosophy." The latter was organized in 1925 in Frankfurt am Main in support

[18]On Alexander Herzberg see Schernus (1994).

[19]See Sect. 1.4, below.

of the journal *Annalen der Philosophie* which already had something of a scientific orientation. Initially, it was launched as a Journal in support of the "as if" philosophy of the Neo-Kantian Hans Vaihinger.

During the Society's first year and a half, Reichenbach remained skeptical about its viability, formally becoming a member only in October 1928. Ironically enough, it was Neurath who prompted Reichenbach to become a full member of the Berlin Society and eventually to change its agenda. Neurath's idea was for a reformed Berlin Society that would be a counterpart—not necessarily a satelite organization—of the Ernst Mach Association.[20]

It so happened that at just this time Petzoldt fell ill and resigned in May 1929. Reichenbach, Dubislav and Herzberg were thereupon elected to the Board of the Society: Reichenbach as a President (*Vorsitzender*), Dubislav as a Manager (*Geschäftsführer*). On June 30, 1929 Reichenbach wrote to Carnap: "Recently Dubislav and I were integrated into the Board, where we, together with Herzberg, have the real power" [HR 013-39-34]. The Berlin Group's interest in the Society's work grew during the ensuing 2 months, particularly after the Vienna Circle published its manifesto in August. The next couple of years saw Reichenbach and his colleagues transform the structure of the Society, such that by the end of 1931 the "Society for Empirical Philosophy" was rechristened the "Society for Scientific Philosophy."[21]

The Society's membership largely represented the scientific elite of Berlin but also of other scientific centers in Germany. Most were seasoned researchers and respected authorities in their fields, many of them holding lead positions at prestigious academic departments and institutes (Hoffmann 1994, 27). Boasting more resident Nobel Prize winners than any other city on the planet,[22] the Berlin of that period was a world-class centre of scientific research; and the lineup of lecturers hosted by the Society included no less than three Nobel Prize laureates: Max von Laue, Otto Meyerhoff, and Wilhelm Oswald.

Besides providing renowned conventional scientists the opportunity to disseminate their findings to some of their most distinguished colleagues in other scientific disciplines, the Society was a forum for *innovative* scientists, like the founder of Gestalt psychology Wolfgang Köhler and the brain researcher Oskar Vogt. Not surprisingly, the Society attracted talented up-and-coming *interdisciplinary* scholar-scientists, such as the biologist and systems theorist Ludwig von Bertalanffy from Vienna. As a means of furthering their own original research programs, these scientists sought out precisely the sort of stimulus to innovative thinking that the Society's philosophically keyed interdisciplinary discussions fostered.[23]

Finally, the Society also exerted influence on the wider cultural environment. Its list of members included leading avant-garde intellectuals such as Bertolt Brecht

---

[20]Cf. Reichenbach's letter to Philipp Frank of 1.05.29 [HR 014-06-31].

[21]Cf. Sect. 1.7, (ii), below.

[22]Cf. Leitko (1998, 154).

[23]Cf. Sect. 1.7, (ii), below.

and Robert Musil,[24] both of whom often attended its sessions. The leftist social philosopher Karl Korsch, a close friend of Dubislav, was twice the Society's featured lecturer.

1.4 Intellectual Background

Besides divergent programs and organizational formats, the differences between the Vienna Circle and the Berlin Group can be also reconstructed in terms of their intellectual pedigrees. While the leading figures of the Vienna Circle emphasized its mission in continuing the work of Ernst Mach, Reichenbach resisted associating his Group with Mach's name. He categorically repudiated Mach's practice of dismissing for positivistic reasons fruitful scientific theories—the atomic theory in physics, for example. Reichenbach saw the philosophy of the Berlin Group as historically related to "Kantianism and Friesianism," with particular intellectual debts to Ernst Cassirer and Leonard Nelson (Neurath 1930, 312).[25]

This suggests the Berlin Group's deep roots in German philosophy of the "big nineteenth century" (1789–1914) which featured two major currents of thought. The more widely known of the two, especially outside Germany, was German Idealism. Less well known was the nineteenth-century German scientific philosophy associated with Fechner, Fries, Herbart, Lotze, Hertz, and Helmholtz.[26] Even the Neo-Kantians, whom the Logical Empiricists sought so forcefully to disprove, were mainly interested in the epistemology of *science*—in both its natural and humanistic forms (i.e., in both the *Naturwissenschaften* and the *Geisteswissenschaften*). What distinguished the scientifically oriented German philosophers of the *fin de siècle* from the later Logical Empiricists was that the former persisted in the belief that philosophy has its own discrete realm of knowledge. The most influential proponents of this old dogma were Cassirer, the Marburg Neo-Kantian, and Nelson, the Göttingen Neo-Friesian.[27]

Leonard Nelson (1881–1927) played a particularly significant role in this story. Nelson's professional friendship with David Hilbert and his group in Göttingen marked a new kind of interdisciplinary collaboration between philosophy and

[24]The Viennese Musil was 1931–33 in Berlin.

[25]This sentence in Neurath's "Remarks" was written by Reichenbach. Cf. Reichenbach's letter to Otto Neurath of 24.04.1930 [HR 013-41-70].

[26]How little this period of the German philosophical thought is known today is clear when we glance in the Routledge *Philosophy of Science Encyclopedia* in which we read: "What is called philosophy of science today has its roots in both the British and the Austrian tradition ... (with Bolzano, Mach, and others)" (Sarkar and Pfeifer 2006, xi).

[27]On the influence of Ernst Cassirer on the Berlin Group see the last paragraph of Sect. 1.7, (ii), as well as Chap. 4. It deserves notice that Nelson and Cassirer were engaged in a heated dispute. In 1906 Nelson published a very negative review of Hermann Cohen's book *Logik der reinen Erkenntnis* (1902). Cassirer's answer to Nelson was reciprocally antagonistic.

mathematics (Peckhaus 1990). In 1904 Nelson founded the journal *Abhandlungen der Fries'sche Schule* (n.s.), which by 1937 produced six volumes, with four carefully prepared issues per volume. Notable among the many distinguished papers that first saw print in the journal is Grelling's exposition of the already mentioned paradox named in his honor and four contributions by Paul Bernays.

In 1913 Nelson founded the Jakob Friedrich Fries Society, which met regularly until 1921. This Society's charter, drafted as early as 1908,[28] clearly shows it to be a predecessor of the Society for Empirical/Scientific Philosophy. Like the Berlin Society, the Göttingen Fries Society was an interdisciplinary forum where philosophers, scientists, and mathematicians together mooted various philosophical problems related, most importantly, to the latest results of scientific and mathematical research, presented in their original, technically articulated form—not abridged or simplified in any way.

Thanks mainly to the good offices of David Hilbert and Felix Klein, in 1919 the University of Göttingen appointed Nelson an Extraordinary Professor (*Extraordinarius*) of "systematic philosophy of the exact sciences." Symbolically enough, Moritz Schlick was the second on the Göttingen short-list. The appointment committee selected Nelson "because at the time he was the author of more influential work and had greater impact on his colleagues than Schlick" (Franke 1991, 136). Unfortunately, Nelson's appointment came on the heels of the Great War and, dashing Hilbert's hopes, he precipitately threw himself into the fight for (far left) political causes, dying exhausted at the age of 45 in 1927.

As we shall see in Sect. 1.5, clear lines of succession extended from the Jakob Friedrich Fries Society to the Berlin Group. Except for 3 years spent in Munich, Kurt Grelling worked together with Nelson in Göttingen from 1905 till 1922. The mathematician and David Hilbert's assistant Paul Bernays, Nelson's close friend, was an active member of the Fries Society and later became affiliated with the Berlin Group. Hilbert himself found the work of the Society for Empirical Philosophy so important that he took care to convince Reichenbach and his colleagues to change the Society's name, which toward the end of 1931 became, as we have noted, the "Society for Scientific Philosophy" (Joergensen 1951, 48). Indeed, this title much more accurately reflected the Society's character.[29] Interestingly, one of the last papers read before the Society for Scientific Philosophy in 1934 was presented by Grete Hermann, who had been Nelson's academic assistant in the last few years of his professorship.[30]

The historical roots of the Berlin Society for Empirical Philosophy, by contrast with the Berlin Group, go back to 1912 with the founding of the Society for Positivist Philosophy by Joseph Petzoldt in Berlin. Petzoldt's Society briefly

---

[28]Cf. Peckhaus (1990, 152 f).

[29]See Sect. 1.7, (ii), below.

[30]Grete Hermann was also active in earlier sessions of the Society. In one of them she claimed that quantum physics can be easily made to agree with determinism; Werner Heisenberg found this idea very interesting (Danneberg and Schernus 1994, 396–7, n. 26).

published its own journal, *Zeitschrift für positivistische Philosophie* (1913–1915) and was ultimately absorbed by the Kant Society in 1921, its members forming the "positivist group" within that association (Danneberg and Schernus 1994, 401). Another antecedent of the Berlin Society for Empirical Philosophy was the German Union of Monists, with Ernst Haekel as its poster child. Out of this group came such active members of the Berlin Society as Count Georg von Arco, Max Deri, and Reichenbach's intimate friend Alexander Herzberg, who later became a full member of the Berlin Group. From 1920 till 1931 the Union published the journal *Monistische Monatschrifte*.[31]

Walter Dubislav, who became a member of the Society in May 1927, played a pivotal mediating role in establishing its affiliation with the Berlin Group. Petzoldt came to befriend Dubislav (Dubislav 1929b) and proved instrumental to the younger man's successful Habilitation in 1928.

1.5 The "First Berlin Group"

As already noted, between 1910 and 1913 Kurt Grelling studied political economy in Munich. Reichenbach might well have first gotten to know Grelling while he himself was in Munich in 1912 and 1913. As Flavia Padovani reports, "they both were very actively engaged in the *Freistudentenschaft*. [. . . So] they could have met at one of the Free Student Body meetings in Munich" (Padovani 2008, 37).

Be that as it may, while in Göttingen in the spring of 1914, Reichenbach befriended members of Leonard Nelson's group of neo-Friesians, the central figure of which was Grelling. Later Reichenbach reported that already in 1914 Grelling criticized in discussions his attempt "to base probability claim on a claim of certainty" (Eberhardt and Glymour 2008, 23). Between April and September 1914 Dubislav was also at the University of Göttingen, renting the house at 59 Nikolausberger Weg, which was next door to that of Leonard Nelson who lived at 61. It is an established fact that during his stay in Göttingen in 1914, Dubislav developed a serious interest in the philosophy of Fries and Nelson, an abiding interest that later became clearly evident in Dubislav's *Die Fries'sche Lehre von der Begründung* (1926a), *Über den sogenannten analytischen und synthetischen Urteile* (1926b), and *Zur Methodenlehre des Kritizismus* (1929a).

It is thus more than likely that the three future members of the Berlin Group discussed philosophy as early as the hot summer of 1914. Ten years later Reichenbach recollected in a letter to Erich Regener: In 1914 "I was befriended by some members of the Nelson Circle who like myself were interested in problems of natural philosophy" [HR 016-16-03]. This friendship of the young Reichenbach gained prominence at the vegetarian restaurant in Göttingen on one of the first days of the general mobilization in August 1914, where he got involved in a brawl with

[31] On the history of the Monist Group cf. Herzberg (1928).

a nationalist group of students who were harassing foreign nationals. This incident led to talk that Reichenbach was a "Nelsonian." Indeed, the vegetarian restaurant in Göttingen was the local hangout of the neo-Friesians, Nelson being perhaps the first philosopher ever rigorously to defend animal rights with philosophical arguments.

Around 1925, when Reichenbach began seeking a professorship, the persistent rumors about his affiliation with Nelson became a problem for him. To be sure, Nelson's Group was (rightly enough) seen in Weimar Germany as politically far left and as such inappropriate to associate with if one were a prospective servant of the state (as the position of state university professor was, and still is, in Germany). This explains why Reichenbach tried to play down his connection with the neo-Friesians. In the previously cited 1925 letter to Erich Regener, Reichenbach insisted that he was never a member of the Nelson circle.

There was, however, also another reason why Reichenbach kept his distance from Leonard Nelson: this was a dispute with Nelson that occurred in the summer of 1914 again.[32] Interestingly enough, the disagreement did not concern the philosophy of science but rather the philosophy of education, about which both men nurtured passionately held positions at the time. While Nelson insisted that all "progressive people" in Germany ought to adopt Hermann Lietz's educational reforms, Reichenbach argued that the "movement of the young" that he championed was much more ambitious than Lietz's program. The theory of education that Reichenbach himself embraced originated with Gustav Wyneken, a sharp critic of Lietz.[33]

Worth noting here is also that Reichenbach, who earned his PhD in 1915, had Paul Hensel as his dissertation director in Erlangen. Hensel, it turns out, was close to Nelson (who called him "honorary uncle" [*Nennonkel*]). The neo-Friesians were evidently instrumental in establishing the relation between Reichenbach and Hensel. Given that Reichenbach never studied in Erlangen, his move to select Hensel as his dissertation director would remain rather a puzzle if not for his contact with the neo-Friesians.[34]

Hensel came to play an important role in the pioneering Conference on Exact Philosophy held at Erlangen in March of 1923.[35] The Conference convened in the villa of the newly established Philosophical Academy, which had been founded by Hensel's former doctoral student Rolf Hoffmann.[36] In 1925 the Academy launched

---

[32]Cf. Reichenbach (1914). The character of this dispute is to be perhaps better understood with reference to the fact that, in general, Reichenbach had problems with persons that purposivelly strived to influence the public opinion. Typical examples are Otto Neurath and Carl Popper. Leonard Nelson was at least as resolute to exercise influence on society as these two. (I am indebted for this remark to Andreas Kamlah.)

[33]Additional information on the conflict between Hermann Lietz and Gustav Wyneken is to be found in Chap. 7.

[34]Flavia Padovani, for example, deplores: "The reason why Reichenbach finally veered off to Hensel is not clear" (Padovani 2008, 39).

[35]Cf. Carnap (1936, 14).

[36]Cf. Thiel (1993).

the journal *Symposion*. Although only a single volume ever appeared (in four Issues, the fourth published in 1927), Reichenbach, Schlick and Kurt Lewin contributed important articles. In 1928, the successor to the Academy's publishing house, der Weltkreis-Verlag, issued Carnap's *Aufbau*.

We can trace both the subject and content of Reichenbach's dissertation to the influence of Ernst von Aster, under whom Reichenbach studied at the University of Munich in 1912 and 1913–later Reichenbach repeatedly said that von Aster was the person he was most indebted to in philosophy. While Reichenbach was still in Munich, von Aster published his *Prinzipien der Erkenntnislehre* (von Aster 1913). It contains a section (Chapter V, § 5, 290–9) on probability in which von Aster refers to two works only: Carl Stumpf's paper "On the Concept of Mathematical Probability" (1892) and Kurt Grelling's 1910 paper "Philosophical Foundations of the Calculation of Probability." Grelling's piece was mainly a review paper that defended three ideas: (i) objective ("ontological") interpretation of probability against Carl Stumpf's subjectivism; (ii) discrimination between mathematical and philosophical probability introduced by Jacob Friedrich Fries (in his *System of Logic*, 1811); (iii) the coupling of probability with induction. It deserves notice that Reichenbach followed (i) and (iii) till the end of his days.[37,38]

The positions substantiated in the present section discredit the so called Neurath–Haller thesis. The latter contends that the Vienna Circle was a product of the "Vienna liberal enlightenment", by which Otto Neurath and Rudolf Haller referred to Austrian philosophers who putatively had notable affinities with British empiricism and distrusted the obscure German Idealism. By contrast, the German philosophers of the time, or so argued Neurath and Haller, followed above all "Kant and the Kantians, together with Fichte, Hegel and Schelling" (Neurath 1936, 687)—understood, of course, as enemies of science and experience.[39] It is on the strength of this claim that Neurath and Haller sought to explain why we can regard the philosophical events in Austria as "a chapter of an intellectual development in Europe, which had no success in Germany" (ibid., 676).

This assessment is mistaken. First of all, as we have seen, nineteenth-century German philosophy was also scientifically oriented. Second, the Vienna Circle's influence in Vienna itself was rather limited. The majority of the professional philosophers in the Austrian capital around 1930 were idealists (Stadler 1991). Finally, the Berlin Group's history as we have reviewed it in the present chapter makes it clear that the German scientific philosophy was also developed in the Weimar Republic.

---

[37] In his Dissertation Reichenbach also discussed Ernst Friedrich Apelt's *Theory of Induction* (1854), which appears on Reichenbach's bibliography. Apelt was student and friend of Fries. In the draft of the dissertation, Reichenbach refers as well to Fries' *Essay in a Critique of the Principles of Calculus of Probability* (1842).

[38] Cf. Eberhardt and Glymour (2008, 15 ff).

[39] Today, this claim is controversial: cf. Friedman and Nordmann (2006).

New studies in the history of philosophy of science adduce further evidence against the Neurath–Haller Thesis. Above all, they reveal that the Neo-Kantians as well as Husserl exerted a formative influence on the early Carnap (Friedman 1999; Mayer 1991). Reichenbach also started out as a Neo-Kantian, counting Ernst Cassirer and Alois Riehl among his teachers. What's more, Kant's philosophy itself was clearly scientific in orientation (Friedman 1992). Indeed, Kant is now recognized as having originated a universe of ideas that can be seen as sponsoring both of idealism *and* of scientific philosophy. In other words, a philosopher could be Kantian and at the same time orient his thinking by appeal to science and mathematics—precisely what the Neurath–Haller Thesis denies.

1.6 Realms of Joint Work

Unfortunately, it is difficult to reconstruct the joint work of the members of the Berlin Group proper (1926–1933). The original sources are scant and there is practically no secondary literature on the Group's collaborative activities. What we can discern, however, is that the Berliners concentrated their efforts in three areas: logic, epistemology, and ethics.

(a) *Logic*. The collaboration of Reichenbach with Grelling and Dubislav concentrated principally on logic and meta-mathematics. It is evident that Reichenbach deliberately sought collaborators proficient in this realm—they would be of help in his effort to elaborate his own program for "logical (axiomatic) analysis of science." This point is supported by the fact that while in the late 1920s and the early 1930s Berlin hosted other scientifically oriented philosophers, they were not invited to join the Berlin Group. One such thinker was the physicist Paul Hertz, a friend of Reichenbach's from the time of the 1923 Erlangen Workshop.

    Collaboration with Dubislav on logic proved especially valuable for Reichenbach. Dubislav's work on definitions[40] helped Reichenbach to clarify his position on coordinative definitions. A product of this collaboration was Dubislav's 1929 paper *Elementarer Nachweis der Widerspruchslosigkeit des Logik-Kalküls* (Dubislav 1929c). Appearing in the *Crelles Journal*, this essay features Dubislav's "quasi truth-tables"[41]; Reichenbach himself pursued work along the same lines. Three years later he employed Dubislav's tables in his paper *Wahrscheinlichkeitslogik*.[42] It propped Reichenbach's theory of probability according to which propositions have three predicates: true, false, and their prediction-value or weight (Reichenbach 1938, 28).

(b) *Epistemology*. In the already mentioned paper *Logistic Empiricism in Germany and the Present State of Its Problems* (1936a), Reichenbach recalled that

[40]Cf. Dubislav (1926c, 1927, 1931).

[41]Cf. Chap. 8.

[42]See also Reichenbach (1947, 127 n).

instead of investigating the principle of verification, or getting entangled in the protocol-sentence debate, the members of the Berlin Group concentrated on probability, which they treated as "theory of propositions about the future." As Reichenbach put it, "there were years of work in Berlin [on this subject], filled with fresh starts and tentative solutions, proposed in ardent discussions, before a definitive theory was reached" (p. 152). Reichenbach here supplies indispensable, firsthand insight into the Group's preoccupations, character, and modus operandi.

We can reconstruct Grelling's position on probability and induction during these years mainly from a text published in *Erkenntnis* which preserves for us his contribution to the discussion on probability at the 1929 Prague Conference.[43] This is a programmatic document in which Grelling and Reichenbach attacked Waismann's Wittgensteinian position and that of Carnap as well. Grelling, in particular, insisted that science is possible only when based on the principle of induction—the principle that endows science with its predictive power. However puzzling it might be deemed, the inductive principle is in any case neither empirical nor tautological. "If I were a Friesian," declares Grelling, "I would say that this principle is synthetic a priori" (p. 278). But he was no longer a Friesian. Grelling admitted that he had no solution to this problem; but he was content to have laid it out in its most clear and compelling terms.

In September 1929, Reichenbach shared this problem with Grelling: the only non-empirical point in his epistemology was the principle of induction, or "Hume's problem". After years of joint effort, in 1932/1933 Reichenbach reached a solution to this problem: Induction is based on conjectures, or posits, that are a result of our assessments of the facts. Posits are ultimately a product of our free will and so are purely empirical. Reichenbach called this position *radical empiricism*, and put it to work in his most influential books: *Experience and Prediction* (1938), and *The Rise of Scientific Philosophy* (1951).

Dubislav took up the problem of probability and induction in a lecture before the Society for Empirical Philosophy on December 10, 1929. The text was published almost *mot-à-mot* in chapter 4.7 ("Induction") of his book *Natural Philosophy* (Dubislav 1933, 99–114) in which Dubislav concurs with Reichenbach's analysis in virtually every respect.

Grelling started to disagree with Reichenbach's new theory of induction only in 1936 when he adopted Carnap's position that there are two kinds of probability—philosophical and statistical[44]; the difference between them being a matter of convention, of "syntax."[45]

(c) *Ethics*. Dubislav and Reichenbach also shared a joint position in ethics, one that opposed the Vienna Circle's doctrine on the subject. Although both schools

[43]"Diskussion über Wahrscheinlichkeit", *Erkenntnis* 1 (1930): 260–85 (Grelling's contribution is on p. 278).

[44]In fact, this was his position in Grelling (1910). Cf. Sect. 1.5, above.

[45]Cf. Grelling's letter to Reichenbach of 28.01.1936 [HR 013-14-04].

took anti-cognitivist stands in ethics, the Vienna philosophers championed a form of emotivism: they maintained that value judgments are expressions of our emotions. This position distinguished two forms of understanding, knowledge and emotions, the problem with it being that, as a matter of fact, this position was based on a conception of the German (Dilthey's) "life-philosophy" (cf. Gabriel 2004) that the Vienna Circle officially radically opposed.

In contrast, Reichenbach and Dubislav regarded all ethical propositions as implicit commands.[46] Thus as with scientific propositions, which are posits, the propositions of ethics are, according to Reichenbach and Dubislav, products of the free will.[47] The two philosophers saw this position as most radically empiricist.

1.7 Autonomy of the Berlin Group

The Vienna Circle and the Berlin Group were unquestionably closely related academic communities. While studying at the University of Vienna in the Winter Term 1929/1930, the Berliner Carl Hempel took part in sessions of the Vienna Circle; and Martin Strauss was from 1934 through 1938 Philipp Frank's post-doctoral student in Prague. On the other hand, Carnap and Neurath lectured at the Berlin Society for Empirical/Scientific Philosophy. Richard von Mises, in his turn, served as a bridge between the Berlin Society and the Ernst Mach Association.

At the same time, each of the schools had its own distinctive character. Indeed, as we already have remarked and as we shall see further in the present section, the two groups proceeded on grounds that were clearly divergent from each other—the scientifically oriented philosophy in Vienna took as its points d'appui philosophy of language and philosophical logic, while the Berlin Group's activities centered on logical analysis of the newest scientific discoveries.[48] This difference in orientation is clear in the two "Introductions" to *Erkenntnis*, one by Reichenbach and the other by Schlick (Reichenbach 1930; Schlick 1930), which outlined two different programs of scientific philosophy. Understandably, in view of the considerable dissimilarity of their programs, Schlick declined several invitations to lecture before the Berlin Society, while the Vienna Circle never invited Reichenbach to read a paper—this despite several requests on his part to do so[49] and the fact that the Circle did host other international philosophers and logicians, such as Eino Kaila and Alfred Tarski.

---

[46]Cf. Dubislav (1937), Reichenbach (1947, 344), Reichenbach (1951, 280 ff). Cf. Chap. 7.

[47]Nicholas Rescher, who considers himself one of the successors of the "Berlin Group in America" (Rescher 2005), developed this stance in his *The Logic of Commands* (1966).

[48]Cf. Sect. (ii), below.

[49]Cf. Reichenbach's letter to Schlick from 2.01.1933 [HR 013–30–13].

Apparently, Vienna and Berlin treated their collaboration as a marriage of convenience.[50] Ostensibly their shared objective was to show the educated public of the time that *together* they made a front of scientifically oriented thinkers that opposed the traditional philosophy, with its claim to be an autonomous discipline featuring its own truths. In fact, however, there were considerable differences between Vienna Circle from the Berlin Group. Three in particular are of interest here:

(i) *Different Masterminds.* As we have previously remarked, while Vienna, in a general sense, took Ernst Mach as its guiding spirit, Berlin was post-Friesian. With respect to active participants in the two groups, Wittgenstein exerted a formative influence upon scientific philosophers in Vienna. Indeed, as early as the academic years 1923 through 1925, the *Tractatus* was discussed *mot-à-mot* in Schlick's seminars. Starting in 1926, the nucleus of the Vienna Circle systematically studied Wittgenstein's book (Stadler 1997, 227 f.). Crediting the seminal power that Wittgenstein's ideas had in the evolution of Vienna Circle philosophy helps to make comprehensible the Circle's preoccupation with such matters as the discrimination of metaphysics from science, the elimination of metaphysics (that is why they were called "logical *positivists*") and the principle of verification. These topics were scarcely discussed in Berlin, where Wittgenstein's impact was comparatively limited.[51]

The Berlin Group's philosophical hero was not Wittgenstein but Bertrand Russell. This is evidenced by, among other things, Kurt Grelling's translation into German of four of Russell's books in the 1920s (Milkov 2005). Grelling's translations no doubt aided Reichenbach in mastering Russell's philosophy, something evident in two brief but highly informed and laudatory essays titled, "Bertrand Russell" (Reichenbach 1928b, 1929).

Russell, however, was a complex philosopher whose views were subject to radical shifts and whose thought cut across the interests of both the Vienna Circle and the Berlin Group. On the one side, especially in the teens of the last century, he largely subscribed to Frege's and Wittgenstein's philosophy of language. This bears on Russell's doctrine of the logical construction of the world out of elements of the given (snese-data), by way of the logic of relations; for he developed this line of thought under Wittgenstein's influence (Milkov 2002). On the other side, Russell remained deeply interested in the latest scientific developments and discoveries, specifically with an eye toward subjecting them to philosophical analysis and to assessing their value for philosophy. It was this second, scientifically attuned Russell that most interested the members of the Berlin Group.

[50] See the motto to this chapter.

[51] In fact, the only message of Wittgenstein assimilated in Berlin was the thesis that logic is tautological in character.

The Berlin Group had closest contacts with two of the lead figures in the science and the mathematics of the day: Einstein and Hilbert. Of course, these two men also influenced the Vienna Circle, albeit in a somewhat different way. The Circle construed the formative findings of Hilbert and Einstein as, first and foremost, definitive refutations of the idea of apriori truths. The neo-positivists held that the sciences proceed on the basis of material that we perceive through our senses. This position was embraced in Berlin as well (in the form of physicalism), in particular after Reichenbach introduced his "radical empiricism" thesis in 1932/1933. Prior to that development, scientific philosophers in Berlin were more interested in specific philosophical aspects of the newest scientific and mathematical theories. This explains, among other things, why at that period Reichenbach, unlike Neurath, did not hesitate to call himself a "philosopher."[52] Indeed, the former's main objective was to cultivate the *philosophical* insights engendered by the most recent discoveries and theories in science and mathematics.

In the 1920s Reichenbach won wide recognition as the leader of the "defense belt around Einstein".[53] To be sure, some members of the Vienna Circle were penetrating philosophical interpreters of the Theory of Relativity. Schlick in particular, who trained under Max Planck, was the first to propose it an empirical interpretation (Schlick 1918). As we have observed, however, after 1924 he turned to the intensive study of Wittgenstein, which precipitated a dramatic shift in his philosophical interests (Ferrari 2008).

Especially noteworthy here is that between 1926 and 1929, the defining years of the Vienna Circle and the Berlin Group, Reichenbach led deeply probing Group discussions with Einstein,[54] while Schlick and Carnap (Carnap as from 1927) hosted intensive meetings with Wittgenstein as the star guest. These engagements proved formative for the two groups.

Also highly influential among the members of the Berlin Group was David Hilbert. As we have seen, all three of the leading Berliners—Grelling, Reichenbach and Dubislav—studied under Hilbert in Göttingen (Hilbert was Grelling's *Doktorvater*). For Hempel, too, "the atmosphere in Göttingen was very, very stimulating. The two terms there influenced [him] very strongly" (Hempel 2000, 5). This was not the case with the three principal figures of the Vienna Circle: Schlick, Neurath, and Carnap. The latter two had never studied in Göttingen, and when Schlick was there as a graduate student in 1904 and 1905, he was totally preoccupied with experimental work in physics and never sought out Hilbert (Iven 2008, 108 ff.). It can come as no surprise, then, that

[52]Moreover, Reichenbach showed willingness for "peaceful debates" with "speculative", or idealistic, philosophers, such like Oskar Becker. See Reichenbach (1931a).

[53]Stölzner (2001, 108). Reichenbach's papers in apology of the Theory of Relativity were recently published in Reichenbach (2006).

[54]One piece of evidence is Einstein's review of Reichenbach's *Philosophy of Space and Time*. Cf. Einstein (1928).

while, for example, Walter Dubislav arduously defended Hilbert's formalism against the logicism of Frege (Dubislav 1930), Carnap was and remained a logicist who considered axiomatic simply an "applied logistic."[55]

Hilbert's greatest influence on the Berlin Group flowed from his project to axiomatize the sciences, something clearly related to Reichenbach's program for logical analysis of science. To be more exact, Hilbert claimed that in order to set out the foundations of a given science, we have to "set up a system of axioms which contains an exact and complete description of the relations subsisting between the elementary ideas of that science."[56] Reichenbach's objective in the 1920s was precisely this: to axiomatize the philosophy of space and time and to formulate it in strict logical order. It was realized in three of his books: Reichenbach 1920, 1924 and 1928. In this way he synthesized two lines of influence, that of Hilbert and that of Einstein.

Scarcely anything demonstrates more clearly the different orientations of Vienna and Berlin (under Göttingen influence) than the criticism that Hans Hahn and Philipp Frank leveled at Hilbert's notion of axiomatization as a general theory of science. Taking the lead from Ernst Mach's positivistic postulate for strict demarcation between mathematics and physics, Hahn and Frank argued that the axiomatic method, which strives at "deepening the foundations" of science, is infected with metaphysics (Stölzner 2002). In this, they followed the motto of the Vienna Circle: "In science there are no 'depths': there is surface everywhere" (Neurath 1929).

(ii) *The Berlin Program for Logical Analysis of Science.* We have already seen indications that the empiricism of the Berlin Group differed substantially from the empiricism that members of the Vienna Circle embraced. At its height, the Circle advanced the neo-positivist shibboleth that "the meaning of every statement of science must be statable by reduction to a statement about the given"—i.e., about sense-data or experiences (ibid., 309). Its guiding positivist theme was roughly speaking an Ernst Machian interpretation of Wittgenstein's *Tractatus*.

By contrast, the Berlin Group propounded the view that the only sources of knowledge are scientific observations, experiments, and theories—not some apriori judgments, as Kant believed, nor our sense-data. Moreover, it contended that this view "*is the very first condition of empiricism*" (Reichenbach 1936a, 152).[57]

Apparently, the schools of Vienna and Berlin subscribed to different versions of empiricism. Interestingly enough, Kurt Lewin insisted that instead of "empiricism," the position of Reichenbach should be called "observatism"

[55]Cf. Chap. 13. It deserves notice that around 1930, Carnap's interest in Hilbert radically increased (arguably, under Dubislav's and Reichenbach's influence), a development that found expression in his *Logical Syntax* (1934).

[56]Hilbert (1900, 447). Cf. Peckhaus (2003).

[57]Italics mine—N. M.

(Lewin 1925, 91). More so since the Berlin Group and the Society were interested not only in problems of empirical science but also in those of theoretical physics, theoretical biology, mathematics, and technique—and these are scarcely "empirical."

Reichenbach agreed with Kant that science functions thanks to certain principles and axioms. He maintained, however, that these come and go with every significant scientific discovery. Reichenbach's conclusion was that the task of philosophy is to "logically analyze" the newest scientific theories: it "penetrates the results of the special sciences to philosophical question-formulation" (Anonymous 1930, 72). This it does in order: (i) to distill their axioms and principles; (ii) to explicate the connection of our experience with the conceptual systems of scientific theories. Philosophy realizes these aims, according to Reichenbach, with the help of coordinating definitions that constitute the objects of science.

In this way Reichenbach replaced Kant's apriori judgments with relativised and dynamic constitutive principles that change from theory to theory such that every well-developed theory has its own constitutive principles (Friedman 2005, 125). Moreover, pace Kant, the principles of different sciences do not on this account diverge radically one from another: they are not different in type. It follows that the task of scientific philosophers is to explore the everchanging fundamental principles that sponsor all of the sciences.

This conclusion played a seminal role in the formation of the *Society for Scientific Philosophy*.[58] It makes clear, in particular, why once Reichenbach and his friends acquired control over the "Society for Empirical Philosophy" after June 1929 they renamed it by the end of 1931 into the "Society for Scientific Philosophy."

Plainly the ultimate, if merely adumbrated objective animating the program for logical analysis of science was to compare the principles of different scientific disciplines in order to stimulate the further development and articulation of those principles (Milkov 2011, 151 n. 14). Such was the rationale for gathering together scientists of different disciplines. Apparently, the hope was that these interdisciplinary studies would lead to the birth of new disciplines, such as the now-established fields of cybernetics, game theory, and systems theory. The program also sought to foster interdisciplinary influences similar to that which some 20 years later the physicist Erwin Schrödinger exerted on the biologists Francis Crick and Maurice Wilkins, and which lent impetus to the discovery of the structure of DNA. And while in the latter instance the cross-disciplinary interaction and the discoveries did not occur in the setting of the Society for Scientific Philosophy, the principals had connections with it: Schrödinger's assistant, Fritz London, for example, was a quondam member of the Berlin Group; as already mentioned, the Society for Scientific Philosophy hosted Ludwig von Bertalanffy as lecturer three times and members of the

[58]Cf. Sect. 1.3, above.

Berlin Group published extended reviews of his writings over the years (see Hempel 1951). Finally, Reichenbach expressed the conviction that his own work in "natural philosophy" anticipated Werner Heisenberg's Principle of Indeterminacy (cf. Reichenbach 1931b, 40 f.) (Cf. Chap. 6).

The Vienna Circle's interdisciplinary program exhibited considerably more of an ideological cast than did that of the Berlin Group. In particular, the Circle sought to demonstrate the unity of science in terms of a system of related concepts, the ultimate aim being to show that the humanities do not follow a method fundamentally different from that of the natural sciences. In this connection it deserves notice that comparing the papers read before the Ernst Mach Association with those presented at the Society for Scientific Philosophy reveals that the Society hosted lecturers which were more in line of the leading science of the time (cf. Danneberg and Schernus 1994, 478–81; Stadler 1997, 379–81).

Reichenbach was convinced that there was an acute need for a logical analysis of science such as his. Scientists prefer, as he noted, to concentrate on discovering new facts and constructing new theories, rather than on making their theories logically—and epistemologically—more coherent: this is the task of scientific philosopher. "Scientific research does not leave a man time enough to do the work of logical analysis," observed Reichenbach, "and … conversely logical analysis demands a concentration which does not leave time for scientific work" (Reichenbach 1951, 123). The two academic disciplines—science and scientific philosophy—share the same subject-matter and so *both* should be harnessed in tandem to pull the "car of knowledge." Reichenbach urged, moreover, that scientific philosophy is as technically sophisticated as the sciences themselves, and that it frequently corrects scientific theories when they are initially framed. The logical analysis of science is thus anything, then, but mere scientific journalism.

Carnap openly criticized Reichenbach on this score, insisting that "the investigation of facts is the task of the natural-scientific, empirical research, the investigation of the language forms is the task of the logical, syntactical analysis" (Carnap 1936, 265). The latter task, asserted Carnap, is the defining project of scientific philosophy.

When speaking of the Berlin Group's work in "logical analysis" one needs to bear in mind that in early analytic philosophy the term was understood in two different senses. On the one hand, for the likes Schlick and Carnap "logical analysis" referred above all to philosophical logic and philosophy of the language of science; on the other, thinkers such as Reichenbach used the term in conjunction with distilling the new principles of science. In fact, "logical analysis," "conceptual analysis," "clarifying of concepts," "conceptual confusion," "logical forms," "analysis of the language of science," and "rational reconstruction" all meant significantly different things for Reichenbach than they did for Schlick and Carnap. When, for example, Reichenbach spoke about the "logical analysis" of science, what he had in mind was closer to what we think of as "axiomatic analysis." He first evinced serious interest in logic only after 1928, under the influence of Grelling and Dubislav.

From a historical perspective, logical analysis of science, such as the Berlin Group conceived it, exhibited close methodological affinities with the Marburg Neo-Kantian analysis (principally Ernst Cassirer's analysis) of the structure (the "logic") of science.[59] Relative to its context in the scientific philosophy of the time, Reichenbach's program, in particular, shares elements of Kurt Lewin's interdisciplinary variant of Cassirer's method (Lewin 1925).[60] In fact, when Reichenbach initially elaborated his method of the logical analysis of science in his *Relativitätstheorie und Erkenntnis apriori*, Lewin (1920) was the only published work to which he referred in its support (Reichenbach 1920, n. 20).[61] From that period forward and till the end of the 1920s, Reichenbach and Lewin regularly cited each other's work in their publications (Wittmann 1998, 184).

(iii) *Social Work*. Besides the differences in theory, the Vienna Circle and the Berlin Group pursued divergent social aims. Whereas Neurath's goal was to reorganize the social and economic life of his country (and eventually the world), Reichenbach nurtured the more modest social ambition of elevating the status of science in society and, more particularly, in philosophy. A "great evangelist of science" (van Fraassen 2002, 224), Reichenbach sought among other things to set up professorships throughout Germany in the scientific philosophy of nature. To this end, in 1931 Reichenbach composed a sixty-page petition that he submitted to the Ministry of Science, Art and Education of the Weimar Republic. The cover letter was signed by a host of luminaries, including Einstein and Hilbert. Reichenbach hoped to publish the petition in *Erkenntnis*, but this plan fell through when his Vienna Circle colleagues objected (Danneberg and Schernus 1994, 404 n. 56).

Roughly at the same time, at the request of Wolfgang Windelband, the Prussian Minister of Science, Art and Education, Moritz Schlick wrote a formal assessment of Reichenbach's achievements as scientific philosopher. Schlick's findings were unequivocal: "I consider [Reichenbach's] main ideas in analysis of causality and probability false. It appears that in this realm a peculiar, inflexible adherence to certain ideas prevents him from reaching the ultimate depth."[62] Ironically enough, Wolfgang Windelband was the son of Wilhelm Windelband, the head of the Southwest Neo-Kantians, whom both the Viennese and Berlin schools of scientific philosophers fiercely opposed.

---

[59] This point is confirmed by the fact that Reichenbach referred to Ernst Cassirer (his professor at the University of Berlin), when he spoke about the historical roots of the Berlin Group in Neurath (1930, 312) (cf. n. 25).

[60] Cf. Chaps. 4, 5 and 14.

[61] The 1920 program of Lewin–Reichenbach was most closely followed by Paul Oppenheim (Oppenheim 1926). Cf. Chap. 13.

[62] Moritz Schlick's letter to Wolfgang Wildenband, 15.03.1931.

1.8 Logical Positivism and the Rise of Logical Empiricism

The story of the Berlin Group in its relation to the Vienna Circle is of special interest also because of what it brings to light in the history of analytic philosophy in North America. It is well established that the Group exerted a profound influence in the USA through those of its members who, unlike Dubislav and Grelling, managed to immigrate to America after January 1933. Reichenbach and Hempel in particular contributed signally to a radical change of philosophical climate in the United States, as did Olaf Helmer and Paul Oppenheim (Cf. Chap. 14).

Ronald Giere asserts that after Carnap and Reichenbach resettled in North America, they made themselves over from scientific philosophers (*wissenschaftliche Philosophen*) into philosophers of science as a means of adapting to their new academic milieu:

> They realized, quite rightly, that works like the *Aufbau* and *Relativitätstheorie*, which were written in the context of a cultural, scientific, and philosophical tradition that did not then exist in North America, would not be much appreciated in the North American context. So they put their efforts into other projects, ones better suited to their new intellectual and cultural environment. (Giere 1996, 337)

Giere's account, here, is unconvincing. In the first place, by referring only to Reichenbach's early *Relativitätstheorie* (1920) he obscures the fact that in the later and more mature *Philosophie der Raum–Zeit Lehre* (1928) Reichenbach produced what is in effect the manifesto of the program for scientific philosophy as it is also practiced today. In fact, elements of philosophy of science had been generally in play in Europe since the late 1920s. In large part this was due, as we have noted (see Sect. 1.7, above), to participants of the Berlin Group, for whom the nascent program evolved pretty much along the same lines that would distinguish the philosophy of science as a autonomous sub-discipline in the USA.

In more recent years Wesley Salmon substantiated these points. A student of Reichenbach's at UCLA Berkeley, Salmon repeatedly asserted what Reichenbach had underscored back in the mid-1930s (Reichenbach 1936a, 151 f.), namely that while the Vienna Circle advanced a doctrine of *logical positivism*, one distinguished by the sharply anti-metaphysical stance shared by each of its members, the Berlin Group championed a program of *logical empiricism* that logically analyzed the latest findings and theories of the sciences and mathematics. Despite real affinities, the Vienna Circle and the Berlin Group were associations whose structure and proceedings reflected different agendas. True enough, it was the Circle's Otto Neurath who in 1931 coined the term "logical empiricism" (Neurath 1931, 297), one which he came to employ regularly. Nonetheless, Reichenbach more accurately captured the history of the Germanophone scientific philosophy when in 1936 he divided it into Vienna logical positivism and Berlin logical empiricism.

Salmon declared that the second movement the Berlin logical empiricism—"completely superseded" the first one: the Viennese logical positivism (Salmon 1999, 333). But what the evidence indicates more specifically is that this development signaled the historical triumph of the Berlin Group's program and the concomitant eclipse of the Vienna Circle's positivist agenda in the realm of philosophy of science. Soon, the logical positivism, the doctrine once shared by all the leading members of the Vienna Circle, was abandoned, whereas the program for the logical analysis of science persisted and remains viable even today (Salmon 2001, 233 f.). In this connection, one should bear in mind that leading members of the Vienna Circle abandoned logical positivism prior to the decimation of the Vienna and Berlin groups under the Nazis.

Too many authors today conflate these developments, and simply identify the Vienna Circle philosophy as "logical empiricism."[63] Typically, the move to represent the Circle in this light gets justified along some such lines as the following: "Logical empiricism is really the story of the development of themes articulated within logical positivism. [... It is] the reexamination, modification, and (alternatively) rejection and endorsement of the themes of logical positivism" (Hardcastle 2006, 458, 464).

If the evidence adduced in the preceding discussion has made anything clear, it is that such a reading is a mistake. The two movements formulated and pursued two different, albeit related, programs of scientific philosophy. To ignore this difference is to obscure the historical record.

1.9 Philosophy of Science versus Analytic Philosophy of Language

Recently, Nicholas Rescher published an account, somewhat similar to the present one, of the Berlin Group's scientific philosophy and its influence in the United States (cf. Rescher 2005). Rescher traces back to Berlin the philosophy of science developed over the last four decades at the University of Pittsburgh, as well as at other major research universities across the USA—from Berkeley, Bloomington, and Boston, to Minneapolis and Princeton. He numbers among the luminaries in this American extension and development of the Berlin legacy Carl Hempel, Adolf Grünbaum, Wesley Salmon, Baas van Fraassen, Alberto Coffa, Larry Laudan, and John Earman, among others.

Accurate as Rescher's picture may be, it is essential to bear in mind that the program of the Berlin Group did not totally overshadow the legacy of the Vienna Circle in American analytic philosophy. Indeed, W.V. Quine and his followers at Harvard and elsewhere by and large hewed to Circle's project, continually developing and often correcting it. Moreover, Quinean exact philosophy has proven to be more influential than the philosophy of science inspired by Reichenbach and Hempel.[64]

---

[63]Among the authors that are against such conflation are Philipp Kitcher (2001, 148), and Peter Godfrey-Smith (2003, 22).

[64]This point has become especially prominent during recent decades (cf. Howard 2000, 75 f.).

While Quine himself understood his philosophy to be closely tied to science, in truth, it was not. Being a very general "science," Quine's doctrine differs perceptually from Reichenbach's real (concrete), or "actual" science. More conceptual in nature than scientific are (i) Quine's discussions of the dichotomy between analytic and synthetic propositions; (ii) his criticism of the sense-data theory, of the "radical reductionism" thesis, of the correspondence theory of truth, etc.; (iii) his defense of the ontological commitment of language; (iv) his interest in "what there is" and which are "the ultimate constituents of the Universe." Quine's philosophy of science is more of an analytic philosophy of language, broadly conceived—one that includes analytic ontology and analytic epistemology. While these sub-disciplines probe the methods and conditions of scientific praxis, they do not address recent scientific discoveries and theories and their immediate philosophical importance.

Analytic philosophy of language (or philosophical logic) originated, as we have noted, with Frege. It saw its most significant early development in some of Russell's works from the period of 1903–1919, reaching its pinnacle in Wittgenstein's *Tractatus logico-philosophicus* and in the work by the Vienna Circle (cf. Sect. 1.7, (i)). It is true that Quine is known to have shown little interest in Wittgenstein's book[65]; but he closely followed the Carnap of 1926–1935 who did draw largely upon Wittgenstein, and ultimately Frege.

This and similar approaches were later named "external philosophy of science." The "external" aspect that marks this current of thought has to do with the fact that it is not developed in terms of any "inspection of the procedures actually followed by scientists" (McMullin 1970, 24). The "internal philosophy of science," by contrast, "relies for its warrant upon a careful ... description of how scientists actually proceed" (ibid., 26). It was this philosophy that Reichenbach and his colleagues initially formulated around 1930.

Hilary Putnam supplies a telling personal record of the experience of jettisoning analytic philosophy of language of science for Reichenbach's logical empiricism in a way that reflects the difference between the two programs:

> I did a year of graduate work at Harvard in 1948–1949, where I came under influence of Quine's views on ontology and his scepticism concerning the analytic/synthetic distinction. At that point, I was in a mood that is well known to philosophy teachers today: it seemed to me that the great problems of philosophy had turned out to be pseudoproblems. [...]
>
> Within a few months of my arrival in Los Angeles in the fall of 1949 these philosophical "blahs" had totally vanished. What overcame my "philosophy is over" mood, what made the field come alive for me, made it more exciting and more challenging than I had been able to imagine, was Reichenbach's seminar, and his lecture course on the philosophy of space and time. (Putnam 1991, 61)

Lamentably, for decades philosophers cast a blind eye to the all-important contrasts that differentiated these two philosophical currents. Russell, in particular, had insisted that they were mutually complementary. He discovered the serious disagreements between the two only in his later years (in particular, in his *Human*

[65]In contrast to his friend Burton Dreben and his acolytes at Harvard.

Knowledge, 1948), when he returned to his old ambition to pursue philosophical inquiries into science. Likely enough it was Reichenbach who awoke Russell from his philosophy-of-language slumbers, for the two shared an office when Russell was at the University of California at Berkeley in 1940.

1.10 Epilogue

The foregoing sections enable us to recast, in a general way, our picture of the history of analytic philosophy in post-World War II America. During the second half of the twentieth century analytic philosophy in the United States evolved along lines that reflected the contrasting currents of scientific philosophy that took their definitive form in Berlin and Vienna around 1930. Roughly speaking, whereas Carnap and his student Quine, as well as the latter's follower, Donald Davidson, were engaged mainly with problems of analytic philosophy of language, in conjunction with philosophical logic, Reichenbach and his student Hempel (along with Hempel's students Putnam, Grünbaum, Salmon, and their followers van Fraassen and John Earman) devoted themselves, by contrast, to internal philosophy of science and hence pursued the philosophy of science that analyzes the facts of concrete scientific practice.[66]

References

Anonymous. 1930. Gesellschaft für empirische philosophie: Berlin. *Erkenntnis* 1: 72–73.
Ayer, Alfred. 1936. *Language, truth, and logic*. London: Gollancz.
Blumberg, Albert E., and Helmut Feigl. 1931. Logical positivism. A new movement in European philosophy. *Journal of Philosophy* 28: 281–296.
Carnap, Rudolf. 1936. Von der Erkenntnistheorie zur Wissenschaftslogik. In *Wiener Kreis*, ed. Michael Stoeltzner and Thomas Uebel, 260–266. Hamburg: Felix Meiner.
Cohen, Hermann. 1902. *Logik der reinen Erkenntnis*. Berlin: Cassirer.
Danneberg, Lutz, and Schernus Wilhelm. 1994. Die Gesellschaft für wissenschaftliche Philosophie. In *Hans Reichenbach und die Berliner Gruppe*, ed. Lutz Danneberg et al., 391–481. Braunschweig: Vieweg.
Danneberg, Lutz, et al. (eds.). 1994. *Hans Reichenbach und die Berliner Gruppe*. Braunschweig: Vieweg.
Dubislav, Walter. 1926a. *Die Fries'sche Lehre von der Begründung: Darstellung und Kritik*. Dömitz: Mattig.
Dubislav, Walter. 1926b. *Über den sogenannten analytischen und synthetischen Urteile*. Berlin: Weiss.
Dubislav, Walter. 1926c. *Über die definition*. Berlin: Weiss.
Dubislav, Walter. 1927. *Über die definition*, 2nd ed. Berlin: Weiss.

[66]Preliminary versions of this Chapter were read at the Universities of Bochum, Graz, Pittsburgh and Vancouver. I am grateful for stimulating discussions.

Dubislav, Walter. 1929a. *Zur Methodenlehre des Kritizismus*. Langensalza: Beyer & Söhne.

Dubislav, Walter. 1929b. Joseph Petzoldt in memoriam. *Annalen der Philosophie* 8: 289–295.

Dubislav, Walter. 1929c. Elementarer Nachweis der Widerspruchslosigkeit des Logik-Kalküls. *Journal für die reine und angewandte Mathematik* 161: 107–112.

Dubislav, Walter. 1930. Über den sogenannten Gegenstand der Mathematik. *Erkenntnis* 1: 27–48.

Dubislav, Walter. 1931. *Die definition*, 3rd ed. Leipzig: Felix Meiner.

Dubislav, Walter. 1933. *Naturphilosophie*. Berlin: Junker und Dünnhaupt.

Dubislav, Walter. 1937. Zur Unbegründlichkeit der Forderungssätze. *Theoria* 3: 330–342.

Eberhardt, F., and C. Glymour. 2008. Introduction. In *The concept of probability in the mathematical representation of reality*, ed. Hans Reichenbach, and trans. Frederick Eberhardt, and Clark Glymour, 1–36. Chicago: Open Court.

Einstein, Albert. 1928. Hans Reichenbach: *Philosophie der Raum-Zeit-Lehre. Deutsche Literaturzeitung* 1: 19–20.

Ferrari, Massimo. 2008. Moritz Schlick in Wien: Die Wende der Philosophie. In *Moritz Schlick: Leben, Werk und Wirkung*, ed. Fynn O. Engler and Mathias Iven, 91–113. Berlin: Parerga.

Franke, Holger. 1991. *Leonard Nelson*. Ammersbek bei Hamburg: Verlag an der Lottbek.

Friedman, Michael. 1992. *Kant and the exact sciences*. Cambridge: Cambridge University Press.

Friedman, Michael. 1999. *Reconsidering logical positivism*. Cambridge: Cambridge University Press.

Friedman, Michael. 2005. Ernst Cassirer and contemporary philosophy of science. *Angelaki* 10: 119–128.

Friedman, Michael, and Alfred Nordmann (eds.). 2006. *The Kantian legacy in nineteenth-century science*. Cambridge, MA: MIT Press.

Gabriel, Gottfried. 2004. Introduction: Carnap brought home. In *Carnap brought home. The view from Jena*, ed. S. Awodey and C. Klein, 3–23. Chicago: Open Court.

Gerner, Karin. 1997. *Hans Reichenbach: Sein Leben und Wirken*. Osnabrück: Autorenpress.

Giere, Ronald. 1996. From *wissenschaftliche philosophie* to philosophy of science. *Minnesota Studies in the Philosophy of Science* 16: 335–354.

Godfrey-Smith, Peter. 2003. *Theory and reality: An introduction to the philosophy of science*. Chicago: The University of Chicago Press.

Grelling, Kurt. 1910. Die philosophische Grundlagen der Wahrscheinlichkeitsrechnung. *Abhandlungen der Fries'schen Schule* 3: 439–478.

Grelling, Kurt. 1929. Realism and logic: An investigation in Russell's metaphysics. *The Monist* 39: 501–520.

Grelling, Kurt. 1930. Die Philosophie der Raum–Zeit-Lehre. *Philosophischer Anzeiger* 4: 101–128.

Grelling, Kurt. 1933. Bemerkungen zu Dubislav, *Die Definition. Erkenntnis* 3: 189–200.

Grelling, Kurt. 1939. A logical theory of dependence. In *Foundations of gestalt theory*, ed. Barry Smith 1988, 217–228. München: Philosophia.

Grelling, Kurt, and Paul Oppenheim. 1937/1938. Der Gestaltbegriff im Lichte der neuen Logik. *Erkenntnis* 7: 211–225. 357–359.

Grelling, Kurt, and Paul Oppenheim. 1939. Logical analysis of 'Gestalt' as 'functional Whole'. In *Foundations of gestalt theory*, ed. Barry Smith, 210–216. Philosophia: München.

Haller, Rudolf, and Stadler Friedrich (eds.). 1993. *Wien–Berlin–Prag. Der Aufstieg der wissenschaftlichen Philosophie*. Wien: Hölder–Pichler–Tempsky.

Hardcastle, Gary. 2006. Logical empiricism. In *The philosophy of science: An encyclopedia*, ed. Sahotra Sarkar and Jessica Pfeifer, 458–465. London: Routledge.

Hempel, Carl. 1951. General system theory: A new approach to unity of science. *Human Biology* 23: 313–322.

Hempel, Carl. 1991. Hans Reichenbach remembered. *Erkenntnis* 35: 5–10.

Hempel, Carl. 2000. Intellectual autobiography—The interview with Richard Nollan. In *Science, explanation, and rationality*, ed. James H. Fetzer, 3–35. Oxford: Oxford University Press.

Hentschel, Klaus. 1991. *Die Korrespondenz Petzold-Reichenbach*. Berlin: ERS.

Herzberg, Lily. 1928. Die philosophischen Hauptströmungen im Monistenbund. *Annalen der Philosophie* 7: 113–135. 177–199.

Hilbert, David. 1900. Mathematical problems. *Bulletin of the American Mathematical Society* 8: 437–479.

Hoffmann, D. 1994. Zur Geschichte der Berliner 'Gesellschaft für empirische/wissenschaftliche Philosophie'. In *Hans Reichenbach und die Berliner Gruppe*, ed. Lutz Danneberg et al., 21–31. Braunschweig: Vieweg.

Hoffmann, D. 2007. The society for empirical/scientific philosophy. In *Cambridge companion to logical empiricism*, ed. Alan Richardson and Thomas Uebel. Cambridge: Cambridge University Press.

Howard, Don. 2000. Two left turns make a right: On the curious political career of North American philosophy of science at midcentury. *Minnesota Studies in the Philosophy of Science* 18: 25–93.

Iven, Mathias. 2008. *Moritz Schlick. Die frühen Jahre (1881–1907)*. Berlin: Paregra.

Joergensen, Joergen. 1951. *The development of logical empiricism*. Chicago: University of Chicago Press.

Kaila, Eino S. 1930. *Der logistische neupositivismus. Eine kritische studie*. Turku: Annales Universitatis Fennicae Aboensis.

Kitcher, Philipp. 2001. Carl Hempel. In *A companion to analytic philosophy*, ed. A.P. Martinich and D. Sosa, 148–159. Oxford: Blackwell.

Leitko, Hubert. 1987. *Wissenschaft in Berlin*. Berlin: Dietz.

Leitko, Hubert. 1998. Wissenschaft in Berlin um 1930. In *Hans Reichenbach: Philosophie im Umkreis der Physik*, ed. Hans Poser and Ulrich Dirks, 139–155. Berlin: Akademie Verlag.

Lewin, Kurt. 1920. *Die verwandtschaftsbegriffe in biologie und physik und die darstellung vollständiger stammbäume*. Berlin: Bornträger.

Lewin, Kurt. 1925. Über Idee und Aufgabe der Vergleichenden Wissenschaftstheorie. *Symposion* 1: 61–93.

Luchins, Abraham S., and Edith H. Luchins. 2000. Kurt Grelling: Steadfast scholar in a time of madness. *Gestalt Theory* 22: 228–281.

Mayer, Verena. 1991. Die Konstruktion der Erfahrungswelt—Carnap und Husserl. *Erkenntnis* 35: 287–304.

McMullin, Ernan. 1970. The history and philosophy of science: A taxonomy. *Minnesota Studies in the Philosophy of Science* 5: 12–67.

Milkov, Nikolay. 2002. The joint philosophical program of Russell and Wittgenstein (March–November 1912) and its downfall. *Contributions of the Austrian Wittgenstein Society* 10: 60–62.

Milkov, Nikolay. 2005. Russell studies in Germany today. *The Bertrand Russell Society Quarterly* 125(6): 35–47.

Milkov, Nikolay. 2011. Anmerkungen des Herausgebers. In *Ziele und Wege der heutigen Naturphilosophie: Fünf Aufsätze zur Wissenschaftstheorie*, ed. Hans Reichenbach and N. Milkov, 147–158. Hamburg: Felix Meiner.

Moran, Dermot (ed.). 2008. *The Routledge companion to twentieth century philosophy*. London: Routledge.

Mulder, Henk. 1968. Wissenschaftliche Weltauffassung: Der Wiener Kreis. *Journal of the History of Philosophy* 6: 386–390.

Neurath, Otto. 1929. The Vienna circle of the scientific conception of the world. In *Empiricism and sociology*, ed. Marie Neurath and Robert S. Cohen, 301–318. Dordrecht: Reidel.

Neurath, Otto. 1930. Historische Anmerkungen. *Erkenntnis* 1: 311–314.

Neurath, Otto. 1931. Physikalismus. *Scientia* 50: 297–303.

Neurath, Otto. 1932. Protocol sentences. In *Logical positivism*, ed. Alfred J. Ayer, 199–208. New York: Free Press.

Neurath, Otto. 1936. Die Entwicklung des Wiener Kreises und die Zukunft des logischen Empirismus. In *ibid.*, *Gesammelte philosophische Schriften*, 2 vols., ed. Rudolf Haller et al., 673–702. Wien: Hölder–Pichler–Tempsky.

Oppenheim, Paul. 1926. *Die natürliche Ordnung der Wissenschaften: Grundgesetze der vergleichenden Wissenschaftslehre*. Jena: Fischer.
Padovani, Flavia. 2008. Probability and causality in the early works of Hans Reichenbach, Ph.D. dissertation, Geneva: University of Geneva.
Peckhaus, Volker. 1990. *Hilbertprogramm und kritische Philosophie: das Göttinger Modell interdisziplinärer Zusammenarbeit zwischen Mathematik und Philosophie*. Göttingen: Vandenhoeck & Ruprecht.
Peckhaus, Volker. 1994. Von Nelson zu Reichenbach: Kurt Grelling in Göttingen und Berlin. In *Hans Reichenbach und die Berliner Gruppe*, ed. Lutz Danneberg et al., 53–86. Braunschweig: Vieweg.
Peckhaus, Volker. 2003. The pragmatism of Hilbert's programme. *Synthese* 137: 141–156.
Petzäll, Åke. 1931. Logistischer Positivismus: Versuch einer Darstellung und Würdigung der philosophischen Grundanschauungen des sog. Wiener Kreises der wissenschaftlichen Weltauffassung. *Göteborgs högskolas årsskrift* 37: 3. 36 pages.
Poser, Hans, and Ulrich Dirks (eds.). 1998. *Hans Reichenbach: Philosophie im Umkreis der Physik*. Berlin: Akademie Verlag.
Putnam, Hilary. 1991. Reichenbach's Metaphysical picture. *Erkenntnis* 35: 61–75.
Reichenbach, Hans. 1914. Zum Lietzschen Vortragsabend. *Göttinger Akademische Wochenschau* 10:5 (12.06.), 38.
Reichenbach, Hans. 1920. *The theory of relativity and a priori knowledge*. Trans. Maria Reichenbach. Berkeley: University of California Press.
Reichenbach, Hans. 1928a. *Philosophie der Raum–Zeit Lehre*. Berlin: de Gruyter.
Reichenbach, Hans. 1928b. [Russell:] An early appreciation. In *Bertrand Russell: Philosopher of the century*, ed. Ralph Schoenman, 129–133. London: Allen & Unwin.
Reichenbach, Hans. 1929. Bertrand Russell. In *Obelisk Almanach*, 82–92. Berlin and Munich: Drei-Masken Verlag.
Reichenbach, Hans. 1930. Zur Einführung. *Erkenntnis* 1: 1–3.
Reichenbach, Hans. 1931a. Zum Anschaulichkeitsproblem der Geometrie. *Erkenntnis* 2: 61–72.
Reichenbach, Hans. 1931b. *Ziele und Wege der heutigen Naturphilosophie*. Leipzig: Felix Meiner.
Reichenbach, Hans. 1936a. Logical empiricism in Germany and the present state of its problems. *The Journal of Philosophy* 33: 141–160.
Reichenbach, Hans. 1936b. Ansprache bei der Begrüßung der Pariser Kongresses. In *Actes de congrès international de philosophie scientifique. Paris 1035, Tome I: Philosophie scientifique et empirisme logique*, 16–18. Paris: Hermann.
Reichenbach, Hans. 1938. *Experience and prediction*. Chicago: University of Chicago Press.
Reichenbach, Hans. 1947. *Elements of symbolic logic*. New York: The Free Press.
Reichenbach, Hans. 1951. *The rise of scientific philosophy*. Berkeley: University of California Press.
Reichenbach, Hans. 1953. *Der Aufstieg der wissenschaftlichen Philosophie*. Trans. Maria Reichenbach. Berlin: Herbig.
Reichenbach, Hans. 2006. *Defending Einstein: Hans Reichenbach's writings on space, time, and motion*, ed. Steven Gimbel and Anke Walz. Cambridge: Cambridge University Press.
Rescher, Nicholas. 1966. *The logic of commands*. London: Routledge.
Rescher, Nicholas. 2005. The Berlin school of logical empiricism and its legacy. In *Studies in 20th century philosophy*, ed. Nicholas Rescher, 119–148. Ontos: Frankfurt.
Salmon, Wesley. 1999. The spirit of logical empiricism: Carl G. Hempel's role in twentieth-century philosophy of science. *Philosophy of Science* 66: 333–350.
Salmon, Wesley. 2001. Logical empiricism. In *A companion to the philosophy of science*, ed. W.H. Newton-Smith, 233–251. Oxford: Blackwell.
Sarkar, Sahotra, and Pfeifer Jessica. 2006. The philosophy of science: An introduction. In *The philosophy of science: An encyclopedia*, ed. Sahotra Sarkar and Pfeifer Jessica, xi–xxvi. London: Routledge.
Sarkar, Sahotra, and Jessica Pfeifer (eds.). 2006. *The philosophy of science: An encyclopedia*. London: Routledge.

Schernus, Wilhelm. 1994. Alexander herzberg: Psychologie, medizin und wissenschaftliche philosophie. In *Hans Reichenbach und die Berliner Gruppe*, ed. Lutz Danneberg et al., 33–51. Braunschweig: Vieweg.
Schlick, Moritz. 1918. *Allgemeine Erkenntnislehre*. Berlin: Springer.
Schlick, Moritz. 1930. Die Wende der Philosophie. *Erkenntnis* 1: 4–11.
Simons, Peter. 1997. Review of Kush, *Psychologism. The British Journal for the Philosophy of Science* 48: 439–443.
Stadler, Friedrich. 1991. Aspects of the social background and position of the Vienna circle at the University of Vienna. In *Rediscovering the forgotten Vienna circle*, ed. Thomas Uebel, 51–77. Dordrecht: Kluwer.
Stadler, Friedrich. 1997. *Studien zum Wiener Kreis*. Frankfurt: Suhrkamp.
Stölzner, Michael. 2001. Die Kausalitätsdebatte in den Naturwissenschaften. Zu einem Milieuproblem in Formans These. In *Wissenschaft: Transformationen im Verhältnis von Wissenschaft und Alltag*, ed. H. Franz, 85–128. Bielefeld: Institut für Wissenschafts- und Technikforschung.
Stölzner, Michael. 2002. How metaphysical is 'deepening the foundations'?—hahn and frank on Hilbert's axiomatic method. *Vienna Circle Institute Yearbook* 9: 245–262.
Stumpf, Carl. 1892. Über den Begriff der mathematischen Wahrscheinlichkeit. *Sitzungsberichte der philosophisch-philologischen und historischen Classe der Königlich Bayerischen Akademie der Wissenschaften* 20: 37–120.
Thiel, Christian. 1993. Carnap und die wissenschaftliche Philosophie auf der Erlanger Tagung 1923. In *Wien–Berlin–Prag Der Aufstieg der wissenschaftlichen Philosophie*, ed. Rudolf Haller and Stadler Friedrich, 175–188. Wien: Hölder–Pichler–Tempsky.
van Fraassen, Bas. 2002. *The empirical stance*. New Haven: Yale University Press.
von Aster, Ernst. 1913. *Prinzipien der Erkenntnislehre*. München: Quelle & Meyer.
Wittmann, Simone. 1998. *Das Frühwerk Kurt Lewins*. Frankfurt: Peter Lang.

Chapter 2
The Berlin Group and the USA: A Narrative of Personal Interactions

Nicholas Rescher

I am grateful for this opportunity to survey the history of my interactions and collaborations with members of the Berlin Group. The contacts I developed in this context had a significant formative influence on course of my work and on the development of my career.

As a freshman at Queens College in the spring of 1946 I enrolled in a course in Philosophy of Science with Professor C.G. Hempel of whom I already knew something via a recent student of his, Charlotte Knag, my geometry teacher at Flushing High School. During my Queens College days I took all the courses with Hempel that I could and formed a friendly relationship with him.

Upon completing undergraduate studies at Queens, I went on to do graduate work at Princeton. Here Hempel had put me in touch with his good friend and collaborator Paul Oppenheim, who had been settled there for about a decade. At Hempel's suggestion, Oppenheim enlisted me as a collaborator. He had long been interested in the concept of Gestalt—i.e., in the theory and application of the concept of structure in psychology, cognitive theory, and the theory of science. Years before, Oppenheim had cooperated with Kurt Grelling in investigating this topic but their research had been interrupted by the disaster being visited upon all concerned in Europe by the Nazis. My own collaboration with Oppenheim issued in a joint piece on "Logical Analysis of Gestalt Concepts" published in the *British Journal for the Philosophy of Science* in 1955 (Rescher and Oppenheim 1955). However, further collaboration with Oppenheim was aborted by my brief period of military service during the Korean War.

Upon discharge from military service in 1954 I went to work in the Mathematics Division at the RAND Cooperation in Santa Monica, California. They hired me at Hempel's suggestion and I was assigned to a group lead by his longtime friend and

N. Rescher (✉)
Department of Philosophy, University of Pittsburgh, Pittsburgh, PA, USA
e-mail: rescher@pitt.edu

N. Milkov and V. Peckhaus (eds.), *The Berlin Group and the Philosophy of Logical Empiricism*, Boston Studies in the Philosophy and History of Science 273, DOI 10.1007/978-94-007-5485-0_2,

collaborator Olaf Helmer. Equipped with a lively sense of humor and an inquisitive, open-minded spirit, Helmer was an easy-going individual. Notwithstanding his seniority to me in position and in age (by 18 years), we got on smoothly as equal partners working in cooperation.

While my service at RAND was largely oriented in other directions, I continued to work at some problems that had been engaging the attention of the Berlin Group. In particular, during my time at RAND I was drawn under Helmer's influence to issues of prediction and futurology. I worked with him and our colleague Norman Dalkey in developing the so-called Delphi-Method of expert-interactive prediction.[1] (All three of us held Ph.D.'s in philosophy which may have increased our sympathy for eccentric approaches.)[2]

Because our interests in futuristics went beyond relevancy to RAND's prime concerns, Helmer and I also pursued our collaboration by periodic evening work sessions held alternatively at our homes. During 1955 we met periodically in the evening or on weekends, either at Helmer's house on Mandeville Canyon Road or at my house on Bestor Boulevard in Pacific Palisades. Eventually published in 1958 as RAND paper P-1513 entitled "On the Epistemology of the Inexact Sciences," this paper was re-issued in 1960 as RAND Report R-353 having meantime been published under the same title in *Management Sciences*, vol. 6 (1959), pp. 25–52. In their book on *The Delphi Method: Techniques and Applications* (Reading, MA: Addison Wesley, 1975), the editors Harold A. Linstone and Murray Turoff characterized this Helmer-Rescher publication as "a classic paper which was very adequate for the typical technology forecasting applications for which Delphi has been popular" (p. 15).[3] This line of work inaugurated a longstanding interest in forecasting on my part, which then resulted in article publications and ultimately in my 1998 book on *Predicting the Future* (Rescher 1957, 1998).

During this period I also carried on some investigations on inductive reasoning and the theory of confirmation—a topic which, under the influence of Hans Reichenbach, had preoccupied Hempel and Helmer since the early 1940s. This interest led in the short run to my paper on "A Theory of Evidence" (Rescher 1958) and in the long run to my 1980 book on *Induction* (Rescher 1980). My interest in matters of confirmation and induction were further stimulated by occasional meetings at Rudolf Carnap's Santa Monica house where Helmer and I and the mathematician L.J. Savage discussed issues relating to the theory of probability and

[1] "The Methodology of the Inexact Science." initially circulated as a RAND paper in the mid 1950s, it was subsequently published as Helmer and Rescher (1959).

[2] The method drew inspiration from the study by Kaplan et al. (1950). It was initially explained in Helmer and Rescher (1959) (This article reprints an internal Rand Corporation paper of 1958, and was the earliest discussion on Delphi published in the open literature.)

[3] A comprehensive bibliography of Delphi-relevant writings is given in Linstone and Turoff (*op. cit.*), 575–605. Of the six items published prior to 1963 specifically dealing with Delphi, I am the author or co-author of three—that is, half of them. For further references see p. 299 ff. of Cooke (1991) (Chapter 11, entitled "Combining Expert Opinion" is particularly relevant). Good discussions of Delphi are also found in Martino (1972).

induction. Carnap was by far the most senior of us, but I was impressed by the extent to which he too was open-minded and undogmatic, willing to exchange ideas with his juniors and even prepared to learn things from them.

I want to seize this opportunity to set the record straight on one point. In the course of the Vietnam War, the RAND Corp. became one of the bogeymen to the Liberal Left in America—a symbol of all that was reprehensible about intellectuals in the service of a war-mongering establishment. (When I lectured in Oxford in the spring of 1974, one young participant told me of an American colleague who refused to attend my lectures because he heard that I had once worked for RAND!) The fact, however, is that what RAND's studies in those days primarily showed was the effective impracticability of waging a nuclear war. All those war-gaming simulations pointed towards the desirability of war-avoidance. The prime thrust of RAND's work in those days was entirely defensive, the prime object being security of the US through an effective defense rather than encompassing any aggressive measures. To the best of my knowledge and recollection, not a single study done at RAND in those days was devoted to matters of aggrandizement and power projection. The large and important contributions that RAND mad to American military preparedness in those critical times of the 1950s were entirely concentrated on matters of defense.

And RAND was also unjustly charged with further malfeasance. In 2005 George A. Reisch published a book entitled *How the Cold War Transformed Philosophy of Science* (Reisch 2005. See also Mirowski 2005; Abella 2008; Hounshell 1997; Steinmetz 2005). Its principal theses include three contentions.

- "An unlikely combination of intellectual and political forces taking root in Cold War anticommunism shaped ... the research undertaken by leading philosophers." (Author's descriptive statement.)
- "These intellectual and political forces were concentrated in and around the RAND Corporation. The significance of these personal connections among RAND, operations research, and logical empiricism ... is that they shaped the [philosophical] profession's view of itself and its public profile during and after the 1950s." (p. 351)
- The science-oriented world-view of the RAND-involved philosophers impelled the subject and practitioners away from humanistic concerns—with unfortunate effects. "Had the profession ... encouraged its brightest lights to supplement their technical work in philosophy with analysis of public issues and debates ... one cannot but wonder whether ... a more ... informed public and possibly a more peaceful, economically stable, and just world would not seem as naïve and dreamlike as they do today." (p. 388)

Like most conspiracy theories, this view of RAND as an evil democracy-undermining malignly anticommunist octopus sending its ideologically corrupting tentacles out into the wider society overlooks some crucial thesis-contravening facts. Granted, RAND employed a number of philosophy-trained individuals in those days—rare among government-sponsored think-tanks. But most of them were, like myself, refugee immigrants and—

- It was not cold-war anticommunism that impelled those philosophers drawn to what Reichenbach called "logical empiricism" to prioritize science, but the prominence of science in the Weimar Germany in which most of them had their cultural foundation.
- Their opposition to and distrust of communism was not the product of the McCarthyite hysteria of the 1950s but antedated it as part of their revulsion to totalitarianism in general as product of their experience in Nazi Germany.
- Their emphasis on value-free science was not the product of an opposition to values but rather to the infusion of cognitively extraneous values into scientific inquiry. They saw the projects of Hitler's "Aryan Science" and Stalin's "Communist Science" as corruptions against which science proper—ideologically unfettered objective inquiry—should be protected.
- These philosophers at RAND were not drawn there by the military-geared mission of the organization, but through a personal relationship. For Olaf Helmer came to RAND by chance (via his wartime encounter with John Williams, who directed the Mathematics Division), and the others became drawn to RAND through longstanding connections with Helmer, who saw this as a mode of mutual aid to fellow refugees as well as access to a posse of unusually talented theoreticians.
- The RAND connection did not impel these philosophers into scientistically antihumanistic allegiance. Insofar as they has such views—and mostly did not—these long antedated the RAND connection and were formed independently of it.
- As a mere drop in a large ocean, those RAND-connected philosophers—some dozen in all out of a total of six thousand or so—did not and could not determine the thought-orientation of the profession at large. Apart from the inherent diversity of the group itself, there is the fact that it was but a dozen individuals in a profession of several thousand idiosyncratic and independent thinkers. They were only one tree in a vast forest of speculation. Moreover—
- Given the cultural isolation of academic philosophy in the U.S., it is clear that even if (per impossible) the philosophical profession at large had prioritized social reform over rational inquiry—adopting Marx's injunction that the task of philosophy is not to understand the world but to change it—this would not and could not produce "a more peaceful, economically stable, and just world." Given the marginal status of philosophy in American life—or for that matter the world at large—the idea of such a massive and monumental potential is ludicrous.

But in any case, what rendered the philosophy of science of the 1950s and 1960s a technical, apolitical, and ideologically aseptic enterprise was not anti-Marxism or anti-communism, but a revulsion against totalitarian efforts—mostly by Nazis fascists and Stalinist communists—to enlist science in their totalitarian cause. These people felt that science should be done for the sake of understanding and not of politics. "Science should be done scientifically, so let's keep political predilections and personal ideologies out of it" was effectively their motto. Rather than rejecting human values, they saw unfettered inquiry itself as a prime value.

I myself left RAND at the end of 1956 to take up a professional post at Lehigh University, drawn there by another former Hempel student, Adolf Grünbaum. And after some 20 years at RAND, Helmer left in 1968 to join with Theodore Gordon and Randites Paul Baran and Arnold Kramish in founding The Institute for the Future—a futurology think-tank—whose Eastern branch, located in Middletown Connecticut, he headed for a time. (I was offered a post there, but could not see my way clear to leave my philosophy professorship.) However, neither of us left RAND because of any ideological estrangement.

After his mandatory retirement at Princeton in 1976 Hempel, took up a post-retirement appointment at the University of Pittsburgh, whose Philosophy Department faculty I had joined in 1961. For a decade, until his second and final retirement in 1985 we were colleagues. While we never conducted any active collaboration we were on the best of collegial terms—indeed I edited a Festschrift for his 75th anniversary. And we worked in active cooperation to sustain the life of the Department. My own interests during these years included issues in metaphysics and perspectives on philosophical pragmatism, and I am not sure that Hempel altogether approved such radical departures from the tenor of his own earlier sympathies. It is one of the regrets of my life that I did not during these busy years take steps to bring this matter to a more decisive resolution. The reminder of most opportunities is the disadvantage of retrospection. His unusual longevity has made it possible to keep up an occasional friendly contact with Helmer over the years—unfortunately only via the mails, given the extent of geographical separation.

What impressed me deeply throughout my interaction with Oppenheim, Hempel, and Helmer was the inclination of these émigré members of the Berlin Group to look on one another in a profoundly collegial perspective. Our interaction was not just that of investigators sharing a common interest but that of members of a family concerned to support each other in their careers and their professional lives. We were all refugees from Nazi Germany and having shared a difficult past were for this reason (so I think), inclined to the idea of making the present as smooth as possible for one another.

In any case, the confluence of former Hempel students at Pittsburgh and the prominence of its Center for Philosophy of Science can be seen as the prime legacy of the Berlin Group in the USA.[4]

References

Abella, Alex. 2008. *Soldiers of reason: The RAND corporation and the rise of the American Empire*. New York: Harvard.

Cooke, Roger M. 1991. *Experts in uncertainty*. New York: Oxford University Press.

---

[4]The Center members who had some training under Hempel include: John Earman, Adolf Grünbaum, Gerald Massey, and Nicholas Rescher. Earman apart, each of us served as Center director for a period of years.

Helmer, Olaf, and Nicholas Rescher. 1959. On the epistemology of the inexact sciences. *Management Science* 6: 25–52.
Hounshell, David A. 1997. The Cold War, RAND, and the generation of knowledge, 1946–1962. *Historical Studies in the Physical and Biological Sciences* 27: 237–267.
Kaplan, Abraham, A.L. Skogstad, and M.A. Girshik. 1950. The prediction of social and technological events. *Public Opinion Quarterly* 14: 93–110.
Martino, Joseph P. 1972. *Technological forecasting for decision making*. New York: American Elzevier.
Mirowski, P. 2005. How positivism made a pact with the postwar social sciences in the United States. In *The politics of the politics of method in the human sciences: Positivism and its epistemological others*, ed. George Steinmetz, 142–172. Durham: Duke University Press.
Reisch, George A. 2005. *How the Cold War transformed philosophy of science*. Cambridge: Cambridge University Press.
Rescher, Nicholas, and Paul Oppenheim. 1955. Logical analysis of Gestalt concepts. *The British Journal for the Philosophy of Science* 6: 89–106.
Rescher, Nicholas. 1957. On prediction and explanation. *The British Journal for the Philosophy of Science* 8: 83–94.
Rescher, Nicholas. 1958. A theory of evidence. *Philosophy of Science* 25: 87–94.
Rescher, Nicholas. 1980. *Induction: An essay on the justification of inductive reasoning*. Oxford: Blackwell (German tr. 1987. *Induktion*: *Zur Rechtfertigung des Induktiven Schliessens*. München: Philosophia Verlag).
Rescher, Nicholas. 1998. *Predicting the future: An introduction to the theory of forecasting*. Albany: State University of New York Press.
Steinmetz, George. 2005. *The politics of method in the human sciences: Positivism and its epistemological others*. Durham: Duke University Press.

Olaf Helmer with Dorothy and Nicholas Rescher at the University of Pittsburgh, 1963

Part II
Historical–Theoretical Context

Chapter 3
J. F. Fries' Philosophy of Science, the New Friesian School and the Berlin Group: On Divergent Scientific Philosophies, Difficult Relations and Missed Opportunities

Helmut Pulte

Vor dem Irren aber, so glauben wir,
schützt einzig und allein das Nichtdenken.

(Walter Dubislav, *1922*)

3.1 Fries' Development of Kant's Philosophy of Science

Fries never shied from admitting his indebtedness to Kant's approach and explicitly subordinated his own thought to the core elements of that framework. Specifically, "Kant's distinction of analytic and synthetic judgements, the fundamental question of how synthetic judgements a priori are possible, the discovery of the transcendental guideline and the system of categories and ideas, the discovery of pure intuition, and finally the implementation of the doctrines in his critiques" (Fries 1967–2011, vol. 29, 808).

If one aims at characterising Fries' own philosophical work—especially with regard to the New Friesian School and the Berlin Group—one is well advised to distinguish between two of its key facets: The first aspect, although inventive, is highly contested with respect to its philosophical method. However, it is without serious implications for the general understanding and estimation of science. The second aspect while having been widely neglected during Fries' lifetime, is also

---

This paper is the largely extended version of a talk given at the workshop "Die Berliner Gruppe" (Paderborn, September 3–5, 2009). I would like to thank the participants for constructive discussions and Janelle Pötzsch for polishing the English of this paper.

H. Pulte (✉)
Institute of Philosophy, University of Bochum, Bochum, Germany
e-mail: Helmut.Pulte@rub.de

N. Milkov and V. Peckhaus (eds.), *The Berlin Group and the Philosophy of Logical Empiricism*, Boston Studies in the Philosophy and History of Science 273, DOI 10.1007/978-94-007-5485-0_3,

highly inventive. What is more, it is quite progressive as regards the philosophy of science and mathematics.

The first aspect of Fries' work regards his anthropological criticism of reason (Fries 1828–1831). Herein, he aimed to dispel what he called Kant's 'transcendental prejudice,' i.e. the view that even our *a priori* knowledge is in need of proof (which Kant tried to provide via a 'transcendental deduction' concerning the categories). According to Fries, we can justify the basic judgements of our cognition neither by transcendental or logical deductions, nor by demonstrations based on pure intuition. Instead, we have to make them explicit via a reflective introspection of reason. In order to achieve this 'demonstration' (*Aufweisung*), he suggested a regressive method of analysis of inner experience via reason, which is said to lead to (and at the same time, make aware) our basic judgements. Somewhat misleadingly, he called this procedure 'deduction,' and demarcated it from both the proof via first principles in propositional form as well as from demonstrations by intuition.

Since demonstrations are psychological procedures of introspection, Fries was often criticised for defending psychologism, in the sense of a reduction of philosophical judgements to empirical psychology. Kuno Fischer, for instance, famously phrased it like this: "Whatever is *a priori* can never be recognized *a posteriori*" (Fischer 1862, 99). But, in fact, Fries aimed at a psychological method of demonstration, not at empirical justification of a priori knowledge. As such, it is quite misleading to label him a psychologist (Sachs–Hombach 1999).

Fries' theory of justification by proof, demonstration, and deduction became pivotal for the science-orientated New Friesian School,[1] though it had no direct consequences for foundational issues of the 'exact' sciences. For, his psychological demonstrations did not develop any modification as regards the synthetic principles that are a priori of mathematics and the theory of motion. Moreover, he considered both Euclidian geometry and Newtonian mechanics to be sufficiently substantiated by these principles, though he gave them a methodological meaning that offered some opportunities for the later development of physics.[2]

Now I would like to explore the second aspect of Fries' philosophy, mentioned above, which often seems remarkably modern and is to be found 'below' the indicated level of a priori foundation. One might describe this project as further developing Kant's philosophy of science in a methodological and empirical direction. Such thoughts are less prominent in his major philosophical works than in his *Mathematische Naturphilosophie* (Fries 1822), in his books on logic (e.g. Fries 1837) and in several of his textbooks on the natural sciences. Here, Fries took significant steps to develop Kantian theory, out of a desire to reconcile it with the sciences of his times.

[1]See esp. Dubislav (1926a, 1929), Eggeling (1904), Grelling (1907), Kastil (1918), Nelson (1904, 1962).

[2]This is an important aspect with respect to special relativity to which I will come back later (see Sect. 3.4).

Fries was a philosopher with an excellent knowledge of mathematics and the natural sciences,[3] and he knew very well that Kant's *First Critique* and his *Metaphysical Foundations of Natural Science* only provided a philosophical foundation for a small area of mathematics and 'science proper.' For example, Kant never seriously undertook the philosophical analysis or justification of calculus, of formal algebra, of the theory of probability, or of analytical mechanics. Indeed, as is well known, he even relinquished the idea that chemistry could acquire the status of a proper science.

Fries' strategy was to extend Kant's approach to these 'new' sciences in two different manners. On the one hand, he developed a methodology of the empirical sciences that cast Kant's synthetic principles as a priori heuristic guidelines (*Maximen*) of empirical investigation, in areas where their constitutive character was by no means obvious. Here, he could tie in with Kant's analogies of experience of the first *Critique* and in the *Critique of Judgement*. On the other hand, he 'stretched' Kant's idea of science as a deductive *system* by disentangling the concepts of 'system' and 'theory.' For, while he held that there is only *one* system of scientific knowledge that stands as a regulative ideal, in Kant's sense, Fries thought that different empirical theories (sciences) governed by different 'local' principles are possible. In his 'philosophy of mathematics'—a term seemingly introduced by Fries[4]—he likewise extended the area of 'proper knowledge' gained by reason from the construction of concepts: He broadened Kant's understanding of mathematical knowledge by introducing 'productive imagination' (*productive Einbildungskraft*) as a foundational instance. Consequently, he asserted that syntax, i.e. the theory of pure laws of arrangement, should be considered as part of mathematics on equal footing with arithmetic (Fries 1822, 64–65; cf. Bernays 1933, 109).

Both these facets of Fries' new architecture of the philosophy of mathematics were representative of the actual mathematical developments of his time, which were coined not so much by geometry or (synthetic) mechanics as by formal arithmetic, algebra and 'analytical' mathematical physics. This is all the more important as Fries not only aimed to supply a broader foundation for 'pure' mathematics. He also thought that such a general foundation (i.e. beyond Euclidean geometry and elementary arithmetic) could stand as a source for fruitful hypothesis-building in the realm of the empirical sciences.

In what follows I will elucidate a number of the specific achievements of Fries' philosophy of science and mathematics. However, as these accomplishments are

[3]Besides the favorable statements on his abilities by the mathematician Carl Friedrich Gauß, the theoretical physicist Wilhelm Weber and others (cf. König and Geldsetzer 1979) one can appeal also to the naturalist Alexander von Humboldt: "Fries, in his mathematical–philosophical orientation, is a beneficence for Germany" (Henke 1937, 256).

[4]For a detailed historical report see König and Geldsetzer (1979, 45), and Pulte (1999a, 71–76).

described and analysed elsewhere in some detail,[5] I shall confine myself to those results and consequences for the 'exact sciences' which I consider relevant for the work of the Neo-Friesian School, and for their relationship to the Berlin Group:

(i) On the basis of an objective conception of probability, Fries offered the first philosophical analysis that sets out the legitimate area of application for probability statements (Fries 1842; see Fischer 2004). Via E. F. Apelt, J. von Kries and others, this approach gained some influence on the later discussion on probability (Grelling 1910; Reichenbach 1916, 1932). Experts of that time saw in Fries "the most consistent moulder" of objective probability (Sterzinger 1911, 52).

(ii) According to Fries, an indispensable task of any philosophy of mathematics is what has come to be described as 'critical mathematics.' This endeavor became an integral part of Hilbert's program of meta-mathematics: a philosophical justification (*Deduktion*) of the first mathematical principles or axioms. Without any doubt, this part of Fries' program—perpetuated by L. Nelson, G. Hessenberg, O. Meyerhof and others—was the most important one with respect to acceptance in the philosophical–mathematical community. Its influence on Hilbert's axiomatics—irrespective of manifest divergences—is obvious and well documented (Peckhaus 1990, 1999). Within the New Friesian School this topic probably allowed the most direct and intense recourse to Fries' original approach (see esp. Hessenberg 1904, 1907; Nelson 1905b, 1906, 1927; Grelling and Nelson 1908; Bernays 1930).

(iii) In his theory of rational induction, Fries relinquishes Kant's ideal of a *system* of experience in favour of a multiplicity of theories. A system continues to exist as a *synthetic a priori* foundation for mechanics. However, a multitude of theories is possible within this system, whose heuristic maxims may have a *constitutive* function (see Pulte 1999b). The theory of electricity or magnetism, for example, may have its own maxims that can gain constitutive relevance. This means that those maxims are—as candidates for general laws of nature—related to the mechanical laws of motion only in a weak sense of compatibility. As such, separate scientific theories serve as theoretical backgrounds for the

[5] See Pulte (1999a, 2005a (esp. Ch. IV), and 2006). For Fries' conception of 'theory' and 'system' as well as for foundational aspects of his methodology, the *Grundriß der Logik* (Fries 1827) is most important. His philosophy of mathematics and the more applied aspects of his methodology can be found in his *Mathematische Naturphilosophie nach philosophischer Methode bearbeitet* (Fries 1822). A general estimation of his achievements in both respect is given by the excellent introduction of the Editors (König and Geldsetzer 1979). For Fries' philosophy of pure mathematics see also Schubring (1999) and Herrmann (2000, Ch. 3). His contribution to the theory of probability is analyzed in Fischer (2004). The heuristic dimension of Fries' concept of probability is meticulously analyzed in van Zantwijk (2009, esp. Ch. 5). Some philosophical implications of his perception and interpretation of analytical mechanics are investigated in Pulte (2005b). A more general evaluation of Fries' philosophy of science and the broader 'aprioristic tradition' is intended in Herrmann (2012). A comprehensive analysis of German philosophies of nature in the early nineteenth century, including Fries' approach, is Bonsiepen (1997).

acquisitions of further experience: Observation always depends on 'guiding maxims.' While this theory of rational induction played an important role in the first Friesian School (see esp. Apelt 1854), it was of minor importance for the New Friesian School.

(iv) (Limited) Fallibilism and Conventionalism: 'Below' the level of synthetic principles a priori, empirical laws can basically be revised by new experiences. New hypotheses, however, must not contradict any *a priori* principles and are to be formulated in such a way that they can be "refuted for certain by experience" (Fries 1822, 21). In addition to this 'Popperian' element, Fries also introduces a conventional element at the same level. Specifically, he holds that for a fixed sets of phenomena, several empirically equivalent explanatory laws are possible. Between those, neither experience nor reason can decide, but only considerations of simplicity and convenience. Moreover, conflicting observation *never* challenges a single law, but *all* theoretical assumptions on which the deductive explanation of this observation is based (cf. Pulte 1999b for a more detailed discussion).

(v) Theory of space and motion: Regardless of the 'modern' elements of philosophy of science, described above, Fries was a 'Kantian conservative' as regards Euclidean geometry. Other geometries deserving of this name, i. e. axiomatized theories of pure space, were out of his ken. As such, he attempted to prove Euclid's parallel axiom in order to solve the ongoing public discussion about it in favour of a 'unique' Euclidean geometry (Herrmann 2000, 132–136 and 222–232). This 'Euclidean fixation' had a lasting impact on the New Friesian School, especially on Nelson (see his 1905b, 1906, 1927), which will be discussed later. However, Fries was quite aware that using Euclidean geometry to elaborate a theory of motion is problematic. Specifically, he noted that the distinction of a straight line as the trajectory of an inertial motion is in need of merely conventional fixations (Fries 1822, 413–418). Moreover, motion in general is basically relative: "We always have to talk about *relative* spaces, which are movable und which we may find moving, without ever coming to an absolute space as, so to speak, a fixed basic form of the world" (Fries 1822, 422). In order to deal with this *problem of relativity*, we have to *postulate* certain rules, under which the construction of motion is possible (Fries 1822, 423–424). His follower, E. F. Apelt, maintains likewise that "there is no absolute space [. . .] for assessments, in experience we have to take space as *comparative* (relative)" (Apelt 1910, 554–555). As far as I can see, these considerations on space remained unnoticed in the New Friesian School, and played no role for the Berlin Group either. They are, however, interesting for their discussion about the theory of relativity to which I will come back later (see Sect. 3.4).

To sum up, Fries' achievements are considerable, but only certain aspects of his philosophy of mathematics, (i) and (ii), have received attention, while interesting aspects of his philosophy of science, (iii)–(v), remained largely unnoticed. As such, it makes sense to take a look at the reception of his philosophy from a more general point of view in order to yield a better understanding of these findings, before we discuss their implications for the relationship of the New Friesian School and the Berlin Group in more detail.

3.2 Fries Reception and Deflation: Historiographical Remarks with Regard to Berlin

While Fries' efforts to reconcile philosophy, mathematics and the sciences received positive feedback with his contemporaries, the later reception of his work was less favourable. First of all, mainly because of a politically motivated interdiction to teach, Fries himself failed to set up a philosophical school. What is more, his most eminent disciple, E. F. Apelt (1812–1859), suffered an untimely death. Therefore the (first) 'Friesian school,' spearheaded by that latter scholar, was a philosophical flash in the pan. In addition, the reception of Fries' work within academic philosophy suffered from the dominance of German Idealism (especially Hegel and his adherents), to which his philosophy was opposed. Later, Neo-Kantianism and its imperative of going straight 'Back to Kant' led to a disregard of post-Kantian developments, even if they stood in close relation to his work. For these reasons and others, mainly rooted in the problematic German historiography of philosophy and the sciences (see Pulte 1999a), Fries' attempt to bring philosophy and science together was poorly received in the later nineteenth and early twentieth century, outside of the New Friesian School. Given this background, it is hardly surprising that direct references by the Berlin Group to the work of *Fries* are—apart from Dubislav and Grelling—rare exceptions. But, even beyond those considerations, the height of that alliance (1927–1933) was a century removed from the publication of Fries' most relevant contributions to the philosophy of science, and its disinterestedness in (or even hostility to) historical research (cf. Hentschel 1991, 34) made such a reach back in time out of the question.

Reichenbach's early leanings towards Kant's a-priorism are well known, and his perspective of the post-Kantian development is quite similar to that of many Neo-Kantians. Namely, that it is a period of philosophical degeneration and misunderstanding of science. This attitude is still visible in his late book on *The Rise of Scientific Philosophy* (more a book of historical fairytales than a serious historical investigation). Therein, Fichte, Schelling, Hegel and others are disqualified as "as if philosophers" (Reichenbach 1969, 142) with no affiliation to science. Whereas, Fries is not even mentioned.

On Reichenbach's approach, the legitimate follower of Kant is not the 'Kantianism' of academic philosophy, but philosophy following a "method of analysing science" (Reichenbach 1920, 71) that is applied to the latest achievements of science. As he states, "(o)ne should proceed with the history of philosophy, which attired herself in systems until Kant, not with the pseudo-systems of epigones, but with a new philosophy which originated from the *science* of the nineteenth century and has been further developed in the twentieth century."[6] Thus, he simply did not consider Fries a congenial philosopher with closely related aims and interests. Rather, it seems that he referred to Fries only once, albeit positively. In his *Elements of Symbolic Logic* Reichenbach stated that with respect to Fries' *New Critique of Pure Reason*: "(t)he fact that a proposition stating that a formula is logically necessary is in itself an empirical statement seems to have been first pointed out by J. F. Fries [. . .]" (Reichenbach 1947, 188). Also, in his dissertation on probability, he did refer at least to the objectivistic concept of probability of E. F. Apelt, J. von Kries and K. Grelling, who again referred to Fries (Reichenbach 1916, 215–223).

The consultation of the works of other members of the Berlin Group like Carl Gustav Hempel, Alexander Herzberg, Wolfgang Köhler or Kurt Lewin yields an equally disillusioning picture. At least Richard von Mises, in his *Kleines Lehrbuch des Positivismus*, allowed Fries an earnest endeavour of advancing Kant's theory "in a scientific sense." However, he surprisingly asserts that Fries "tried to constitute the Apriori psychologically by some sort of analysis of feelings of evidence—which is very close to our viewpoint" (von Mises 1939, 391). This is startling, since von Mises was not even too close to his own viewpoint in this systematic misjudgement.

As already mentioned, Dubislav and Grelling had a different attitude towards Fries. Grelling left the New Friesian School in 1922, after an argument with Nelson about Einstein's theory of relativity (Peckhaus 1990, 148; cf. Sect. 3.4). Whereas Dubislav had probably come into contact with Fries' and Nelson's philosophy during his studies of mathematics (inter alia with Hilbert) in Göttingen from 1914 onwards. Given their exposure, Dubislav (see sep. his 1926a, b, 1929) and Grelling (see esp. 1906, 1907, 1910) published on Fries and Nelson. Indeed, Grelling even published *with* the latter, on the topic of logic (e.g. Grelling and Nelson 1908).

Both Dubislav and Nelson later belonged to the "founding generation" (Rescher 2006, 282) of the Berlin Group, and were quite active members (Danneberg and Schernus 1994; Hoffmann 1994; Peckhaus 1994). Otto Neurath's short description of the Berlin Group mentions that Reichenbach, Dubislav and Grelling "focused primarily on logical and physical problems as starting points of epistemological critique (toeholds in Kantianism and Friesianism, influence of Cassirer and Nelson)" (Neurath 1930, 390; cf. Hentschel 1991, 30). Grelling, Dubislav and (the early) Reichenbach, from 1927 onwards, counter-balanced to a certain extent the strong positivistic leanings of the group, emanating from the Mach-orientated subgroup around Joseph Petzoldt. However, Reichenbach's subsequent departure from Kant's a priori was already terminated when he got into closer contacts to Dubislav and

[6] Reichenbach (1969, 142). For him, scientific philosophy after Kant is simply a kind of 'Science as Philosophy.' Herbert Schnädelbach describes under this heading the changing relation of both areas after 1831 in a quite adequate manner: "Philosophy deserts to science to a degree that threatens its identity." (Schnädelbach 1991, 113) This strategy of defense, which can be detected in different philosophical movements of the nineteenth century, develops in Reichenbach's systematic turn of this historical development to the only legitimate form of philosophy at all. Ironically enough, Fries called for a philosophy that itself is "rigorous science (*strenge Wissenschaft*)" (Fries 1828–1831, vol. 3, 169).

Grelling. Thus, their philosophical influence on him—in terms of the mediating 'Friesian elements' described in the first section—was obviously very limited.

As such, apart from Grelling and Dubislav, the relationship between the Berlin Group and the work of Fries is mainly a history of *missed chances* (cf. also Sect. 3.3). For, while Reichenbach is more or less right in maintaining that one should not forget that the history of philosophy is "history and not philosophy" (Reichenbach 1969, 364); equally correct is the idea that a serious study of the history of philosophy can lead to interesting, maybe continuative or—to complete the augmentation in Reichenbach's sense—even original 'scientific philosophy.' But Reichenbach obviously stuck to the assumption—hardly justifiable by logic or experience—that even the most basic and seminal ideas of this philosophy depend on *present* scientific research: "He who contributes to the new philosophy does not look back, because his work would not profit from historical considerations" (Reichenbach 1969, 364).

3.3 Divergent Scientific Philosophies: The New Friesian School and the Berlin Group

It is clear that Fries' work largely failed to attract the attention of the Berlin Group, but what was the relation of its members to the New Friesian School, and what are their distinctive features? Freely adapted from Viktor Kraft, one might say that neither the Berlin Group nor the New Friesian School are 'unambiguous units' (cf. Haller 1993, 61). That is, they were not philosophically homogenous groups that can be characterized and differentiated via some rare common convictions. However, both groups were manifestations of a discontent with the academic philosophy of their time. In addition, both groups were concerned with a close collaboration of the different sciences and philosophy. In both groups one encountered, not only philosophers, but also mathematicians, natural scientists and other academics. However, this is where the similarities end. A closer look at the New Friesian School reveals substantive differences:

Nelson founded this school in 1903, when he was still a student in Göttingen. The founding members from philosophy, mathematics and other disciplines (Blencke 1978) were committed to the basic philosophical theorems of the Kant–Friesian philosophy as they were passed over by the (first) Friesian School around E. F. Apelt. From the beginning, Nelson laid claim to the philosophical and organizational leadership of the new school (Franke 1991, 66–71). Indeed, by 1904 he had already launched the *Abhandlungen der Fries'schen Schule, Neue Folge* as the mouthpiece of the new foundation. Co-edited by L. Nelson, G. Hessenberg and G. Kaiser, the *Abhandlungen* appeared with interruptions from 1904 to 1936. From the beginning it was meant to spread and develop the 'true' Kantian philosophy in the tradition of Fries and Apelt and to counter-balance the strong influence of Neo-Kantianism in the German philosophical journals of that time. In 1913, Nelson backed up the New Friesian School—a more or less informal group without institutional setting—with the *Jakob Friedrich Fries Gesellschaft*. It organized conferences and gained influential members like D. Hilbert (Peckhaus 1990, 152–154). The programmatic statements of the *Abhandlungen* and the discussions about the aims of the *Gesellschaft* allow for a rather precise appreciation of the New Friesian School and a demarcation of the Berlin Group.

In order to see how this is so, it is helpful to begin by disposing of a (possible) misunderstanding: To begin with the disposal of a (possible) misunderstanding: The attitude towards the history of philosophy seems *prima facie* quite similar and does not mark a criterion of demarcation. The commitment of the New Friesian School to Kant, Fries and Apelt[7] should be understood as a *systematic* one, not as an appeal to extensive historical research. Nelson starts his first contribution to the *Abhandlungen* with the motto: "There are scholars who hold the opinion that the history of philosophy (both old and new) itself is philosophy; these Prolegomena are not written for them" (Nelson 1904, 1). Whereas the Berlin Group states in its appeal from 1927 that it feels compelled to an empirical philosophy "on [the] basis of the experiences of the single sciences" (Hentschel 1991, 25), the systematic primacy of this earlier school of 'scientific philosophy' is the critical method in the line of Fries, especially the idea of an empirical-psychological self-introspection of human reason in order to uncover apriori-knowledge without transcendental deduction.

It has to be stressed, however, that even Nelson and other members of his school did not analyse and exhaust Fries' contributions to the philosophy of science with the accurateness it deserves: They strongly focused on his 'new' *Vernunftkritik* (cf. Sect. 3.1) and extensively analysed its epistemological implications. As such, they appreciated and developed his philosophy of (pure) mathematics (e. g. Hessenberg 1904, 1907; Nelson 1905b, 1906, 1927), and they also discussed his theory of rational induction and deduction in some detail (e.g. Nelson 1904, 1905a). However, neither Nelson nor other members of the group broached the issue of the conventionalist and fallibilist elements in Fries' philosophy of science, nor did they fully grasp his theory of space (cf. Sect. 3.1, (iii)–(v)). Because these innovative aspects of Fries' philosophy of the empirical sciences were not really reflected in the New Friesian School, their general attitude regarding the foundation of physics remained radically *conservative*, as I will subsequently show. It is this conservatism that I consider to be the main obstacle for a fruitful relation of the Berlin Group to the philosophy of the empirical sciences.

Nelson's dogmatism, which has no intellectual roots in Fries' philosophy, reveals the ambivalent role the empirical sciences played within the New Friesian School. Typifying this issue is the fact that he issued two prefaces within the first issue of the *Abhandlungen*. He begins with the manifesto of the First Friesian School, dating

---

[7]This commitment becomes most obvious from the Editor's foreword of the first issue of the *Abhandlungen* and is accompanied by a strong rejection to any other forms of Kantianism, which are charged of abandoning Kant's true critical method, being unscientific and obscurantism. They are philosophical sects which the history of philosophy will overcome as present science overcame "Patricius, Robert Fludd and Jakob Böhme. *Kant*, *Fries* and *Apelt*, however, will continue to stay next to *Keppler*, *Galilei* and *Newton*" (Hessenberg et al. 1904, xii).

from 1847, on which he then elaborates the second preface without uncovering any time-boundedness (cf. Pulte 2005a, Ch. V and VI) of this nearly 60 year old document. A few sentences from the 'old new' preface will highlight the fundamental relationship between scientific philosophy and empirical sciences to which Nelson recommitted the New Friesian School from the beginning[8]:

(a) Any philosophy which is in accordance with the exact sciences can be true, any one which is conflict with them must necessarily be wrong. [...]
(b) All knowledge of nature is inductive, it does not stem from philosophical concepts, but from experimentation and observation. [...]
(c) Induction alone would not lead to any fixed results, if it were not aided by philosophy of nature. Such philosophy of nature is and can be only the one whose mathematical principles have been developed by Neuton [sic!] and whose metaphysical basis has been clarified by Kant. Such mathematical philosophy of nature forms the background of all inductions and regulates their processes. [...] It is therefore nothing more than a delusion to believe that the inductive sciences exist independently of philosophy.

If we take these statements at face value—and the comments of the school on 'mathematical philosophy of nature' provide no reason to do otherwise—it is clear that the relation between scientific philosophy and empirical science is marked by a strong, almost necessary mutual dependence, which becomes obvious from the three points made above. *First*, scientific philosophy must not clash with the 'exact sciences'—if she does, it is to her disadvantage. So far, this is in line with the empiricist program of the Berlin Group. However, Nelson states very clearly at this early point—not yet occupied with their program, but with Positivism and Neokantianism—that according to this criterion, only the philosophy of Kant and Fries will remain due to its "*scientific method*" (Hessenberg et al. 1904, viii). *Second*, All empirical sciences are in need of observation and experimentation, and all their proper knowledge depends on rational induction. *Third*, the Kant-Friesian philosophy solely identifies the principles of Newton as the most general principles of rational induction. Accordingly, the inductive sciences, if they are to be considered as scientific, are dependent on the Kant-Friesian metaphysics of nature for justification.

This, of course, is a decisive point of demarcation between Nelson's view—the 'official doctrine' of the New Friesian School, with respect to the foundation of the empirical sciences—and the later position of their Berlin Group, mainly fixed by Reichenbach in his analysis of space and time in the succession of Einstein's theories of relativity. Nelson never revised his position from 1904—the year before the special theory of relativity emerged—in his later career. Rather, he integrated the 'double link' between scientific philosophy and a supposedly infallible science described above by means of Fries' theory of non-intuitive immediate knowledge

[8]Hessenberg et al. (1904, iv–vi); numbers added by me. The heading of the foreword is: "Vorwort der alten Folge, zugleich Vorwort der neuen Folge." See also Apelt et al. (1847, 3–5).

in a *certistic* theory of scientific knowledge. The synthetic principles of the natural sciences are to be justified by a synthetic a priori principle of rational induction. While, that principle is itself rooted in immediate a priori knowledge. Karl Popper—obviously not aware of Dubislav's relevant analysis of the foundational problem in Fries' philosophy (Dubislav 1926a)—perceptively criticised Nelson's circular reasoning in his early work *Die beiden Grundprobleme der Erkenntnistheorie* (Popper 1994, 110–114). I will not discuss the philosophical ambiguity of Popper's criticism,[9] but confine myself to what might be regarded as its 'moral' with respect to the foundations of the empirical sciences from a Friesian point of view. Specifically, it is untenable to establish the ultimate philosophical foundation of a *unique* system of knowledge by a fixed set of synthetic principles a priori—be they determined by a transcendental deduction or by empirical introspection. However, it does make sense to strive for the uncovering of first synthetic principles, by Fries' method of 'regressive abstraction,' on the basis of present scientific knowledge as a whole. The principles gained by this method are not 'absolute' but 'relative' a priori. That is, they can change in the course of the successive development of our scientific knowledge. As such, they act as heuristic directives for the application of the basic (or constitutive) concepts involved. I claim that such a 'liberalisation' follows the genuine intellectual tradition of Fries' philosophy of science, which aims indeed at a dynamical synthesis of Kantian apriorism and scientific development (cf. Sect. 3.1). Therefore, it is not by accident that philosophers from the Neo-Friesian tradition like Paul Bernays (1953, 125–131) or Stephan Körner (1979, 6–13; cf. 1970, 1984) later veered in this direction. As regards the mathematical philosophy of nature (or mechanics), this broadening fits even better with Fries' original approach, as the pure intuition of space and time does not amount to immediate knowledge in his sense (Bernays 1953, 119) and as his construction of motion does not rely on Newton's absolute space, but on relative spaces (cf. Sect. 3.1, (v)).

Reichenbach's early *Relativitätstheorie und Erkenntnis a priori* is certainly affine to this broadened Friesianism (cf. Reichenbach 1920, 1–5, 46–58), as well as—to some extent—the early discussion of the theories of relativity in the Berlin Group. However, the New Friesian School did not indicate in its *official* statements up to 1927 (the year when Nelson died and the Berlin Group was founded) any sympathy for such a course of liberalisation. Quite contrary, Nelson unflinchingly adhered to his *certism*, as regards his mathematical philosophy of nature, after the emergence of special relativity and, as far I can see, nearly until his death (cf. Sect. 3.4). In 1908 he opposed Ernst Mach's view on mechanics, as follows:

[9] On the one hand, Popper's charge of either circularity or infinite regress—in the context of his well-known trilemma of justification—falls short of the Friesian claim to achieve an demonstration (*Aufweisung*) bei introspection and *not* by a quasi-logical justification of a priori knowledge. On the other hand, the Friesians have to admit that this demonstration serves for a certain kind of justification—Nelson's claims above do make this quite obvious. However, contrary to the logical structure of Popper's criticism this justification does not aim at the truth of special propositions a priori, but at the *whole* of the transcendental perception (cf. Fries 1828–1831, vol. 2, 99–100). See Sachs-Hombach (1999) for a closer examination of Popper's criticism and why it does not do justice to Fries' method of demonstration.

"As the principles of mechanics do not stem from experience, it is only consequent when those who want to proceed empirically are converting the fundamental laws of mechanics into arbitrary assumptions, because the practicability of which is a matter of larger or lesser convenience only. However, with these [laws] they abandon any objective criteria of scientific truth and return to a pre-Galilean level of science." (Nelson 1908, 298) At the core of his adherence to (what he regards as) a 'Newtonian' foundation of the empirical sciences is his advocacy of metaphysics as an integral part of science itself. On this understanding, it is the task of true scientific philosophy to unveil this metaphysics and its fundamental role, in order to keep, so to speak, 'science itself scientific.' As Nelson writes, "(h)e who wants to eliminate metaphysics from science hands science over to a metaphysics *outside* of science—as without metaphysics no judgements are possible at all,—i.e. he unwittingly and unconsciously pays science over to mysticism. This should be considered in due time by those who regard the matter of science and enlightenment with passion" (Nelson 1908, 299). Popper's later warning addressed to Wittgenstein and the Vienna Circle sounds similar, though *he* insisted on a demarcation of metaphysics and science: "Positivistic radicalism annihilates metaphysics and along with it science" (Popper 1982, 11). And indeed, though the Berlin Group did not accentuate its anti-metaphysical bias as strongly as the Vienna Circle, Nelson's conception of scientific philosophy is quite different at this point. That is, scientific philosophy, for him, is not only about logical and methodological analysis of existent science, but also about its ineradicable metaphysics and its legitimate fundamental claims. Quite contrary, the "method of analysing science (*wissenschaftsanalytische Methode*)" of the Berlin Group was meant "to oppose consciously all claims of a philosophy which affirms an autonomous right of reason and which would like to establish a priori valid propositions which are not subject to scientific criticism" (Anonymous 1930, 72). Here we find developed the basic point of demarcation between the two scientific philosophies, the New Friesian School and the Berlin Group developed. All affinities in the areas of logic and the philosophy of (pure) mathematics notwithstanding, they had basically incompatible ideas about how the foundations of the empirical sciences should look like. This divergence takes a concrete shape and becomes most virulent with the rise of Einstein's theories of relativity—even more so as Reichenbach from 1920 to 1929 was their "busiest and most persistent defender against the most varied forms of contradictions and attacks" (Hentschel 1990, 178).

3.4 Relativity and Geometry in the New Friesian School

Basically, Nelson's adherence to Newton's mathematical principles of natural philosophy constitutes an *a priori* fixation on the space/time structure of classical mechanics. It is therefore hardly surprising that the New Friesian School's examinations of the special theory of relativity (SRT) in the *Abhandlungen* are rare and rather critical. Indeed, the general theory is ever only mentioned once, in an article of the *Abhandlungen* published after Nelson's death (Bernays 1933).

Otto Berg's paper "Das Relativitätsprinzip in der Elektrodynamik" from 1912 and Paul Bernay's paper "Über die Bedenklichkeiten der neueren Relativitätstheorie" from 1911 (published in revised form in 1918) deal with Einstein's SRT in a competent, fair and critical manner. Both accept the empirical findings and consider the technical apparatus of the special theory in some detail (Berg 1912, 336–375; Bernays 1918, 463–474). Both, however, are sceptical about to what extent the principles of Einstein's new theory really solve the fundamental problems of classical mechanics, or whether they are even mandatory in order to do so. The new concept of simultaneity poses special problems for both (Berg 1912, 376–378; Bernays 1918, 475–478). Additionally, they refer independently of each other to Walther Ritz's emission theory of light as a possible alternative with respect to Einstein's principle of the constancy of light velocity in vacuum (Berg 1912, 379; Bernays 1918, 479–481), in order to show that SRT is not a necessary consequence of the relevant empirical findings. Most importantly, both explicitly reject that philosophy has to admit basically new intuitions of space and time. Berg maintains that Einstein's principle of relativity exceeds empirical evidence and is, therefore, "a proposition that still can be confirmed or rejected. [...] The view that one has to adhere to the principle of relativity in any case cannot be derived from experience, but corresponds to a metaphysical need the warrant of which we would not like to discuss here" (Berg 1912, 382). Here, the strong suspicion becomes obvious that SRT entails 'bad metaphysics' disguised as empirical science. Bernays underlines the matter of principle in an even stronger manner and questions the legitimacy of using physical research to casting doubt on the "a priori given (*das a priori Gegebene)*" properties of space and time by a combination of experiment and new theory-building (Bernays 1918, 475). With respect to simultaneity, he develops an argument that is reminiscent of Kant's third analogy of experience (cf. Pulte 2010, 243–244), later picked up by Nelson (1962, 684–687). It concludes: "These disquisitions should be sufficient to show that, for the pure philosophical standpoint, the view that the theory of relativity entails new insights about the relation of space and time depends on a mere delusion" (Bernays 1918, 478). Though Bernays considered SRT to have significant explanatory power, especially with respect to electrodynamics, he thought that its basic principle had to be rejected, "because for the decision about the acceptability of a theory its explanatory value can only be regarded after its apriori (i. e. basically methodological) admissibility is guaranteed." As such, his basic message to the Friesian philosophers is that they need not be worried about Einstein's SRT: "The main result of these considerations is that there is no sufficient reason to doubt the hitherto existing conceptions of time and space" (Bernays 1918, 482). Other statements of the *Abhandlungen* at that time (1905–1918) are more or less mere echoes of Nelson's conservatism in this regard.[10]

Independent of the discussion on Einstein's theories, though systematically linked to his new physics, was the attitude of the New Friesian School towards non-Euclidean geometries. Nelson picked up the ongoing fundamental debate in geometry in 1905 and linked it—as did Hessenberg in the year before (Hessenberg 1904)—to Hilbert's axiomatics, in order to engross his program for the revival of Fries' critical mathematics (cf. Peckhaus 1990, 158–168). Hilbert's criteria for axiomatic systems (consistency, independence, completeness) are utilised for the 'critical project.' Nelson pursues his aim to demonstrate the superiority of a Kant-Friesian approach to geometry over other (i. e. empiricist and logicist) approaches, by and large turning the tables on his opponents. For, Kant's thesis that the axioms of mathematics have non-logical origin and that their validity does not depend on experience is *best proven* by the possibility of consistent, non-Euclidian geometries (cf. Nelson 1905b, 388 and 392). While the axioms of both Euclidean geometry and of non-Euclidean geometries are consistent, *only* the axioms of Euclidean geometry are additionally rooted in the pure intuition of space and, therefore, are apriori *and* synthetic. The consistency of axiomatic systems is neither sufficient for the truth of their axioms, nor for the existence of the matters they are meant to represent. Therefore, the main difference between the Euclidian geometry and its rivals is epistemological in nature. That is, the axioms of the former have a privileged origin in pure intuition, whereas the axioms of the latter do not.

This argumentation, in favour of a 'two-tier geometry,' was widely accepted among the remaining members of the New Friesian School approximately until Nelson's death. As such, it backed their rejection of Einstein's physics temporarily. However, after the general theory of relativity proved to be of remarkable success, such lines of argumentation became difficult to defend. In short, the employing of a non-Euclidian, i.e., epistemic inferior geometry would lead to such spectacular empirical successes was something the Neo-Friesians could hardly cope with. For the group's conservatism, as regards the discussion of space and time, became a problematic confinement for those members (or friends) of the Friesian School best versed in the 'exact sciences,' i. e. for Dubislav, Grelling and Bernays.

The year 1920 marked a turning point in the development of the group, due to the publication of Reichenbach's book *Relativitätstheorie und Erkenntnis a priori*. For, that work was dedicated to explaining how Kant's a priori might be conserved, even in the light of the theories of Einstein (Kamlah 1979, 475–477; cf. also Sect. 3.5).

---

[10]Kurt Grelling, for example, in 1907 did not doubt the philosophical justification of Newtonian mechanics (Grelling 1907, 169–171), but later changed his view. Alfred Kastil's presentation of Fries' theory of knowledge is equally 'conservative' with respect to the theory of space and time (Kastil 1918), as is Kowalewsky's analysis of Kant's treatment of the antinomies of pure reason (Kowalewsky 1918). Other references from the *Abhandlungen* might be added, though most of them are marginal as regards space and time. In general, mathematical philosophy of nature played no important role in this organ of the New Friesian School, and to a certain extent the later volumes reflect Nelson's turn to ethics and political philosophy (cf. Franke 1991).

As such, Reichenbach's approach was very appealing to several Neo-Friesians. The old demand of the Friesian tradition to disparage a philosophy which contradicts science (cf. quotation 11, point [1.] above) had to be taken seriously in the light of Einstein's challenge, and the need for a 'Kantian' philosophy that met both the old demand and the new challenge became pressing. At a meeting of the *Jacob Fries Gesellschaft*, Grelling gave a talk on the "Theory of Relativity and Critical Philosophy" in which he sided with Reichenbach; the minutes reveal that he was blamed because of his sharp antithesis to critical philosophy.[11] He ultimately fell out with Nelson and joined the Berlin Group. Furthermore, Dubislav, Bernays and others were impressed by Reichenbach's new analysis. Although he did not belong to Nelson's circle, Dubislav was a sagacious and, on principle, also a favourable critic of Fries' theory of justification. With regard to the general theory of relativity, he admonished not so much Fries (who could not have known about such a theory) but the Neo-Friesians Nelson and Hessenberg for holding a philosophy of geometry which "stands in complete contrast to the methodological proceeding of the modern physicist."[12] He also reproached the Kant-Friesian philosophy of mathematics for misusing pure intuition as an "asylum for sluggish reason" (Dubislav 1926a, 73). Anyone familiar with Kant's understanding of *ignava ratio* knows what a serious offence against the Kantian ideal of scientific philosophy Dubislav charged the most important representatives of the New Fries School with.

As is well known, such criticism of the 'renegades' from the Neo-Friesian camp accords quite well with the attitudes of the later Berlin Group towards 'critical' theories of space and time. Although this group did not formally constitute itself until 1927, its 'predecessor,' the *Gesellschaft für positivistische Philosophie*, was a forum where Einstein's SRT was discussed affirmatively and defended against philosophical criticism (Hentschel 1991) from 1912 onwards. Its leading figure, Joseph Petzoldt, belonged next to Reichenbach as amongst the most active supporters of Einstein; with both forming a philosophical stronghold around

[11] Minutes of the meeting of the *Fries-Gesellschaft* from August 15 and 16, 1921 (Nachlass Nelson, Bll. 243–253). I did not see these minutes and refer for further details to Peckhaus (1990, 148, n. 437). Peckhaus makes quite clear that Grelling later dissociated himself from Fries' philosophy, especially from its theorem that mathematics and 'science proper' is based on synthetic principles a priori.

[12] Dubislav (1926a, 72; cf. 71). Dubislav's sharp rejection of an alleged epistemological superiority of the Euclidean geometry deserves to be quoted more extensive, because it reveals the role of physics quite exact: "He who claims that Euclidean geometry would not only rest on consistent principles, but be also a mathematical discipline that can raise a claim to truth par excellence (*Wahrheit schlechthin*), which accordant to its truth character (*Wahrheitscharakter*) be alone applicable to real objects with success, stands in complete contrast to the methodological procedure of the modern physicist, because he [the physicist] does not appeal to pure intuition and does not dogmatically distinguish with its help one special geometry, but he takes that geometry as a basis of his geometry, which serves best to derive time, position and type of future events from present empirical knowledge. These will, when they actually take place, corroborate the suitability of the geometry in question. This means that he is prepared in principle to apply under all consistent geometries a different one in case that this allows for a more exact prediction" (Dubislav 1926a, 72–73). For his discussion of the theory of relativity, see also Dubislav 1933, 144–150.

Einstein's physics (Hentschel 1990). Reichenbach's later conventionalist answer to the problem of how geometry and physics are to be coordinated, emerging from his early Kantianism from 1920 onwards, gained broad support in logical empiricism and beyond (e.g. Grünbaum 1973; Friedman 1983).

I would like to close this section with a note on Nelson. In the second half of his short academic career he was more interested in 'practical' philosophy in a broad sense than in philosophy of science. Due to internal disputes (Franke 1991, 143–150) and the developments sketched above, the *Gesellschaft* lost a couple of experts in the philosophy of science. As a consequence, the activities of the New Friesian School in the field of *scientific philosophy* decreased dramatically after 1921. However, there are at least some short published statements which evince what Nelson's philosophical position was after this defeat. Significant, in this respect, are his posthumously published Göttingen lectures from 1919 to 1926 on *Fortschritte und Rückschritte der Metaphysik* (cf. Kraft 1962, 728). Within a defense of Fries' *hylologische Weltansicht* as a philosophically well-founded form of mechanism, he casually admits that classical mechanics is—mainly due to Einstein's theory of relativity—in a critical stage of its development and might perhaps collapse (Nelson 1962, 682–684). However, he warns against the "empirical dogma" and criticises any attempt to "draw premature metaphysical conclusions" from the present unclear states of physics because science is not entitled to do so. In fact, his advice is: "In view of this situation it seems not only justified [...] but it is the only position compatible with critical natural philosophy (*kritische Naturphilosophie*) to abstain from such metaphysical claims and to limit oneself to the conventionalist point of view which demands from physical theories only that they have a heuristic meaning."[13] This 'conventionalist-heuristic retreat' is vague and expectant, and obviously has no consequences for Nelson's discipleship to Kant's and Fries' foundation of classical mechanics. My hunch is that Nelson did not go further (and, in a way, *could* not go further) because of the epistemological consequences a full acceptance of Reichenbach's interpretation—even at its early, 'Kantian' stage from 1920—would have had. I will now give the main reason for this interpretation.

[13]Nelson (1962, 684). In the following, the 'heuristic meaning' of this intermediate physical theory is linked, again, to Fries' heuristic interpretation of mechanical principles. Nelson here also discusses the concept of simultaneity with regard to Kant's postulates of empirical thought in general in the Transcendental Analytic in order to show that the modern physicist is "in complete agreement with critical metaphysics" (Nelson 1962, 684–685). Already in 1921 he stated against Oswald Spengler that Einstein's theory should not to be understood as a symptom of decline of physics; with a reference to Hilbert he positively appraises its axiomatic form (Nelson 1921, 520–521).

3.5 Reichenbach in 1920 and Nelson: The Basic Epistemological Difference in a Nutshell

In his book *Relativitätstheorie und Erkenntnis a priori* (1920), Reichenbach focused on the question of what kind of a philosophy of space and time could do justice to both theories of relativity. As further conditions, he sought a position that would do without the synthetic aprioris of Kant and without drifting into an epistemologically untenable empirical conception of space and time. Reichenbach's answer consists mainly of an introduction of and a strict differentiation between two kinds of principles: the *axioms of coordination (Zuordnung)* and the *axioms of connection (Verknüpfung)*. The first ones are the principles of physical geometry which are not empirical, i. e. not subject to confirmation or refutation by certain observations or experiments. The second group, however, consists of empirical laws which can only be gained by observation and experiment. This split corresponds to the two different meanings Reichenbach finds in Kant's synthetic a priori judgements: the first being 'apodictically valid or valid for all times,' the second being '*constitutive* of the objects of experience'[14] (Reichenbach 1920, 46–58). For his part, Reichenbach adheres *only to the second* meaning; that the axioms of coordination are *a priori* in the Kantian sense of being constitutive for experience. Accordingly, he thought that we must define axioms of coordination before we can gain empirical knowledge by connecting sense data to points in space–time. This means that we cannot discuss the truth of any proposition that refers to experience without a priori fixation of coordinating principles.

However, Reichenbach dismisses the far-reaching first meaning of a priori, i. e. being necessary and immutable, which was also prominent in the writings of Kant, Fries, and Nelson. While the axioms of coordination do not depend on concrete experience, they do depend on the empirical state of knowledge of their time. Seen against this background, Einstein had good reasons to introduce *other* coordinating principles than Newton and Maxwell. The axioms of coordination can generally be revised according to new evidence, despite their being constitutive for experience. In short, they are not *apodictic* in the Kantian sense (Reichenbach 1920, 53).

With this 'bisection' of Kant's a priori, Reichenbach got rid of the problems that arose from the synthetic a priori in Kant's theory of space and time. He thus realized the remarkable 'gain' of aligning this position with the theories of relativity. However, from a Neo-Kantian (as well as a Neo-Friesian) perspective this gain is offset by a serious 'loss'. Specifically, because of the bisection of the *a priori*, our scientific knowledge can never be demonstrated to be *certain*, i. e. any change of the axioms of coordination will change the conceptual framework of physics and therefore its objects. To put it in 'Reichenbach's nutshell': "Here our view

[14]Reichenbach (1920, 46–58); cf. Klein (2000) for Reichenbach's physical geometry in the context of conventionalism and realism. Moritz Schlick pursued a similar approach and corresponded with Reichenbach on physical geometry. For this discussion as well as the changing concept of science at this stage of logical empiricism see Seck (2008).

differs from that of Kant: While for Kant only the determination of the individual concept is an infinite task, the perspective here is that our concepts of the object of science in general, of reality and its determinability, can only be a matter of subsequent specification" (Reichenbach 1920, 84). Even our best established scientific knowledge consists 'only' in a connection of perceptual experience and conceptual relations. In Nelson's terminology, this means that no comprehensive synthetic principle of 'rational induction' does exist, and all law-like propositions, based on experience, are only statements of probability and thus are, in principle, fallible. This consequence was later summed up by Reichenbach in these words: "There is no certainty at all remaining—all that we know can be maintained with probability only. There is no Archimedian point of absolute certainty left to which to attach our knowledge of the world [. . .]."[15] Ironically enough, this 'Popper-like' statement is directed against the 'absolutism' of positivism, even though it also fits quite well with regard to the work of Nelson who criticised positivism for "destroying not only itself, but also true science" insofar as it intended to eliminate metaphysics (Nelson 1914, 206; cf. 1908) as a warrantor of epistemic certainty. Another irony of the rise of fallibilism, at that time, is that Popper's fallibilism is rooted in his critical analysis of Fries' and Nelson's theory of justification (Popper 1994). By contrast, Nelson and his school considered the epistemological conclusion Reichenbach drew from the theory of relativity unacceptable. During Nelson's lifetime, no one who wanted to remain a member of the New Friesian School crossed this watershed.[16] For them, the acceptance or rejection of Reichenbach's interpretation of Einstein's *Relativitätstheorie und Erkenntnis a priori* was not a mere philosophical subtlety, but a matter of philosophical identity. I think that neither Nelson nor one of the remaining members of his school really faced Einstein's challenge for this reason.

3.6 Epilogue: 'Fries, Who Will Save You from the Friesians?'

After having defended Einstein repeatedly against some orthodox and undeviating Neo-Kantians, Reichenbach wrote despairingly to A. Berliner in April 1921: "Kant, who will save you from the Kantians?" (Hentschel 1990, 507) Along the same line, one might ask: Why did nobody save Fries from the orthodoxy of the Nelson school?

As in history more generally, the course of philosophy and science does not directly correspond to the merits of its protagonists. Indeed, although Fries'

---

[15]Reichenbach (1938, 192); cf. Reichenbach (1969, 272) about the method of "trial and error" as the only method left for prediction. For Reichenbach's epistemology in the broader context of empiricism see Poser (1998).

[16]This is a conjecture which a more extensive study of the sources would have to corroborate. It is certainly correct for the papers in the *Abhandlungen* of the New Friesian School, though due to the lack of experts in the school from 1920 onwards the subject is widely neglected. Bernays (1933) gives a positive estimation of Einstein's physics (cf. Sect. 3.6), but appeared 6 years after Nelson's death.

philosophy of mathematics eventually gained the appreciation it deserves, via the efforts of Nelson, Hessenberg, Dubislav, Grelling, Bernays and others, in this paper I have tried to show that his original philosophy of empirical science had even more to offer than the membership of the New Friesian School realised. For, instead of developing this philosophy further, according to Fries' constant demand for a close interaction of scientific achievements and philosophical reflection, this group meant only to conserve its mathematical philosophy of nature. They did so by trying to shelter it from theories of relativity, the philosophical impact of which were obviously underestimated. This general attitude was exactly *not* what the Friesian and New Friesian School demanded from scientific philosophy, namely: "Any philosophy that is in accordance with the exact sciences can be true, any one that is in conflict with them is wrong with necessity." Thus, Nelson, in spite of his great achievements with regard to the spreading of Fries' philosophy in general, certainly did not save Fries' mathematical philosophy of nature. In fact, he did Fries—in *this* important respect—a regrettable disservice. For, adherents like Dubislav or Grelling, who might have done better, changed sides, while others like Hessenberg or Rüstow confined their activities to perhaps less controversial subjects like the philosophy of (pure) mathematics, logic or social philosophy. Paul Bernays in 1928—a year after Nelson's death—tried to make good for the omissions of the New Friesian School when he set out the "Basic thoughts of Fries' philosophy in its relation to the current state of science." In light of the immense success of both theories of relativity he demanded a revision of the assumption of Kant and Fries that "geometry and physics are within the frame of our intuitive ideas (*anschauliche Vorstellungen*) of space and time, this being a condition of the possibility of scientific knowledge" (Bernays 1933, 107). Although belated, Bernays indicated that the genuine conception of Fries' philosophy of science actually did offer fruitful connections to the modern development of physics. But his calling for such a revision was not published before 1933 and came far too late to influence the ongoing discussion of relativity.

Fries' mathematical philosophy of nature, combining Kantianism and an early predisposition to conventionalist and fallibilist reasoning and alluding to a relational theory of motion, could have been a stimulating source and a point of systematic orientation for Reichenbach. However, he, too, failed to grasp this opportunity to bridge the gap between Kantian philosophy and scientific development. While Nelson did so because he erroneously believed that scientific philosophy demands the persistence of an orthodox Kantian metaphysics of nature, Reichenbach's conception of scientific philosophy excluded any instruction from *history* to the philosophy of science. Here our story exhibits another (third) irony. For, while Reichenbach refused to pay any attention to history in his scientific philosophy, the idea of a 'relativesed apriori' he introduced in his *Relativitätstheorie und Erkenntnis a priori* gained considerable attention and sympathy in present discussion exactly due to the *historical turn* of philosophy of science, and their confirmation of the *historical relevance* of constitutive principles, and their change in the succession of Thomas Kuhn's *Structure of Revolution.*[17]

Reichenbach could not willingly 'save Fries from the Friesians' because he decided—to Fries', and perhaps also to his, disadvantage—not to study the history of philosophy after Kant. However, his *Relativitätstheorie und Erkenntnis a priori* is a remarkable and fruitful approach, for that time, to bridge the gap between Kantian philosophy and the empirical sciences, by ranging "between transcendental method and the method of analyzing science" (Hecht 1994). In this double sense it is a synthesis in the spirit of Fries' mathematical philosophy of nature. Therefore, the question 'Who saved Fries from the Friesians?' is perhaps best answered by saying: Reichenbach, although unwittingly.

References

Anacker, Michael. 2012. *Unterbestimmtheit und pragmatische Aprioris. Vom Tribunal der Erfahrung zum wissenschaftlichen Prozess*. Paderborn: Mentis.

Anonymous. 1930. Chronik: Gesellschaft für empirische philosophie, Berlin. *Erkenntnis* 1: 72–73.

Apelt, Ernst Friedrich. 1854. *Die theorie der induction*. Leipzig: Engelmann.

Apelt, Ernst Friedrich. 1910. *Metaphysik* (1857). ed. Rudolf Otto. Halle : Hendel.

Apelt, Ernst Friedrich, Matthias Jacob Schleiden, Oskar Schlömilch and Eduard Oskar Schmidt. 1847. [Introduction to] *Abhandlungen der Fries'schen Schule*, Erstes Heft, Leipzig: Engelmann, Repr. Hildesheim 1964: Olms, 3–6.

Berg, Otto. 1912. Das Relativitätsprinzip der Elektrodynamik. *Abhandlungen der Fries'schen Schule. Neue Folge* 3(2): 333–382.

Bernays, Paul. 1918. Über die Bedenklichkeiten der neueren Relativitätstheorie. *Abhandlungen der Fries'schen Schule. Neue Folge* 4(3): 459–482.

Bernays, Paul. 1930. Die Philosophie der Mathematik und die Hilbertsche Beweistheorie. *Blätter für Deutsche Philosophie* 4: 326–367.

Bernays, Paul. 1933. Die Grundgedanken der Fries'schen Philosophie in ihrem Verhältnis zum heutigen Stand der Wissenschaft. *Abhandlungen der Fries'schen Schule. Neue Folge* 5(2): 97–113.

Bernays, Paul. 1953. Über die Fries'sche Annahme einer Wiederbeobachtung der unmittelbaren Erkenntnis. In *Leonard Nelson zum Gedächtnis*, ed. Specht Minna and Eichler Willi, 113–131. Frankfurt a. M/Göttingen: Öffentliches Leben.

Blencke, Ernst. 1978. Zur Geschichte der Neuen Fries'schen Schule und der Jakob Friedrich Fries-Gesellschaft. *Archiv für Geschichte der Philosophie* 60: 199–208.

Bonsiepen, Wolfgang. 1997. *Die Begründung einer Naturphilosophie bei Kant, Schelling, Fries und Hegel. Mathematische versus spekulative Naturphilosophie*. Frankfurt a.M: Klostermann.

Danneberg, Lutz, and W. Schernus. 1994. Die Gesellschaft für wissenschaftliche Philosophie. In *Hans Reichenbach und die Berliner Gruppe*, ed. Lutz Danneberg et al., 391–481. Braunschweig: Vieweg.

[17]Cf. Kuhn (1996, esp. 78–80, 144–159). The intimate relation between the present discussion of the 'relativesed apriori' and the history of science is demonstrated in Friedman (2001); for Reichenbach's role see esp. 72–82. For a topical analysis of the subject with strong references to the tradition of pragmatism see Anacker (2012).

Dubislav, Walter. 1926a. *Die Fries'sche Lehre von der Begründung: Darstellung und Kritik.* Dömitz: Mattig.

Dubislav, Walter. 1926b. *Über die definition.* Berlin: Weiss.

Dubislav, Walter. 1929. *Zur Methodenlehre des Kritizismus.* Bad Langensalza: Beyer & Söhne.

Dubislav, Walter. 1933. *Naturphilosophie.* Berlin: Junker und Dünnhaupt.

Dubislav, Walter. 1922. *Beiträge zur Lehre von der Definition und vom Beweis vom Standpunkt der mathematischen Logik aus.* Dissertation. Berlin: Universität Berlin.

Eggeling, Heinrich. 1904. Kant und Fries. Die anthropologische Wendung der Kritik der Vernunft in ihren wesentlichen Punkten erörtert. *Abhandlungen der Fries'schen Schule. Neue Folge* 1(2): 191–231.

Fischer, Kuno. 1862. Die beiden Kantischen Schulen in Jena. Rede zum Antritt des Prorectorats, den 1. Februar 1862. In *Akademische Reden*, ed. Kuno Fischer, 78–102. Stuttgart: Cotta.

Fischer, H. 2004. Jakob Friedrich Fries und die Grenzen der Wahrscheinlichkeitsrechnung. In *Form, Zahl, Ordnung. Studien zur Wissenschafts- und Technikgeschichte. Ivo Schneider zum 65. Geburtstag*, ed. Rudolf Seising, Menso Folkerts, and Ulf Hashagen, 277–299. Wiesbaden: Steiner.

Franke, Holger. 1991. *Leonard Nelson.* Ammersbek bei Hamburg: Verlag an der Lottbek.

Friedman, Michael. 1983. *Foundations of space-time theories. Relativistic physics and philosophy of science.* Princeton: University Press.

Friedman, Michael. 2001. *Dynamics of reason. The, 1999 Kant lectures at Stanford university.* Stanford: CSLI Publications.

Fries, Jakob Friedrich. 1822. *Die mathematische Naturphilosophie nach philosophischer Methode bearbeitet. Ein Versuch.* Heidelberg: Winter [vol. 13 of Fries 1967–2011].

Fries, Jakob Friedrich. 1827. *Grundriß der Logik. Ein Lehrbuch zum Gebrauch für Schulen und Universitäten.* 3rd ed., Heidelberg: Winter [vol. 7, 29–152 of Fries 1967–2011].

Fries, Jakob Friedrich. (1828–31). *Neue oder anthropologische Kritik der Vernunft.* 3 vols., 2nd ed., Heidelberg: Winter [vol. 4, 31–479; vol. 5 and vol. 6 of Fries 1967–2011].

Fries, Jakob Friedrich. (1837). *System der Logik. Ein Handbuch für Lehrer und zum Selbstgebrauch.* 3rd ed., Heidelberg: Winter [vol. 7 of Fries 1967–2011].

Fries, Jakob Friedrich. 1842. *Versuch einer Kritik der Principien der Wahrscheinlichkeitsrechnung.* Braunschweig: Vieweg [vol. 14 of Fries 1967–2011].

Fries, Jakob Friedrich. 1967-2011. *Sämtliche Schriften*, ed. Gert König and Lutz Geldsetzer. [Until now:] 29 vols., Aalen: Scientia.

Grelling, Kurt. 1906. Über einige neuere Mißverständnisse der Fries'schen Philosophie und ihres Verhältnisses zur Kantischen. *Abhandlungen der Fries'schen Schule. Neue Folge* 1(4): 743–757.

Grelling, Kurt. 1907. Das gute, klare Recht der Freunde der anthropologischen Vernunftkritik, verteidigt gegen Ernst Cassirer. *Abhandlungen der Fries'schen Schule. Neue Folge* 2(2): 153–190.

Grelling, Kurt. 1910. Die philosophische Grundlagen der Wahrscheinlichkeitsrechnung. *Abhandlungen der Fries'schen Schule. Neue Folge* 3(3): 439–478.

Grelling, Kurt, and Leonard Nelson. 1908. Bemerkungen zu den Paradoxien von Russell und Burali-Forti. *Abhandlungen der Fries'schen Schule. Neue Folge* 2(3): 301–334.

Grünbaum, Adolf. 1973. *Philosophical problems of space and time*, 2nd ed. Dordrecht/Boston: Reidel.

Haller, Rudolf. 1993. *Neopositivismus. Eine historische Einführung in die Philosophie des Wiener Kreises.* Darmstadt: Wissenschaftliche Buchgesellschaft.

Hecht, H. 1994. Hans Reichenbach zwischen transzendentaler und wissenschaftsanalytischer Methode. In *Hans Reichenbach und die Berliner Gruppe*, ed. Lutz Danneberg et al., 219–227. Braunschweig: Vieweg.

Henke, Ernst Ludwig Theodor. 1937. *Jakob Friedrich Fries. Aus seinem handschriftlichen Nachlasse dargestellt*, 2nd ed. Berlin: Verlag Öffentliches Leben.

Hentschel, Klaus. 1990. *Interpretationen und Fehlinterpretationen der speziellen Relativitätstheorie durch Zeitgenossen Albert Einsteins.* Basel: Birkhäuser.

Hentschel, Klaus. 1991. *Die Korrespondenz Petzold-Reichenbach*. Berlin: ERS.

Herrmann, Kay. 2000. *Mathematische Naturphilosophie in der Grundlagendiskussion. Jakob Friedrich Fries und die Wissenschaften*. Göttingen: Vandenhoeck & Ruprecht.

Herrmann, Kay. 2012. *Apriori im Wandel. Für und wider eine kritische Metaphysik der Natur*. Heidelberg: Winter.

Hessenberg, Gerhard. 1904. Über die kritische Mathematik. *Sitzungsberichte der Berliner Mathematischen Gesellschaft* 3: 21–28.

Hessenberg, Gerhard. 1907. Kritik und System in Mathematik und Philosophie. *Abhandlungen der Fries'schen Schule. Neue Folge* 2(1): 77–152.

Hessenberg, Gerhard, Karl Kaiser, and Leonhard Nelson. 1904. Vorwort der alten Folge, zugleich als Vorwort der neuen Folge. *Abhandlungen der Fries'schen Schule. Neue Folge* 1(1): III–XII.

Hoffmann, D. 1994. Zur Geschichte der Berliner 'Gesellschaft für empirische/wissenschaftliche Philosophie'. In *Hans Reichenbach und die Berliner Gruppe*, ed. Lutz Danneberg et al., 21–31. Braunschweig: Vieweg.

Kamlah, A. 1979. Erläuterungen zum Buch: Relativitätstheorie und Erkenntnis a priori. In Hans Reichenbach, *Gesammelte Werke in 9 Bänden, Band 3: Die philosophische Bedeutung der Relativitästheorie*, 475–480.

Kastil, Alfred. 1918. Jakob Friedrich Fries' Lehre von der unmittelbaren Erkenntnis. *Abhandlungen der Fries'schen Schule. Neue Folge* 4(1): 1–336.

Klein, Carsten. 2000. *Konventionalismus und Realismus. Zur erkenntnistheoretischen Relevanz der empirischen Unterbestimmtheit von Theorien*. Paderborn: Mentis.

König, Gert and Lutz Geldsetzer. 1979. Einleitung zur Abteilung [and] Vorbemerkungen der Herausgeber zum 13. Band. In Fries 1967–2011, vol. 13, 17–94.

Körner, Stephan. 1970. *Categorial frameworks*. Oxford: Blackwell.

Körner, Stephan. 1979. Leonard Nelson und der Philosophische Kritizismus. In *Vernunft, Erkenntnis, Sittlichkeit. Internationales philosophisches Symposion Göttingen, vom 27.–29. Oktober 1977 aus Anlaß des 50. Todestages von Leonard Nelson*, ed. Peter Schröder, 1–13. Hamburg: Meiner.

Körner, Stephan. 1984. *Metaphysics. Its structure and function*. Cambridge: University Press.

Kowalewsky, Michael. 1918. Über die Antinomienlehre als Begründung des transzendentalen Idealismus. *Abhandlungen der Fries'schen Schule. Neue Folge* 4(4): 693–764.

Kraft, Julius. 1962. Nachwort. In *Fortschritte und Rückschritte der Philosophie. Von Hume und Kant bis Hegel und Fries*, ed. Leonard Nelson, 727–735. Frankfurt a.M: Öffentliches Leben.

Kuhn, Thomas S. 1996. *The structure of scientific revolutions*, 3rd ed. Chicago/London: University of Chicago Press.

Mises, Richard von. 1939. *Kleines Lehrbuch des Positivismus. Einführung in die empiristische Wissenschaftsauffassung*, ed. Friedrich Stadler, Frankfurt a. M.: Suhrkamp 1990.

Nelson, Leonard. 1904. Die kritische Methode und das Verhältnis der Psychologie zur Philosophie. Ein Kapitel aus der Methodenlehre. *Abhandlungen der Fries'schen Schule. Neue Folge.* 1(1): 1–88.

Nelson, Leonard. 1905a. Jakob Friedrich Fries und seine jüngsten Kritiker. *Abhandlungen der Fries'schen Schule. Neue Folge.* 1(2): 233–319.

Nelson, Leonard. 1905b. Bemerkungen über die nicht-euklidische Geometrie und den Ursprung der mathematischen Gewissheit. *Abhandlungen der Fries'schen Schule. Neue Folge.* 1(2): 373–392 and 1(3): 393–430.

Nelson, Leonard. 1906. Kant und die nicht-euklidische Geometrie. In Nelson, Leonard. 1970–1977.*Gesammelte Schriften*. ed. Paul Bernays et al. 9 vols., Hamburg: Meiner, vol. 3, 55–94.

Nelson, Leonard. 1908. Ist metaphysikfreie Naturwissenschaft möglich? *Abhandlungen der Fries'schen Schule. Neue Folge* 2(3): 241–299.

Nelson, Leonard. 1914. Über die Unhaltbarkeit des wissenschaftlichen Positivismus in der Philosophie. In Nelson, Leonard. 1970–1977. *Gesammelte Schriften*. Hamburg: Meiner, vol. 1, 201–206.

Nelson, Leonard. 1921. Spuk, Einweihung in die Wahrsagerkunst Oswald Spenglers. In Nelson, Leonard. 1970–1977. *Gesammelte Schriften*. Hamburg: Meiner, vol. 3, 349–552.

Nelson, Leonard. 1927. Kritische Philosophie und mathematische Axiomatik. In Nelson, Leonard. 1970–1977. *Gesammelte Schriften*. Hamburg: Meiner, vol. 3, 191–220.

Nelson, Leonard. 1962. *Fortschritte und Rückschritte der Philosophie. Von Hume und Kant bis Hegel und Fries*. ed. J. Kraft. Frankfurt a.M.: Öffentliches Leben [vol. 7 of Nelson, Leonard. 1970–1977. *Gesammelte Schriften*. Hamburg: Meiner].

Neurath, Otto. 1930. Historische Anmerkungen. Zum Bericht über die 1. Tagung für Erkenntnislehre der exakten Wissenschaften in Prag, 15.–17. September 1929. In Neurath, Otto. 1981 *Gesammelte philosophische und methodologische Schriften*. Wien: Hölder–Pichler–Tempsky, vol. 1, 389–391.

Peckhaus, Volker. 1990. *Hilbertprogramm und kritische Philosophie. Das Göttinger Modell interdisziplinärer Zusammenarbeit zwischen Mathematik und Philosophie*. Göttingen: Vandenhoeck & Ruprecht.

Peckhaus, Volker. 1994. Von Nelson zu Reichenbach: Kurt Grelling in Göttingen und Berlin. In *Hans Reichenbach und die Berliner Gruppe*, ed. Lutz Danneberg et al., 53–86. Braunschweig: Vieweg.

Peckhaus, Volker. 1999. Fries in Hilbert's Göttingen: Die Neue Fries'sche Schule. In *Jakob Friedrich Fries—Philosoph, Naturwissenschaftler und Mathematiker: Verhandlungen des Symposions "Probleme und Perspektiven von Jakob Friedrich Fries' Erkenntnislehre und Naturphilosophie" vom 9.–11. Oktober 1997 an der Friedrich–Schiller–Universität Jena*, ed. Hogrebe Wolfram and Herrmann Kay, 353–386. Frankfurt a.M: Lang.

Popper, Karl. 1982. *Logik der Forschung*, 7th ed. Tübingen: Mohr & Siebeck.

Popper, Karl. 1994. *Die beiden Grundprobleme der Erkenntnistheorie*, 2nd ed. Tübingen: Mohr & Siebeck.

Poser, H. 1998. Glanz und Elend des Empirismus. Hans Reichenbachs Theorie der Erkenntnis. In *Hans Reichenbach: Philosophie im Umkreis der Physik*, ed. Hans Poser and Ulrich Dirks, 157–177. Berlin: Akademie Verlag.

Pulte, Helmut. 1999a. ... sondern Empirismus und Spekulation sich verbinden sollen': Historiographische Überlegungen zur bisherigen Rezeption des wissenschaftstheoretischen und naturphilosophischen Werkes von J.F. Fries und einige Gründe für dessen Neubewertung. In *Jakob Friedrich Fries—Philosoph, Naturwissenschaftler und Mathematiker: Verhandlungen des Symposions "Probleme und Perspektiven von Jakob Friedrich Fries' Erkenntnislehre und Naturphilosophie" vom 9.–11. Oktober 1997 an der Friedrich–Schiller–Universität Jena*, ed. Wolfram Hogrebe and Herrmann Kay, 57–94. Frankfurt a.M: Lang.

Pulte, Helmut. 1999b. Von der Physikotheologie zur Methodologie. Eine wissenschaftstheoriegeschichtliche Analyse der Transformation von nomothetischer Teleologie und Systemdenken bei Kant und Fries. In *Jakob Friedrich Fries—Philosoph, Naturwissenschaftler und Mathematiker: Verhandlungen des Symposions "Probleme und Perspektiven von Jakob Friedrich Fries' Erkenntnislehre und Naturphilosophie" vom 9.–11. Oktober 1997 an der Friedrich–Schiller–Universität Jena*, ed. Wolfram Hogrebe and Herrmann Kay, 301–351. Frankfurt a.M: Lang.

Pulte, Helmut. 2005a. *Axiomatik und Empirie. Eine wissenschaftstheoriegeschichtliche Untersuchung zur Mathematischen Naturphilosophie von Newton bis Neumann*. Darmstadt: Wissenschaftliche Buchgesellschaft.

Pulte, Helmut. 2005b. Formale Teleologie und theoretische Vereinheitlichung. Wissenschaftstheoretische und—historische Überlegungen zu ihrer Beziehung bei Kant und Fries, Kitcher und Friedman. In *Formale Teleologie und Kausalität in der Physik. Zur philosophischen Relevanz des Prinzips der kleinsten Wirkung und seiner Geschichte*, ed. Michael Stöltzner and Weingartner Paul, 77–96. Paderborn: Mentis.

Pulte, Helmut. 2006. Kant, Fries and the Expanding Universe of Science. In *The Kantian legacy in nineteenth-century science*, ed. Michael Friedman and Alfred Nordmann, 101–121. Cambridge, MA: MIT Press.

Pulte, Helmut. 2010. Der Kantische Analogiebegriff und die Theorie der modernen Naturwissenschaften. Eine schematisierende Übersicht. In *Analogien in Naturwissenschaft und Medizin*, ed. Klaus Hentschel, 233–253. Stuttgart: Wissenschaftliche Verlags-Gesellschaft.

Reichenbach, Hans. 1916. Der Begriff der Wahrscheinlichkeit für die mathematische Darstellung der Wirklichkeit. *Zeitschrift für Philosophie und philosophische Kritik* 161: 209–239.

Reichenbach, Hans. 1920. *Relativitätstheorie und Erkenntnis a priori*. Berlin: Springer.

Reichenbach, Hans. 1932. Wahrscheinlichkeitslogik. *Sitzungsberichte der Preussischen Akademie der Wissenschaften* 29: 476–488.

Reichenbach, Hans. 1938. *Experience and prediction*. Chicago: University of Chicago Press.

Reichenbach, Hans. 1947. *Elements of symbolic logic*. New York: Macmillan.

Reichenbach, Hans. 1969. *Der Aufstieg der wissenschaftlichen Philosophie*. Trans. Maria Reichenbach, 2nd ed., Braunschweig: Vieweg.

Rescher, Nicholas. 2006. The Berlin school of logical empiricism and its legacy. *Erkenntnis* 64: 281–304.

Sachs-Hombach, K. 1999. Ist Fries' Erkenntnistheorie psychologistisch? In *Jakob Friedrich Fries—Philosoph, Naturwissenschaftler und Mathematiker: Verhandlungen des Symposions "Probleme und Perspektiven von Jakob Friedrich Fries' Erkenntnislehre und Naturphilosophie" vom 9.–11. Oktober 1997 an der Friedrich–Schiller–Universität Jena*, ed. Wolfram Hogrebe and Herrmann Kay, 119–139. Frankfurt a.M: Lang.

Schnädelbach, Herbert. 1991. *Philosophie in Deutschland 1831–1933*, 4th ed. Frankfurt a.M: Suhrkamp.

Schubring, G. 1999. Philosophie der Mathematik bei Fries. In *Jakob Friedrich Fries—Philosoph, Naturwissenschaftler und Mathematiker: Verhandlungen des Symposions "Probleme und Perspektiven von Jakob Friedrich Fries' Erkenntnislehre und Naturphilosophie" vom 9.–11. Oktober 1997 an der Friedrich–Schiller–Universität Jena*, ed. Wolfram Hogrebe and Herrmann Kay, 175–193. Frankfurt a.M: Lang.

Seck, Carsten. 2008. *Theorien und Tatsachen. Eine Untersuchung zur wissenschaftstheoriegeschichtlichen Charakteristik der theoretischen Philosophie des frühen Moritz Schlick.* Paderborn: Mentis.

Sterzinger, Othmar. 1911. *Zur Logik und Naturphilosophie der Wahrscheinlichkeitslehre*. Leipzig: Mejo.

van Zantwijk, Temilo. 2009. *Heuristik und Wahrscheinlichkeit in der logischen Methodenlehre*. Paderborn: Mentis.

Chapter 4
Ernst Cassirer, Kurt Lewin, and Hans Reichenbach

Jeremy Heis

There has recently been an upsurge in interest in Ernst Cassirer, the neo-Kantian-trained German philosopher whose philosophical and scholarly writings spanned the first four decades of the twentieth century. Historians of philosophy have come to recognize the influence that Cassirer and other Neo-Kantians had on early analytic philosophers. Most prominently, Alan Richardson, Michael Friedman, and André Carus have all argued that Cassirer was a significant influence on Carnap's work up through the *Aufbau*.[1] There are good reasons for historians to emphasize the relationship between Cassirer and the early work of Carnap and other members of the Vienna Circle. Carnap wrote his dissertation under a Neo-Kantian, Bruno Bauch, defending the viability of a broadly Kantian philosophy of space. Moreover, Cassirer was one of the most prominent—if not the most prominent—German philosopher of the exact sciences in the opening decades of the twentieth century. Easily the most subtle and mathematically well-informed of the Neo-Kantians, he was among the vanguard of early twentieth century philosophers seeking to understand the philosophical significance of the revolutionary advances made in logic, mathematics, and physics. Not only did Cassirer write some of the earliest philosophical works on general relativity,[2] but he was one of the first German academic philosophers to give serious attention to Russell's logicism and the new

This paper greatly benefited from comments and conversations with Flavia Padovani, Erich Reck, Thomas Ryckman, and Audrey Yap. I also owe a special debt to Nikolay Milkov for first alerting me to Lewin's relationship with Cassirer and the Berlin Group.

[1] Friedman (1999, Ch. 6), Friedman (2000), Richardson (1998), and Carus (2008). For a helpful (though by now a bit out of date) overview of this literature, see Ferrari (1997).

[2] Cassirer (1921/1923).

J. Heis (✉)
Department of Logic and Philosophy of Science, University of California, Irvine, CA, USA
e-mail: jheis@uci.edu

N. Milkov and V. Peckhaus (eds.), *The Berlin Group and the Philosophy of Logical Empiricism*, Boston Studies in the Philosophy and History of Science 273, DOI 10.1007/978-94-007-5485-0_4,

logic,[3] Dedekind's foundations of arithmetic, and to Hilbert's axiomatic foundation of geometry.[4] Cassirer's commitment to a philosophy of science that engaged with cutting edge science ran deep and was widely known. For example, as a letter from Reichenbach makes clear, Cassirer was the only philosopher to sign onto a petition, composed by Reichenbach in 1931 on behalf of the Gesellschaft für empirische Philosophie, petitioning the German government to create a chair in the philosophy of science.[5] It is not surprising, then, that the younger generation of philosophers of science, such as Carnap, would look to Cassirer's work as an inspiration (and target).

In fact, as John Michael Krois has discovered, Cassirer felt intellectually closer to the philosophers of the Vienna Circle than to any other school. In an unpublished work from in the late 1930s, Cassirer writes:

> In "worldview," in what I see as the ethos of philosophy, I believe that I stand closer to the thinkers of the Vienna Circle than to any other "school"—
>
> The striving for determinateness, for exactitude, for the elimination of the merely subjective and the "Philosophy of feeling;" the application of the analytic method, strict conceptual analysis—
>
> These are all demands that I recognize completely—[6]

However, there is good evidence that the historical connection between Cassirer and the Berlin Group is at least as strong, if not stronger, than that between Cassirer, Carnap, and his other Viennese colleagues. In fact, in the short article entitled "Historical Remarks" from the first volume of the journal *Erkenntnis*, Cassirer is given as an influence on the Berlin Group (though he is not mentioned in connection with the Vienna Circle):

---

[3]Cassirer (1907).

[4]Cassirer (1910/1923, Chs. 2–3).

[5]Letter from Reichenbach to Cassirer, 5 June 1931, HR [025-11-04]; 15 June 1931, HR [025-11-03]. These letters are reproduced in the CD-ROM accompanying Cassirer (2009). The other signers were prominent scientists, such as Hilbert and Einstein, and prominent industrialists. The 15 June letter concerns the best way of formulating the petition, with Reichenbach noting that Hilbert wanted him to present the petitioned chair as an oppositional counterweight to the unfortunate trend in German philosophy away from the philosophy of science. Reichenbach then commented to Cassirer: "I believe that you could not imagine how deep and widespread the animosity is among natural scientists to the prevailing trend in philosophy; it is in fact only your name that is excepted from this judgment."

[6]This text is from a document titled "Zur 'Relativität der Bezugssysteme'," housed in the Cassirer papers at Yale University. The text is quoted in Krois (2000). I don't believe that this quotation shows that Cassirer felt a closer affinity to the Vienna Circle than to the philosophers of the Berlin Group. First, it is not clear whether Cassirer is distinguishing Reichenbach from the Viennese philosophers in this quotation. Second, the Berlin philosophers never formed a 'school' in the way that the Vienna Circle did. Third, all of the values that Cassirer attributes to the Vienna Circle and claims for his own appear just as clearly in the work of Reichenbach and Lewin. (Moreover, many of the *doctrines* that are distinctive to the Vienna Circle—contained for instance in Carnap et al's "The Scientific Conception of the World"(1929/1973)—Cassirer rejected out of hand.)

> The [Berlin Group] concentrates above all on problems from logic and physics as a starting point of an epistemological critique (starting points in Kantianism and Friesianism, influence of Cassirer and Nelson).[7]

In this chapter, I will substantiate this thesis by exploring the influence of Cassirer's thought on two Berlin philosophers: Kurt Lewin and Hans Reichenbach. I choose these two figures in particular for two reasons. First, both took courses with Cassirer while students in Berlin—Lewin in 1910,[8] and Reichenbach in 1913.[9] Second, both explicitly discuss Cassirer's philosophy on multiple occasions in their writings.

The chapter will have four sections. After giving some historical and biographical material about Cassirer, Lewin, and Reichenbach, I'll discuss in Sect. 4.2 the ways that their conceptions of philosophy—as giving an analysis of the sciences—overlap. In the third section, I'll address the relationship between Cassirer's philosophy of science and Lewin's own, emphasizing how Cassirer's philosophy of science provided a theoretical framework for Lewin's research program in experimental psychology. In the fourth and final section, I'll look briefly at the relationship between Cassirer's and Reichenbach's evolving conceptions of the *a priori*.

4.1 Some Biographical Material

Cassirer, Reichenbach, and Lewin enjoyed close personal and intellectual relations throughout their careers. Their interactions began at the University of Berlin, where Cassirer—having completed his dissertation under Hermann Cohen in Marburg in 1899—was a Privatdozent from 1906 to 1919. He moved to Hamburg in 1919, when anti-Semitic academic policies were loosened under the Weimar Republic, and he remained there until 1933, when the Nazi take over forced him into exile. In 1910, Kurt Lewin—a first year graduate student interested in biology and the philosophy of science—took a philosophy of science course with Cassirer, who was just that year publishing *Substance and Function*. In reading a course paper Lewin wrote, Cassirer challenged him to check one of his philosophical claims to see if it was true about psychology. This piqued Lewin's interest, and Lewin soon switched to psychology, completing his dissertation in psychology under Carl Stumpf.[10] After the Great War, he became a Privatdozent in Berlin in 1921. (Like Cassirer, he was kept from a Professorship because he was Jewish.) He joined the Psychological Institute in Berlin, where he stayed until 1933, when he went to America. (The Psychological Institute in Berlin was during the 1920s

[7] Neurath (1930, 312).

[8] Marrow (1969, 9).

[9] See the autobiographical remarks in Reichenbach's 1916 dissertation: Reichenbach (2008, 149).

[10] Marrow (1969, 6).

the home of Gestalt Psychology, and its senior members were the leading Gestalt psychologists, Köhler and Wertheimer.) During his Berlin years, Lewin published works both in experimental psychology and in the philosophy of science, and his teaching alternated between psychology and philosophy seminars. Indeed, as we will see shortly, since he was a working psychologist, his philosophy of science was integrated into his experimental work.

Kurt Lewin is known today as one of the twentieth century's most influential and innovative psychologists, commonly regarded as the founder of modern social psychology. Though he wrote widely in the philosophy of science during his Berlin years,[11] he is best known among historians of philosophy because of his interactions with Reichenbach and the other members of his circle. Lewin's relationship to the Berlin Group is a bit complex. He was not a founding member of the Berlin Group; or at least he is not in the list that Reichenbach gives in his 1936 paper "Logical Empiricism in Germany"—a list that includes only Reichenbach, Dubislav, Herzberg, and Grelling.[12] But Lewin was listed in 1931 (though not 1930) as one of six members of the Executive Committee ("Vorstand") of the "Society for Empirical Philosophy."[13] Lewin delivered a paper to the Society in 1930, which was later published in the first volume of *Erkenntnis*.[14]

Lewin always acknowledged his intellectual debt to Cassirer. This influence is clear in Lewin's philosophical writings, beginning with his earliest writings from his student days. It is patent, for instance, in an early manuscript from 1912, "Erhaltung, Identität und Veränderung in Physik und Psychologie." In this work, which contains in germ two of the main ideas of Lewin's mature philosophy of science—the project of comparative philosophy of science and the notion of "genidentity"[15]—Lewin refers freely and repeatedly to Cassirer's texts and ideas. Moreover, Cassirer's influence extended beyond Lewin's overtly philosophical writings to his experimental psychological research as well. In a 1949 paper for Cassirer's Schilpp volume, Lewin wrote

> [S]carcely a year passed when I did not have specific reason to acknowledge the help which Cassirer's views on the nature of science and research offered. The value of Cassirer's philosophy for psychology lies, I feel, less in his treatment of specific problems of psychology—although his contribution in this field and particularly his recent contributions are of great interest—than in his analysis of the methodology and concept-formation of the natural sciences.[16]

Reichenbach's personal and intellectual relationship with Cassirer goes well beyond the fact that his 1920 Habilitation thesis gave a kind of Neo-Kantian

[11]These writings fill two hefty volumes of Lewin's *Werkausgabe*.

[12]Reichenbach (1936, 143).

[13]Neurath et al. (1931, 310).

[14]Lewin (1931a). This paper was translated into English, with some deletions and occasional additions, as Lewin (1931b). (Citations to this paper will be from the reprint in Lewin 1999.)

[15]On "genidentity," see Lewin (1922).

[16]Lewin (1949). (Citations will be from the reprint in Lewin 1999). See p. 23.

philosophical interpretation of General Relativity. Already in 1914, Cassirer recommended Reichenbach (who could not write his dissertation under the Privatdozent Cassirer) to his mentor and fellow Marburg Neo-Kantian Paul Natorp.[17] The extant correspondence between the two, contained principally in the Reichenbach papers at the Archive of Scientific Philosophy at Pittsburgh and reproduced as part of Cassirer's *Nachgelassene Manuskripte und Texte*, shows a personal and professional relationship that lasted throughout their academic lifetimes. The first extant letter, dated 14 June 1915, concerns Cassirer's (unsuccessful) attempt to convince the editors of *Kant Studien* to publish Reichenbach's dissertation.[18] In his last letter to Cassirer, dated 2 April 1945 (11 days before Cassirer's death), Reichenbach—then a professor at the University of California, Los Angeles—tries to convince Cassirer to take up a position with him in the University of California.[19] Letters from the 1920s document Cassirer's repeated (unsuccessful) efforts to help Reichenbach find a professorship in a philosophy department, and in a 30 September 1930 letter, Reichenbach asks Cassirer to consider sending a paper to the new journal *Erkenntnis*.

4.2 Lewin, Reichenbach, and Cassirer on the Logical Analysis of Science

Let's begin with some words about Cassirer's, Reichenbach's, and Lewin's conception of philosophical methodology. In Reichenbach's 1936 paper, he argued that the "sole significant advance" made by the logical empiricists in Germany was the development of a new "philosophical method in the form of an analysis of science."[20] This method, Reichenbach claimed, was first put forward in his 1920 book, *The Theory of Relativity and A Priori Knowledge*. This method, Reichenbach contends, is akin to Kant's regressive method, but aims to give an analysis of our best scientific knowledge instead of an analysis of our reasoning faculty.

> The program for a philosophical method in the form of an analysis of science was first published, within the context of the movement under discussion, by the author in 1920. What he demanded was the introduction of a method of analysis of science (wissenschaftsanalytische Methode) into philosophy. This was opposed to the Kantian conception of philosophy as a method of establishing conclusions by an analysis of "reason." It was maintained that the Kantian method at its best was nothing else than an analysis of Newtonian mechanics in the guise of a system of pure reason. According to the new view put forward, "reason" was to

[17]Frederick Eberhardt and Clark Glymour, introduction to Reichenbach (2008, 2).

[18]Thanks to Simon Huttegger and Sabine Kunrath for helping me to decipher Cassirer's handwriting in this letter.

[19]Reichenbach wrote: "I have here [at UCLA] a group of talented students interested in my ideas, and they would all be pleased to study with you." These letters are reproduced in the CD-ROM accompanying Cassirer (2009).

[20]Reichenbach (1936, 142).

be grasped only in the concrete form of scientific statements [...] Advancing immediately to realize this program, the present writer entered into a detailed analysis of Einstein's theory.[21]

In a footnote from Reichenbach's 1920 book, he claims that a method akin to his own logical analysis of science appears in the work of two other philosophers: Kurt Lewin, whose "scientific orientation" is the same as Reichenbach's own, and Ernst Cassirer.[22]

Not surprisingly, Lewin casts a similar vision for the philosophy of science (what he calls "Wissenschaftslehre" [theory of science]) in his programmatic 1925 essay, "On the Idea and Task of Comparative Wissenschaftslehre." He writes:

The theory of science, as opposed to the theory of knowledge [Erkenntnislehre], is *not* a science of *researching* as such (thus a science of perceiving and proving, of intuition and systematic investigation), but rather of the *sciences* themselves, as systems of sentences [...] The theory of science is not the theory of "the" science as a sum total of research- and cognitive-acts in the sense of the theory of knowledge... It is also not the theory of science as a plurality of cultural-historical events [Gegebenheiten], but rather the theory of the individual sciences as structures of propositions and problems or doctrinal systems.[23]

Like Reichenbach, Lewin advocates a philosophy of science whose initial data—as it were the observational basis of Wissenschaftslehre as a science[24]—are the concrete individual sciences themselves. This approach is opposed to a theory that begins with the psychological acts or mental faculties (reason, intuition, perception) of individual thinkers, and is equally opposed to a merely historical or sociological approach that looks at the genesis of sciences and scientific theories as historical events calling for historical explanations. Each science is understood instead as a system of concepts and propositions, in abstraction from the psychological events or historical events that brought them into being.[25]

Now compare Cassirer's description of Kant's critical project:

Only now do we fully understand Kant's statement, that the torch of the critique of reason does not light up the objects unknown to us beyond the sense world, but rather the shadowy place of our own understanding. The 'understanding' here is not to be taken in the empirical

---

[21]Reichenbach (1936, 142–143); cf. Reichenbach (1920/1965, 72–73).

[22]Reichenbach (1920/1965, 114).

[23]Lewin (1925). I cite from the reprint in Lewin (1981).

[24]Lewin (1925, 53) advocated that philosophers of science focus on description of the various sciences instead of *deduction* (Lewin 1925, 61; Lewin 1927, 279, translated as Lewin 1992). Compare Reichenbach's advocating an "inductive" over a "deductive" method in the philosophy of science (Reichenbach 1920/1965, 75).

[25]Lewin (1949, 25–26): "Doubtless the researcher is deeply influenced by the culture in which he lives and by its technical and economic abilities. Not these problems of cultural history, however, are in question when the social psychologist has to make up his mind whether or not 'experiments with groups' are scientifically meaningful, or what procedure he may follow for developing better concepts of personality, of leadership, or of other aspects of group life. Not historical, but conceptual and methodological problems are to be answered, questions about what is scientifically right or wrong, adequate or inadequate; although this correctness may be specific to a special developmental stage of a science and may not hold for a previous or a later stage. In other words, the term "scientific development" refers to levels of scientific maturity, to levels of concepts and theories in the sense of philosophy rather than of human history or psychology."

> sense, as the psychological power of human thought, but rather in the purely transcendental sense, as the whole of intellectual and spiritual culture. It stands directly for that entity which we designate by the name 'science' and for its axiomatic presuppositions, but further in an extended sense, for all those orders of an intellectual, ethical, or aesthetic kind demonstrable in reason and perfected by it.[26]

A proper Kantianism, then, for Cassirer accords with Reichenbach's and Lewin's projects: it focuses on an analysis of science, considered as an axiomatic system of concepts and propositions, in abstraction from the psychological acts and historical events that brought it into being.

Reichenbach was thus aware that a method similar to the one he described was already put forward by Cassirer and the other members of the Marburg School. Indeed, compare the following two descriptions of the proper philosophical method. The first is from Reichenbach 1920:

> The results discovered by the positive sciences in continuous contact with experience presuppose [coordinating] principles the detection of which by means of logical analysis is the task of philosophy. [...] There is no other method for epistemology than to discover the principles actually employed in knowledge.[27]

Here Reichenbach is arguing that the primary task of the logical analysis of science is to isolate the a priori elements in our current best physical theories. (Reichenbach's conception of the primary task of an analysis of physics was constant even as Reichenbach's conception of the a priori shifted toward conventionalism, and coordinating principles became coordinating definitions.)[28] The second description is from Cassirer 1906:

> The task, which is posed to philosophy in every single phase of its development, consists always anew in this, to single out in a concrete, historical sum total of determinate scientific concepts and principles the general logical functions of cognition in general. This sum total can change and has changed since Newton: but there remains the question whether or not in the new content [*Gehalt*] that now emerges some most general relations, on which alone the critical analysis directs its gaze, present themselves under a different form [*Gestalt*] and covering.[29]

In this latter passage, Cassirer is expressing his commitment to what the Marburg Neo-Kantians called the "transcendental method" or the method of "transcendental logic."[30] According to this approach, the proper object of philosophy is our best current mathematical sciences of nature. These sciences are the "fact" whose preconditions ("the general logical functions") it is the task of philosophy to study.[31]

---

[26]Cassirer (1918/1981, 154–155).

[27]Reichenbach (1920/1965, 74–75).

[28]See, for instance, Reichenbach (1924/1969, pp. xii–iv).

[29]Cassirer (1906/1922, 16).

[30]See Cassirer (1912), and Natorp (1912). A nice recent discussion is Richardson (2006).

[31]On the Marburg reading of Kant, Kant first isolated the transcendental method and applied it to Newtonian science; in fact, he mistakenly thought that the transcendental preconditions of Newtonian science were the fixed preconditions for all scientific cognition in all times (Cassirer 1906/1922, 18).

"Science" here is no abstract generality: it is the particular sciences and particular theories contained in the writings of the scientists themselves. The preconditions of these sciences are not understood psychologically, as primitive acts innate in the human mind. Such a psychologistic approach is doomed to failure.[32] Rather the goal of the analysis of science is to isolate the highest concepts and principles of our sciences.

Reichenbach was correct, then, to identify Lewin and Cassirer as two other proponents of his conception of the philosophy of science. Still, though, there were differences among the three philosophers. First, the three disagreed on the historical question whether Kant himself had anticipated the project of a logical analysis of science. Reichenbach thought that Kant's project was still psychological. Cassirer argued the opposite,[33] while Lewin articulated a middle position.[34] Second, Reichenbach became convinced that logical analysis could reliably distinguish the factual from the logical elements in a theory only if the target physical theory was given a mathematically exact axiomatization. This view contrasted with a "historical method" such as Cassirer's (and to a lesser extent, Lewin's), which analyzed the logical structure of a theory through a historical analysis of the theoretical scientific work that it grew out of.[35] In fact, Lewin, unlike Reichenbach, showed little interest in distinguishing the a priori and empirical elements in empirical theories. Third, Lewin advocated a "comparative description" of the various concrete sciences. While Reichenbach focused during the 1920s on the analysis of one theory in one science—Einstein's theory of relativity—Lewin wrote works comparing the fundamental concepts and principles of psychology, biology, and physics. Though Lewin found inspiration for this work in Cassirer's writing,[36] this emphasis on comparison is unique to Lewin's Wissenschaftslehre. Since productive comparison requires a critical mass of data, Lewin advocated a focus on observing the various

---

[32]Cohen (1902, 17).

[33]Cassirer read Kant as a proponent of the method. Reichenbach, however, thought that Kant's philosophy confusedly mixed together questions about the logical structure of the sciences with psychological questions. See Reichenbach (1920/1965, 55ff.) and Reichenbach (1922/1981, 29). Schlick agreed with Reichenbach; see Schlick (1921/1979, 331): "Kant certainly wanted to purge [pure intuition] of everything psychological—but I shall never be able to persuade myself that he succeeded." Cassirer defended his reading of Kant against Schlick in Cassirer (1921/1923, 451).

[34]See Lewin (1927, 279): "The Copernican Turn, with which Kant changed the question "*Whether* knowledge is possible" into the question "*How* knowledge is possible," is *one* step"—though not the final step!—"in the development of the theory of knowledge from a speculative science into an observational science. Into a science, therefore, that begins with the investigation of the concrete objects lying before us, instead of a few concepts given ahead of time."

[35]See Reichenbach (1924/1969, xiii).

[36]See Lewin (1949, 26): "A ... reason why I feel Cassirer's approach is so valuable to the scientist is his comparative procedure. Although Cassirer has not developed what might be called a systematic *comparative theory of the sciences,* he took important steps in this direction. His treatment of mathematics, physics, and chemistry, of historical and systematic disciplines is essentially of a comparative nature. Cassirer shows an unusual ability to blend the analysis of general characteristics of scientific methodology with the analysis of a specific branch of science."

sciences, rather than theorizing before the data were all in. Kantian philosophies of science, including presumably Cassirer's own, are too quick to introduce a theoretical superstructure on this data.[37]

4.3 Cassirer and Lewin

In this section my goal is to understand how Lewin's philosophy of science, derived in important ways from Cassirer's own, informed his experimental work as a working psychologist. Lewin outlined the theoretical underpinnings of his psychological research program in a 1931 paper from *Erkenntnis*, "Der Übergang von aristotelischen zum galileischen Denken in Biologie und Psychologie"—a work that draws explicitly and repeatedly from Cassirer. This paper was widely read by psychologists, and was quickly translated into English and published in the English language *Journal of General Psychology*. My organization will be a bit non-standard: instead of explaining one at a time Cassirer's philosophy of science, Lewin's philosophy of science, and Lewin's psychological research, I'll follow Lewin's own presentation in his 1931 paper and discuss all three simultaneously. I'll isolate eight features of Lewin's experimental work. The first four features, I hope to show, can profitably be seen in the context of Cassirer's and the Berlin Group's related projects of giving a logical analysis of science. The last four features can profitably be seen in the context of Cassirer's famous contrast between substance-concepts and function-concepts.

4.3.1 Lewin's Psychological Research Program and Cassirer's Transcendental Method

4.3.1.1 Lewin Expanded the Domain of Psychological Research to Include Behavioral and Social Phenomena Commonly Thought to Be Inappropriate Objects of Psychological Research

Lewin's psychological work was dedicated to widening the subject matter of psychology to include a psychology of behavior and a psychology of social

[37]Lewin (1925, 61). "Even Neo-Kantianism has produced works (e.g., Cassirer 1910) that contained descriptions of a concreteness about the relevant objects that were still not sufficiently concrete. Neo-Kantianism remained too bound to an essentialy deductive 'System'; but it still attained a certain level of descriptive work within the frame of a system. With the question of 'possibility' the fundamental point of view of Kantianism remains the point of view of a not-descriptive theorizing; it remains directed toward *generalities*. The examples often carry the character of mere illustrations for thoughts that are derived from one or some few central ideas (above all from the idea of the unity of consciousness or of knowledge.)"

groups—subjects that many psychologists thought to be illegitimate subjects of psychological research. Indeed, he is today often considered the father of social psychology. In this respect, his work was like Freud's. But Lewin departed from Freud in seeking out experimental (as opposed to therapeutic) methods for testing various theories of individual and group behavior, and again unlike Freud, he wanted to discover mathematical laws for these domains.[38]

4.3.1.2 Lewin Thought That the Expansion of the Domain of Psychological Research Would Require New Concepts

He introduced what he called "Field Theory," expanding the holistic psychology that Gestalt theorists applied to perception to a subject's total place in her environment. A subject's needs and wants form what he calls "Tension systems." Objects in the environment have a "valence," steering the behavior of subjects in their environment. He transferred concepts developed in physics into psychology, introducing talk of "Behavioral Dynamics" and "Group Dynamics," and "vectors." He and his students, in exploring the conditions under which subjects set goals, introduced the concept of a "level of aspiration," a phrase, like his "group dynamics," that has entered the vernacular.[39]

Lewin claimed that his self-conscious conceptual innovation was inspired by Cassirer's philosophy of science. Cassirer recognized, Lewin said, that many of the most important developments in science have been conceptual innovations:

> To proceed beyond the limitations of a given level of knowledge the researcher, as a rule, has to break down methodological taboos that condemn as "unscientific" or "illogical" the very methods or concepts which later on prove to be basic for the next major progress. Cassirer has shown how this step by step revolution of what is "scientifically permissible" dominates the development of mathematics, physics, and chemistry throughout their history.[40]

The Marburg transcendental method involves carrying out anew for each stage in the history of science the project of Kant's first *Critique*. This Neo-Kantianism, unlike Kant's original writings, brings to the center of philosophical reflection the fact that sciences develop, and it seeks to discover how, why, and how it is possible that the sciences develop in the progressive way that they do. (As Lewin put the point, philosophy of science must investigate the Werden—the "becoming"—of the various sciences in their successive developmental stages.)[41] Implicit in the project of carrying out Kant's critical project for each stage in the development of science is the conviction that the advance of science has included more than a further accumulation of more facts or more powerful experimental methods. In each

---

[38]See Lewin (1937).

[39]On these concepts, see the classic papers collected in Part II of Lewin (1999).

[40]Lewin (1949, 26).

[41]Lewin (1925, 75). This phrase echoes Natorp's claim (in Natorp 1912) that science is not just a "faktum," but also a "fieri."

stage of science, the fundamental categories and principles—which delimit the domain of what is thought to be physically possible—are overturned and replaced. As Cassirer's historical research has shown, the progress of science would then be hindered were philosophers (or scientists) to treat the conceptual scheme or fundamental ontology of science as fixed for all time.

4.3.1.3 Lewin Thought That Psychological Experimentation Was Hindered by Adopting the Pose of a Theory-Free "Fact-Collector"; That, Paradoxically, Effective Experimentation Requires Adopting a Theoretical Framework[42]

In fact, Lewin made his international reputation during his Berlin years less because of his experimental results and more because of his willingness to develop (in anticipation, as it were) a new theoretical apparatus for developing and interpreting laboratory experiments.[43]

In this, Cassirer was a clear inspiration. From Kant, Cassirer thinks we should learn not to be embarrassed to admit that we bring to our experiments a set of concepts and principles that make them possible.

> [W]hile a lone sensory perception or mere collection of such perceptions may be able to get along without the guidance of a plan of reason, it is still the latter that first makes experiment precise and possible, 'experience' in the sense of physical knowledge. [...] Before Galileo could measure the magnitude of acceleration in free fall, the conception of acceleration itself, as well as measuring apparatuses, had to exist, and it was this mathematical conception which once for all differentiated his unadorned way of putting the question from that of the medieval scholastic physics. [...] What Galileo laid down in advance, according to the plan of reason, is what initially made it possible for the experiment to be conceived and directed. (Cassirer 1918/1981, 164)[44]

As Lewin noted, Cassirer carried this Kantian thought over into his analysis of natural scientific experimentation. Even the most basic results of physical experimentation—measurements—require instruments whose behavior can only be interpreted through a background theory.[45]

---

[42]Lewin (1949, 28).

[43]On this point, see Brown (1929).

[44]This passage is Cassirer's commentary on Kant, *Critique of Pure Reason (1998)*, Bxii: "[R]eason has insight only into that which it produces after a plan of its own, and that it must not be kept, as it were, in nature's leading strings, but must itself show the way with principles of judgment based upon fixed laws, constraining nature to give answers to questions of reason's own determining. Accidental observations, made in obedience to no previously thought-out plan, can never be made to yield a necessary law, which alone reason is concerned to discover."

[45]Lewin (1949, 27–28) cites with approval Cassirer (1910/1923, 144): "In truth, no physicist experiments and measures with the particular instrument that he has sensibly before his eyes; but he substitutes for it an ideal instrument in thought, from which all accidental defects, such as necessarily belong to the particular instrument, are excluded. ... The corrections, which we make and must necessarily make with the use of every physical instrument, are themselves a work

4.3.1.4 Lewin's Method, Though It Relied on Analogies Between Physics and Psychology, Was Fundamentally Anti-reductionist

Lewin thought explanations could be given of psychological phenomena without reducing the objects, concepts, or laws of psychology to those of physics or physiology.[46] In this respect, Lewin is similar to Cassirer, who thought that each science in each of its phases required its own transcendental analysis. Each of the special sciences 'frames its own questions,' and answers them according to diverse and independent methodologies; none of the methods of the special sciences 'can simply be reduced to, or derived from the others.'[47] Indeed, Cassirer thought that the felt philosophical need for reductionism—even when the practice of the various special scientists did not support it—was motivated by a prior metaphysical conviction that the transcendental method requires us to reject.[48]

In the 1930s, Cassirer associated the metaphysically-motivated reductionism that he rejected with Carnap's claim (in two papers that appeared in *Erkenntnis*: Carnap 1932/1934 and 1932/1959, "Die physikalische Sprache als Universalsprache der Wissenschaft" and "Psychologie in physikalischer Sprache") that the language of physics is a universal language.[49] Lewin had no more sympathy with Carnap's arguments in these papers than Cassirer did. In fact, Reichenbach asked Lewin to write a response to these papers for *Erkenntnis*. Lewin politely declined, and a younger member of the Psychological Institute—Karl Duncker—was deputized instead.[50] Duncker there cited (against Carnap) Lewin's claim that the reducibility of concepts from the special sciences could not be a straightforward 1–1 mapping, but that biological or psychological concepts would at best map into complicated combinations or networks of physical concepts.[51] Elsewhere, Lewin himself argued that science is a unity only in the (comparatively weak) sense that the methods

of mathematical theory; to exclude these latter, is to deprive the observation itself of its meaning and value."

[46]Lewin (1931b, 37).

[47]Cassirer (1923/1955, 76, 77, 78).

[48]See Cassirer (1923/1955, 76): "The object cannot be regarded as a naked thing in itself, independent of the essential categories of natural science: for only within these categories which are required to constitute its form can it be described at all. ... If the object of knowledge can be defined only through the medium of a particular logical and conceptual structure, we are forced to conclude that a variety of media will correspond to various structures of the object, to various meanings for 'objective' relations. The physical object is not the chemical object, nor is it the biological object, because physical, chemical, biological knowledge *frame their questions* each from its own particular standpoint and, in accordance with this standpoint, subject the phenomena to a special interpretation and formation."

[49]See Cassirer (1999 [written 1937], 6–7), and Cassirer (1942/2000, 41).

[50]Ash (1994, 95).

[51]Duncker (1932/1933, 176), citing Lewin (1922).

of individual sciences go through similar developmental stages—for instance, he argues, psychology in his time was passing from a stage akin to Aristotelian physics to a stage akin to Galilean physics.[52]

4.3.2 *Lewin on Substance-Concepts and Function-Concepts*

Lewin argued in his 1931 *Erkenntnis* paper that psychology up that point was reminiscent of Aristotelian physics: it considered only a subset of psychological phenomena to be subject to psychological laws, and in those cases rested content with categorizing subjects into types, with giving generalizations about "normal" cases and eschewing responsibility for explaining why particular cases are the way they are. Contemporary psychological dynamics is thus hindered by a methodology reminiscent of Aristotelian physics, and it needs to move, as physics did, into a Galileian phase.

The language in this paper is unmistakably Cassirerian. The main argument of Cassirer's *Erkenntnisproblem*, vols.1–2, is that the development of modern science and philosophy from the late Renaissance to Kant is a working out of the Galilean ideal of science. Cassirer's *Substance and Function* further argues that modern logic and epistemology have remained wedded to an Aristotelian theory of concepts—grounded in a metaphysics of substances—long after the Aristotelian metaphysics and epistemology were supplanted with the modern science initiated by Galileo. The contrast that forms the main theme of the book—that between substance-concepts and function-concepts—is complex and multi-faceted, and I will try over the course of this chapter to lay out some of the main elements in this distinction.[53] Lewin not only cites Cassirer's book frequently in the essay; he on a few occasions says explicitly that his goal is to initiate a switch in psychology from substance-concepts to function-concepts.[54]

4.3.2.1 Lewin Thought That Psychology Needed to Look for Strict, Exceptionless Laws That Could Unite Psychological Phenomena That Differ Prima Facie

Pre-Galileian physics was hindered by the assumption that only super-lunary phenomena are subject to law and that laws are simply expressions of what happens

---

[52]Lewin (1925, 50–51). There he argues that in the various stages of its historical development one and the same science will require different methods, and that different sciences in the same relative stage of their development will often employ the same method. He concludes: "In view of the fundamental tools [Grundzüge] of the method (also only in this sense) one can speak in the end of a 'unity (better: homogeneity) of all knowledge.'"

[53]For a more detailed discussion of the contrast between "*Substanzbegriff*" and "*Funktionsbegriff*," see Heis (201?).

[54]Lewin (1931b, 40, 44).

in general. Just as post-Galileian physics has done, so too should psychology attempt to find exceptionless laws that apply to psychological phenomena generally.

On Cassirer's view, the Aristotelian notion of laws was intertwined with the Aristotelian view of experience and the Aristotelian metaphysics. For Aristotle, our knowledge of objects of experience was explained by the fact that the forms of objects were perceptible and could be directly transferred from the objects to the mind in perception. Thus, the conceptual repertoire of an empirical scientist is fixed by the nature of empirical objects (substances) and by the nature of our minds to receive these substantial forms. Cassirer thought that Kant had shown that this whole picture is illusory. Even the experience of objects requires a set of concepts, whose meaning is expressed in fundamental laws. No experience is "direct;" it is thus only through the introduction of laws that objects can be experienced at all. And so the domain of law is universal.

4.3.2.2 Lewin Thought That the Development of Psychology Would Require a New Use of Mathematics, and He Thought That the Function of Mathematics Is to Allow Psychologists to Develop General Laws That Can Explain Why a Particular Case Is the Way It Is

Lewin was critical of the way his contemporaries used mathematics in psychology. He thought that the use of precise measurements to determine statistical averages—for instance, in intelligence testing or in determining the properties of the "average 4 year old" —was not very valuable because it did not allow researchers to do as Galileian physicists did—that is, to explain why a particular case has the particular features it does. For instance, compare a simple non-mathematical generalization, like "Every flash of lightning is followed by thunder," with an equation that expresses the temporal interval between seeing the light and hearing the thunder as a function of the distance between the lightning and the observer. In the latter case, but not the former, the law will tell not just that this case of thunder is like all other cases in being accompanied by thunder; it will also say why the temporal interval between lightning and thunder in this particular case differs from the corresponding interval in that case. This is what Lewin wanted. He wanted psychological laws that explained why this 4 year old is doing this now; he did not want to rest content with hearing what an average 4 year old would do—no matter how precisely this average can be measured. A truly Galileian psychology would not just use mathematics—psychologists in 1931 were already doing that—but it would use mathematics for just this purpose.

Again Lewin is sounding a theme from Cassirer. What a mathematical function does—Cassirer argued in *Substance and Function* Ch. 1[55]—is to allow for completely general laws that can capture not only what every event has in common, but also how precisely the individual cases differ from one another. Aristotelian concepts—which express only the common features of a set of objects—do not allow us to recapture the differences among cases.[56]

4.3.2.3 Lewin Thought That the Development of Psychology Was Hindered by a Fear of Introducing "Hidden Variables" That Would Bring Together and Explain a Wide Variety of Psychological Phenomena That, Prima Facie, Are Unrelated

Aristotle thought that the physical world came in fundamental types (each with its own distinctive behavior) and that the type of the object could be ascertained by direct observation. Galileo thought that all physical objects—stars, falling bodies, fluids, etc.—could be explained with the same few laws. But the ultimate success of this program required thinking of objects as composed of homogeneous, microscopic stuff. In the same way, Lewin thought that introducing "tension systems" and psychological vectors would help explain a wide variety of psychological phenomena, even if these tension systems or social forces were not reducible to "directly" observable phenomena.

Cassirer had argued that philosophical scruples about hidden variables and unobservable entities were philosophically on a par with the now discredited Aristotelian metaphysics and epistemology. On the Aristotelian view, philosophers can identify the fundamental forms of objects—the kinds of substances there are—once and for all. And given the Aristotelian theory of perception, it is not surprising that the ontologically privileged set includes only those objects that can be "directly" perceived—that is, not microscopic atoms, nor super-personal psychological entities like groups and social forces. But, given the connection between law and objecthood that Cassirer highlights, the set of objects in our scientific ontologies will develop as our fundamental laws do; and it is a lesson of the transcendental method that we cannot prescribe ahead of time what kind of fundamental laws science might propose.

---

[55]Cassirer (1910/1923, 19–20).

[56]See Lewin (1927, § IV), which cites Cassirer to support the claim that the goal of an experiment is not to find very many equal cases, but rather to find a systematic variation among a sum total of different cases. He argues that, if one thinks of a law as a regularity, a rule, then one thinks that one proves that there is a law by finding the greatest number of equal cases [gleicher Fälle]. But this rests on a faulty theory of induction, refuted already by Cassirer.

4.3.2.4 Lewin Thought That—Unlike Previous Associationist Psychology and Theories of Instincts[57]—A Proper Psychological Dynamics Would Require Viewing Behavior as a Function of Both Environment and the Person[58]

Aristotelian physics explained the dynamics of a physical object by the unchanging nature of the object. Air goes up, no matter its circumstances. In modern physics, though, the direction and velocity of an object depends on its relation to the other objects in its environment. Similarly, in introducing field psychology, Lewin argued that the psychic forces acting on a subject depend crucially on the situation; this dependency is expressed in his now famous equation (often called "Lewin's equation") $B = f(p, e)$.[59] Similarly, in a series of famous experiments from the 1920s he showed that a subject's memory of a fact depends not on past repetition, but on the relevance of the fact to some ongoing task, and he was able to identify the factors in a particular situation that would cause a subject to become angry.[60]

The influence from Cassirer is clear. A metaphysics of substances tries to explain empirical phenomena ultimately in terms of the internal nature of individual objects. Cassirer argued that, in modern physics, empirical phenomena are explained through laws that, through mathematical functions, express the relations between different objects and magnitudes. A physicist then would say that a kind of object exists, not necessarily because we can directly observe it, but because an equation referring to that kind of object allows us to explain the relations among objects that can be more directly perceived. Thus, it is not unchanging objects (or substances) that science is after, but constant laws that express unchanging relations among phenomena.

4.4 Reichenbach and Cassirer on the A Priori

Even this brief survey of Lewin's philosophical writings shows that Lewin—though he shared in broad outlines Reichenbach's goal of a 'logical analysis of science'—did not see his work as supporting empiricism.[61] In fact, answering the question whether natural science presupposes a priori principles—and the accompanying task of systematically sorting the propositions of a theory into the

---

[57] See Lewin (1926) for a criticism of associationist explanations.

[58] Lewin (1931b, 64–65).

[59] This equation, implicit in earlier works, was first introduced in Lewin (1936).

[60] See Marrow (1969, Ch.5). Again, the contrast is with associationism and Freudianism, which try to explain behavior in terms of past experiences rather than through the interaction with the present environment.

[61] Nowhere in Lewin (1931b) does he express any affinity for empiricism. Indeed, the only mention Lewin makes of empiricism is to point out that—paradoxically—the real advance in Galilean physics required introducing unobservable idealized objects like frictionless planes and perfect spheres (Lewin 1931b, 44–45).

a priori and empirical elements—was not part of Lewin's project. It is of course, however, precisely the difference between empiricism and idealism that forms the center of the evolving discussion between Reichenbach and Cassirer over the proper philosophical interpretation of relativity. It is to this topic that I now turn.

4.4.1 Reichenbach on Coordinating Principles

In the first volume of *Erkenntnis*, Reichenbach wrote that with the "method of the analysis of science the Society positions itself in conscious opposition to all pretensions of a philosophy that claims a special right of reason and wants to set up propositions of a priori validity that are not subject to scientific critique."[62] Similarly, in his 1928 *Philosophy of Space and Time*, Reichenbach took himself to have refuted the "philosophy of the a priori,"[63] and he argued that there is a patent contradiction between the theory of space in General Relativity and the philosophy not only of Kant, but also of the various more permissive Neo-Kantians.[64] As is well-known, this was not the rhetorical stance that Reichenbach had taken in his 1920 book, where he argued that Einstein's theory is inconsistent with empiricism and confirms a kind of critical philosophy.[65] This revised Kantianism required separating out two meanings of Kant's a priori principles: as necessarily and unrevisably valid principles, and as principles 'constitutive of the object of knowledge'—that is, at some stage in the history of science.

> Because of the rejection of Kant's analysis of reason, one of its meanings, namely, that the a priori statement is to be eternally true, independently of experience, can no longer be maintained. The more important does its second meaning become: that the a priori principles constitute the world of experience. Indeed there cannot be a single physical judgment that goes beyond the sate of immediate perception unless certain assumptions about the description of the object in terms of a space-time manifold and its functional connection with other objects are made.[66]

Reichenbach rejects the possibility of a priori principles in the first sense, since there is a conflict within General Relativity between the empirical fact that inertial and gravitational mass are equivalent and the purportedly self-evident principle that the metric of physical space is Euclidean.[67] The proper moral of this surprising disconfirmation of a coordinating principle previously thought to be "eternally valid" is not to find a new self-evident principle to take its place, but to reject out of hand the project of identifying apodictic a priori principles.[68]

---

[62]Neurath et al. (1930, 72).

[63]Reichenbach (1928/1957, 67).

[64]Reichenbach (1928/1957, 36).

[65]Reichenbach (1920/1965, ch. 8; 1922/1981, note 21).

[66]Reichenbach (1920/1965, 77).

[67]Reichenbach (1922/1981, 37; 1920/1965, 31).

[68]Reichenbach (1922/1981, 39; 1920/1965, 79)

These considerations, however, do not touch the necessity of a priori principles in Reichenbach's second sense. Indeed, Reichenbach thought that General Relativity well illustrated that special principles are required to coordinate physical objects with the implicitly defined mathematical concepts appearing in the axioms of physics.[69] (Indeed, in 1920/1965 he argues that these principles—which he calls "axioms of coordination"—"constitute the world of experience" (77), since the objects coordinated with our concepts are completely undefined outside of their coordination with our concepts.) In particular, Reichenbach argues throughout the 1920s that the determination of the spatial and temporal metrics presupposes such principles of coordination.

In a famous exchange that has become widely discussed since Alberto Coffa's 1991 book, Schlick argued that what Reichenbach was calling "axioms of coordination" are better understood simply as conventions, and thus as *analytic* a priori principles.[70] Reichenbach was initially unpersuaded, for reasons that he laid out in his 1922 paper *The Present State of the Discussion of Relativity*:

> In the first place, conventionalism does not recognize, as Kant did, that these 'conventions' determine the concept of object, that the particular thing or law is defined only by their help and not by reality alone. Secondly, the term 'convention' overemphasizes the arbitrary elements in the principles of knowledge; as we have shown, their *combination* is no longer arbitrary.[71]

Reichenbach argued throughout the 1920s that coordinating principles function in physical theories only in groups, and that a coordinating principle that allows for a unique or consistent coordination of concepts with things in conjunction with one set of principles might in fact be inconsistent with experience when conjoined with other coordinating principles. In this respect, coordinating principles differ from standard cases of linguistic conventions, whose arbitrariness is unconstrained—a fact patent to any one who has tried to learn a new language as an adult, and run up against the language's frustrating lack of consistency. But considering that Reichenbach's whole point was that these coordinating principles—being relative and not apodictic—do not entirely fit the standard characterization of Kant's "synthetic a priori" principles either, the question whether one should call these coordinating principles "axioms" or "conventions" threatens to become—as Reichenbach seemed to recognize in his 1922 paper[72]—simply a verbal question.

Cassirer himself never consented to labeling the a priori principles of physics as conventions. But given the threat of a terminological draw, it is not clear right away whether Cassirer's opposition to full-fledged conventionalism is itself simply an issue of word choice or rhetorical emphasis. Indeed, both Schlick and Reichenbach worried that the Kantianism in Cassirer's philosophy of physics had been so

---

[69] Reichenbach (1920/1965, 36–37; 54).

[70] Schlick's letter is from November 1920, and is discussed in Coffa (1991, 201–202). Schlick made his criticism in print in Schlick (1921/1979).

[71] Reichenbach (1922/1981, 38–39).

[72] Reichenbach (1922/1981, note 21).

weakened as only to differ verbally from their own empiricism.[73] In the remaining pages of the paper, I will argue that the dispute between Cassirer, Reichenbach, and Schlick was not merely verbal, and that it turned on some of the deepest features of Cassirer's epistemology of science.[74]

4.4.2 Reichenbach's Criticism of Cassirer

Unfortunately, we will find little help in locating the real differences (should there be any) between Cassirer and Reichenbach by looking at Reichenbach's criticisms of Cassirer's interpretation of relativity. In his 1922 review article, Reichenbach praises Cassirer for having "awakened Neo-Kantianism from its 'dogmatic slumber,' while its other adherents carefully tried to shield it from any disturbance by the theory of relativity." But while Cassirer deserves credit for clearly articulating the inconsistency of Einstein's theory with Kant's theory of space, nevertheless Cassirer's "approach is tantamount to a denial of synthetic *a priori* principles, and ... there is no other remedy but to renounce the apodictic character of epistemological statements."[75] This criticism is misplaced (Ferrari 2003, 99ff). According to the Marburg reading, Kant had given a defense of the non-empirical truth of both Euclidean geometry and the principle of causality by showing that they are preconditions of the possibility of Newtonian science. In the same way, Cassirer claims that Riemannian differential geometry and Einstein's principle of general covariance are conditions of the possibility of general relativity.[76] However, Cassirer explicitly denies that these principles, or any other, are apodictic, certain, or self-evident, and he denies that we have any conclusive reason to think that these principles will be constitutive principles in our future physics.[77]

Indeed, many interpreters have claimed that Cassirer's theory of the a priori is an anticipation of the theory of the relativized a priori later articulated by Reichenbach in 1920.[78] Already in 1906, Cassirer wrote:

---

[73]Schlick (1921/1979, 326), and Reichenbach (1928/1957, 36ff).

[74]There is another aspect to Cassirer's resistance to Schlick and Reichenbach's conventionalism, an aspect that I will mention but not further explore. Cassirer argued (against Schlick explicitly) that labeling linguistic meanings as conventions does not explain the prior question How is meaning possible at all? In fact, Schlick can settle for empiricism only because he (mistakenly) thinks that by labeling meanings as conventions he can avoid answering the question altogether. See Cassirer (1927, 136).

[75]Reichenbach (1922/1981, 30).

[76]Cassirer (1921/1923, 415).

[77]See Cassirer (1910/1923, 269).

[78]Richardson (1998, ch.5), Ryckman (2005, ch.2), and Padovani (2011).

> In [science] we find only a *relative* stopping point, [and] we therefore have to treat the *categories*, under which we consider the historical process itself, themselves as variable and capable of change.[79]

Like the relativized a priori program of Reichenbach's *Theory of Relativity and A Priori Knowledge*, the Marburg school's transcendental method requires determining which concepts and laws play the role of Kant's categories and principles at some given stage in the history of science. But there exists within Cassirer's philosophy a second, complementary theory of the a priori as those concepts and principles that remain invariant throughout the entire history of science.

> The goal of critical analysis would be reached, if we succeeded in isolating in this way the ultimate common element of all possible forms of scientific experience; *i.e.*, if we succeeded in conceptually defining these moments, which persist in the advance from theory to theory because they are the conditions of any theory. At no given stage of knowledge can this goal be perfectly achieved; nevertheless, it remains as a demand, and prescribes a fixed direction to the continuous unfolding and evolution of the systems of experience.
>
> From this point of view, the strictly limited meaning of the "*a priori*" is clearly evident. Only those ultimate logical invariants can be called *a priori*, which lie at the basis of any determination of a connection according to natural law. A cognition is called *a priori* not in any sense as if it were *prior* to experience, but because and in so far as it is contained as a necessary premise in every valid judgment concerning facts.[80]

As Cassirer makes clear in surrounding passages, these invariant a priori cognitions include invariant concepts—such as magnitude, number, space, time, and functional correlation—and a priori principles for the formation and selection of theories—such as the principle that physical laws should be simple and of wide scope.

This theory of the a priori is distinct from—and supplementary to—the relativized constitutive a priori that also appears in Reichenbach's early writings. On that conception, a priori principles can change as theories do and a careful logical analysis of our best current theories can isolate these principles successfully. A priori concepts and principles in the second sense of Cassirer's two-part theory of the a priori[81] are absolute, not relative, remaining invariant throughout the history of science. Furthermore, no amount of careful analysis of our current theories will give us anything more than an "educated guess"[82] about what these invariant principles

---

[79]Cassirer (1906/1922, 16).

[80]Cassirer (1910/1923, 269).

[81]Opposed readings of Cassirer's theory of the a priori are given by Friedman—who recognizes only the second, absolute theory of the a priori in Cassirer and denies that he holds to the first (Friedman 2000, 115 ff.),—and by Richardson—who finds both theories in Cassirer's writings but claims that their conjunction is inconsistent (Richardson 1998, ch. 5). In fact, as I argue, the second theory is not inconsistent with the first theory, but necessitated by it.

[82]I owe this phrase to Friedman (2001, 66). This is Friedman's gloss on Cassirer's claim that the philosophical analysis whose goal it is to isolate these a priori elements "at no given stage can be perfectly achieved."

are, for the simple reason that we cannot foresee how science will develop in the future. Perhaps a *metaphysical* "critical analysis" could arrive once for all at a settled list of a priori cognitions, since (on such a view of philosophy) the nature of knowledge or the nature of reality would be presumably fixed and open to philosophical investigation, and its analysis would not have to wait for the results of the projected final science. (Similarly for a *psychological* "critical analysis" of knowledge, since the psychological character of the mind would not evolve as science does.) But forsaking metaphysical and psychological methods for a "logical analysis of science," Cassirer can pretend to no more certainty about these a priori elements than can be offered by our current best (though still fallible!) physical theories.

Why does Cassirer think that we need a second kind of a priori concepts and principles in addition to the first? Cassirer believes that a theory with only relative a priori principles will threaten the objectivity of scientific theory changes. Relativized a priori principles function like Kantian categories: they provide the conditions of the possibility (and thereby also the *limits*) of empirical meaning. Now, if, as Reichenbach claimed in 1920, the objects coordinated with our concepts are *completely undefined* outside of their coordination with our scientific concepts and our scientific concepts are *empty* without being coordinated by a priori principles to experience, then there will be no common stock of meanings no shared concepts between two scientists at different times (or worse yet: between two scientists who disagree on a new theory). But this would undermine the objectivity of theory change, turning the history of science into a sequence of logically and semantically isolated belief systems. But we know that the history of science is a progressive history, moving asymptotically toward the truth about the natural world. A philosophy of science with only relative a priori principles would threaten the fundamental truth that various scientists at various stages in the history of science are all trying to understand one and the same natural world.[83] Given the undeniable changeability of Kantian categories, then, a historically progressive science is possible only if there are some a priori cognitions that remain invariant through all stages of the history of science.[84]

[83]Cassirer (1910/1923, 321–322): 'Going back to such supreme guiding principles [i.e., the '*form* of experience' that persists in all stages of the asymptotic progression toward the fully empirically adequate theory] insures an inner homogeneity of empirical knowledge, by virtue of which all its various phases are combined in the expression of *one* object. The 'object' is thus exactly as true and as necessary as the logical unity of empirical knowledge;—but also no truer or more necessary ... We need, not the objectivity of absolute things, but rather the objective determinateness of the *method of experience*.'

[84]See Cassirer (1906/1922, 16): "The concept of the *history of science* itself already contains in itself the thought of the *maintenance of a general logical structure* in the entire sequence of special conceptual systems."

It is the principles that are a priori in this second sense[85] that Cassirer cannot relabel as conventions. It simply makes no sense to talk of the very same conventions being laid down throughout the history of science, and it makes even less sense to say that there are conventions that we cannot in principle identify for certain. And it is clear that Reichenbach was never tempted to adopt this theory of the a priori. Why not? The answer reveals deep differences between Reichenbach's and Cassirer's epistemologies of physics. Cassirer introduced the theory of the invariant a priori to address concerns about the objectivity of theory changes. Reichenbach himself recognized the epistemological pitfalls introduced by allowing the constitutive principles of physics to change, but he argued in 1920 that these problems could be solved through the "method of successive approximations." On this method, old coordinating principles are to be replaced by new principles such that "for certain approximately realized cases the new principle is to converge toward the old principle with an exactness corresponding to the approximation of these cases"—as Einstein's theory coincides within the limits of observation to the Newtonian theory in small regions. This method "represents the essential point in the refutation of Kant's doctrine of the a priori" because it shows the adoption of new a priori principles follows an objective rule and can be justified in terms that an adherent to the old principles could understand.[86]

A fuller reply to Cassirer's theory of the invariant a priori is outlined four years later in Reichenbach's book *Axiomatization of the Theory of Relativity*. There he argued that General Relativity could be derived from coordinating definitions, mathematical axioms, and what he called "elementary facts." Even these facts—which he calls "objective coincidences"—are not directly perceived but require the imposition of very simple coordinating definitions. Objective coincidences are distinct from "subjective coincidences," which are certain, immediately given, and independent of interpretation.

> Both methods—the observation of subjective and of objective coincidences—are used in physics. But there is a great difference between them. With respect to the second kind, coincidence is inferred, not perceived. Perceptually speaking, a totality of qualities is given, such as dark and bright spots or sound impressions; from these qualities the objective coincidence of things is inferred. When two billiard balls collide with each other, we hear a characteristic noise; but that this experience *noise* constitutes a coincidence of balls is a logical construction. [...] It cannot be maintained, therefore, that the point events of the

[85]There is a further question: Could Cassirer relabel the *relative* a priori principles as conventions (as Reichenbach did)? Again, the answer is No, as Cassirer argues (for instance) at (Cassirer 1910/1923, 186–187) with regard to Newton's principle of inertia. I hope to explain in a future work why Cassirer rejects conventionalism even in the relativized case.

[86]Reichenbach (1920/1965, 69–70). Padovani (2011) argues that Reichenbach in fact does distinguish in 1920 between relative a priori principles and higher level, meta-principles (such as the principle of probability and the principle of genidentity). This reading of Reichenbach brings him closer to Cassirer. Still though, what becomes of these principles after Reichenbach takes his turn to conventionalism? Are they also conventions? As Padovani argues, Reichenbach has no clear, worked out view after 1920.

> theory of relativity are ultimate facts. Only subjective coincidence has the character of the immediately given; subjective coincidence alone is independent of all interpretations and is a necessary condition for all physical observations.[87]

In particular, Reichenbach identifies three kinds of objective coincidences as basic for the theory of relativity: the coincidences of light signals, the readings of a "natural" clock, and the coincidences of a body with the end points of a rigid measuring rod. All of the theorems of the theory can then be derived from coordinating definitions and simple observational facts about lights, clocks, and rods.

By distinguishing objective and subjective coincidences, Reichenbach's immediate goal is to separate his theory from a simple-minded phenomenalist interpretation—like that of Petzoldt—that reads Einstein's dictum that "all our physical experience can be reduced to [space-time] coincidences" as a Machian view that physics can be grounded in the mere co-presence of sensory qualities.[88] A Machian interpretation errs in identifying subjective and objective coincidences. Instead, Reichenbach argues, the objective coincidences involving lights, clocks, and rods are then *inferred* (defeasibly) from subjective coincidences together with some very simple coordinating coincidences. However, though Reichenbach's immediate target is phenomenalism, his distinction allows him to avoid falling into the opposite extreme. After granting—in proper Kantian fashion—that all statements about objective coincidences "contain some measure of theory," Reichenbach considers the following objection:

> Will it still be advantageous, under such circumstances, to start an axiomatization with so-called empirical facts? Are we permitted to consider such particular facts to be more certain than their confirming theory? Does there exist any confirmation other than that of the theory as a whole? Many answer this question negatively, but they are mistaken. There is a way out which is peculiar to all factual knowledge and which rests on the possibility of approximation.[89]

This "way out" has two elements. First, the subjective coincidences are themselves immediately certain, theory-neutral, and form the inductive basis for objective coincidences. The confirmation of these subjective facts is therefore not holistic. Second, though objective coincidences can be inferred from subjective coincidences only with the help of coordinating definitions, the coordinating definitions required to confirm statements about clocks, rods, and lights are themselves *independent* of the differences between relativistic and pre-relativistic physics.[90] These statements

---

[87]Reichenbach (1924/1969, 17–18).

[88]The quotation is from section 3 of Einstein's (1916) "Die Grundlage der allgemeine Relativitätstheorie." Petzold interprets this passage in a Machian way in Petzold 1921, 64. Reichenbach had earlier criticized Petzold's reading in Reichenbach 1922/1981, 17 ff., following Cassirer's criticism from Cassirer 1921/1923, 392–393. Reichenbach refers to Petzold's reading (without giving his name) in Reichenbach 1924/1969, 16. The historical background to these debate is laid out in Ryckman 1992, cf. Ryckman 1994.

[89]Reichenbach (1924/1969, 5–6).

[90]Reichenbach (1924/1969, 6–7, 19).

are then "relatively invariant with respect to a great variety of interpretations." To infer from the presence of sensations of light to the intersection of two light beams requires coordinating definitions; but the particular very simple definitions needed to grasp this "elementary fact" are available to every party in the dispute over Einstein's theory. Granted, there are differences in how the two theories conceive of the travel of a light beam—in straight lines, or along geodesics that are not necessarily straight—but these differences are irrelevant (by the method of successive approximations) in the small regions that are directly perceivable. In this way, Reichenbach claims that the confirmation of these objective coincidences, though mediated by theory in this limited way, is not holistic.

On this view, then, the subjective coincidences are theory-neutral and can thus form a common, shared basis from which our physical theories can be derived using definitions and principles available to all parties. Reichenbach thus secures the objectivity of adopting Einstein's theory without bringing in Cassirer's invariant a priori. Moreover, Cassirer worried that without some constant element in all stages of our science, we would be unable to claim that a new theory presents a better way of understanding the same world as a previous theory. On Reichenbach's view, the theory-neutrality of subjective coincidences explains this: all parties infer their theories ultimately (though defeasibly) from the same, shared intersubjectively available facts. The invariance of the subjective basis and the simple coordinating definitions takes the place of Cassirer's invariant a priori.

Cassirer, however, denied that there could be such subjective coincidences that are both theory neutral and ground our physical theories even in the mediated and defeasible way Reichenbach describes. Along with his teachers Cohen and Natorp, he always denied that there could be "bare impressions" or "simple sensations" that would be available to any subject no matter what concepts or theoretical commitments she possesses.

> Idealism urges that ... all measurement, however, presupposes certain theoretical principles and in the latter certain universal functions of connection, of shaping and coordination. We never measure mere sensations, and we never measure with mere sensations, but in general to gain any sort of relations of measurement we must transcend the "given" of perception and replace it by a conceptual symbol, which possesses no copy in what is immediately sensed.[91]

In fact, when Cassirer criticizes Machian interpretations of Einstein, he draws a crucially stronger conclusion than Reichenbach does:

> [T]he givenness of 'bare' sensations in which abstraction is made in principle from all elements of form and connection, proves to sharper analysis to be a fiction.[92]

Indeed, this rejection is a central component in Cassirer's attack on the Aristotelian substance-concepts that was our concern in Sect. 4.3 of this chapter. The Aristotelian metaphysics of substantial forms underwrote a theory of perception

---

[91]Cassirer (1921/1923, 427).

[92]Cassirer (1921/1923, 390).

according to which the properties of objects could be transferred to a subject simply through causal contact. However, the fact that two subjects are situated in the same world and are receiving impressions from the same physical objects is not sufficient to provide them with a common intersubjective basis. The objectivity of their representations requires not just shared substances in their environment, but common principles of interpretation—what Kantians might call a shared function of synthesis. This requirement drives Cassirer to adopt a confirmation holism that Reichenbach finds objectionable, and it is this requirement that ultimately leads Cassirer, but not Reichenbach, to postulate a non-conventional and non-relativized a priori. It is this difference that fundamentally distinguishes Neo-Kantians like Cassirer from the empiricists of the Berlin Group.

Needless to say, this is not the place to determine the relative merits of Cassirer's and Reichenbach's interpretations of relativity. Still, though, it is clear that the dispute between Reichenbach and Cassirer over whether a priori concepts and principles are simply conventions rests on a prior dispute over the necessity of positing as a transcendental precondition of the objectivity of science a set of invariant a priori cognitions. This dispute itself, then, rests on a fundamental difference over the role of sensations in simple perceptual knowledge and the possibility of shielding off our measurements from our other theoretical beliefs.

This incomplete story of the relations between Lewin, Reichenbach, and Cassirer demonstrates the depth and the diversity of the personal and philosophical interactions between Cassirer and the members of the Berlin Group. Lewin and Reichenbach, though they both studied with Cassirer, drew differently on his work. As a working empirical researcher, Lewin drew from Cassirer's analyses of the history of the sciences morals about the proper methodology for the psychology of his day, and he used these analyses as inspiration for a deeply original and consequential psychological research program. Reichenbach, not himself an empirical researcher, engaged with Cassirer's work in answering a core and perennial philosophical question, the possibility of empiricism. Though all three shared a commitment to a philosophy of science based in the analysis of the actual sciences, they used this method for different purposes and to draw different conclusions. Indeed, it is telling that the very point on which the dispute between Cassirer's and Reichenbach's interpretations of relativity rests—on the possibility of non-holistic confirmation of elementary measurements—Lewin sides with Cassirer and not with Reichenbach. As we saw above, Lewin drew inspiration for his conceptual innovation within experimental psychology from his opposition to the view of the researcher as "fact collector"—an opposition that Lewin drew from Cassirer, citing for support the very passages defending a holistic, Duhemian theory of measurement that Reichenbach ultimately rejected.[93]

---

[93] See note 45.

References

Ash, Mitchell G. 1994. Gestalttheorie und Logischer Empirismus. In *Hans Reichenbach und die Berliner Gruppe*, ed. Lutz Danneberg, Andreas Kamlah, and Lothar Schäfer, 87–100. Braunschweig: Vieweg.

Brown, J.F. 1929. The methods of Kurt Lewin in the psychology of action and affection. *Psychological Review* 36: 200–221.

Carnap, Rudolf. 1932/1934. Die physikalische Sprache als Universalsprache der Wissenschaft. *Erkenntnis* 2:432–465. Trans. M. Black with author's introduction as *The Unity of Science*. London: Kegan, Paul, Trench Teubner & Co.

Carnap, Rudolf. 1932/1959. Psychologie in physikalischer Sprache. *Erkenntnis* 3:107–142. Trans. G. Schick as psychology in physicalist language in *Logical positivism,* ed. Ayer, 165–198. New York: Free Press.

Carnap, Rudolf, Hans Hahn and Otto Neurath. 1929/1973. *Wissenschaftliche Weltauffassung - Der Wiener Kreis*. Vienna. Trans. Artur Wolf as The scientific conception of the world: The Vienna Circle, in Otto Neurath, *Empiricism and Sociology*, ed. M. Neurath and R.S. Cohen, 299–318. Dordrecht: Reidel.

Carus, A.W. 2008. *Carnap and twentieth century thought: Explication as enlightenment.* Cambridge: Cambridge University Press.

Cassirer, Ernst. 1906/1922. *Das Erkenntnisproblem in der Philosophie und Wissenschaft der neuren Zeit. Erster Band*, 3rd ed. Berlin: Bruno Cassirer.

Cassirer, Ernst. 1907. Kant und die moderne Mathematik. *Kant-Studien* 12: 1–40.

Cassirer, Ernst. 1910/1923. *Substanzbegriff und Funktionsbegriff. Untersuchungen über die Grundfragen der Erkenntniskritik.* Berlin: Bruno Cassirer. Trans. William Curtis Swabey, and Marie Collins Swabey in *Substance and function & Einstein's theory of relativity.* Chicago: Open Court.

Cassirer, Ernst. 1912. Hermann Cohen und die Erneuerung der Kantischen Philosophie. *Kant-Studien* 17: 252–273.

Cassirer, Ernst. 1918/1981. *Kants Leben und Lehre.* Berlin: Bruno Cassirer. Trans. James Haden as *Kant's life and thought*. New Haven: Yale University Press.

Cassirer, Ernst. 1921/1923. *Zur Einstein'schen Relativitätstheorie*. Berlin: Bruno Cassirer Verlag. Trans. William Curtis Swabey, and Marie Collins Swabey in *Substance and function & Einstein's theory of relativity*. Chicago: Open Court.

Cassirer, Ernst. 1923/1955. *Philosophie der Symbolischen Formen. Erster Teil: Die Sprache.* Berlin: Bruno Cassirer. Trans. Charles W. Hendel, and Ralph Manheim as *Philosophy of symbolic Forms. Volume One: Language.* New Haven: Yale University Press.

Cassirer, Ernst. 1927. Erkenntnistheorie nebst den Grenzfragen der Logik und Denkpsychologie. Reprinted in *Erkenntnis, Begriff, Kultur*, ed. R. Bast. Hamburg: Felix Meiner Verlag, 1993.

Cassirer, Ernst. 1942/2000. *Zur Logik der Kulturwissenschaften: Fünf Studien.* Göteborg: Wettergren & Kerbers Forlag. Trans Lofts, S.G. as *The logic of the cultural sciences.* New Haven: Yale University Press.

Cassirer, Ernst. 1999. *Ziele und Wege der Wirklichkeitserkenntnis*, ed. Klaus Christian Köhnke and John Michael Krois. Hamburg: Felix Meiner Verlag.

Cassirer, Ernst. 2009. Ausgewählter wissenschaftlicher Briefwechsel. In *Nachgelassene Manuskripte und Texte*, vol. 18, ed. John Michael Krois et al. Hamburg: Meiner.

Coffa, J.Alberto. 1991. *The semantic tradition from Kant to Carnap.* Cambridge: Cambridge University Press.

Cohen, Hermann. 1902. *Logik der reinen Erkenntnis*. Berlin: Bruno Cassirer.

Duncker, Karl. 1932/1933. Behaviorismus und Gestaltspsychologie. Kritische Bemerkungen zu Carnaps 'Psychologie in physikalischer Sprache. *Erkenntnis* 3: 162–176.

Einstein, Alfred. 1916. Die Grundlage der allgemeine Relativitätstheorie. *Annalen der Physik* 49:769–822. Trans. W. Perret, and G. B. Jeffery in *The principle of relativity*, 109–164. New York: Dover Publications, n.d.

Ferrari, Massimo. 1997. Über die Ursprünge des logischen Empirismus, den Neukantianismus und Ernst Cassirer aus der Sicht der neueren Forschung. In *Von der Philosophie zur Wissenschaft. Cassirers Dialog mit der Naturwissenschaft*, ed. Enno Rudolph and Ion O. Stamatescu. Hamburg: Meiner.

Ferrari, Massimo. 2003. *Ernst Cassirer: Stationen einer philosophischen Biographie.* Trans. Marion Lauschke. Hamburg: Felix Meiner Verlag.

Friedman, Michael. 1999. *Reconsidering logical positivism.* Cambridge: Cambridge University Press.

Friedman, Michael. 2000. *A parting of the ways: Carnap, Cassirer, and Heidegger.* Chicago: Open Court Press.

Friedman, Michael. 2001. *Dynamics of reason.* Stanford: CSLI Publications.

Heis, Jeremy. 201?. Ernst Cassirer's *Substanzbegriff und Funktionsbegriff. HOPOS: The Journal of the International Society for the History of Philosophy of Science.*

Kant, Immanuel. 1998. *Critique of pure reason.* Trans. Paul Guyer, and Allen Wood. Cambridge: Cambridge University Press.

Krois, John Michael. 2000. Ernst Cassirer und der Wiener Kreis. In *Elemente moderner Wissenschaftstheorie*, ed. Friedrich Stadler, 105–121. Vienna: Springer.

Lewin, Kurt. 1922. *Der Begriff der Genese in Physik, Biologie und Entwicklungsgeschichte.* Berlin: Bonträger.

Lewin, Kurt. 1925. Über Idee und Aufgabe der vergleichenden Wissenschaftslehre. *Symposion* 1:61–93. Reprinted in *Wissenschaftstheorie I,* ed. Alexandre Métraux as vol.1 of *Werkausgabe*, ed. Carl-Friedrich Graumann. Stuttgart: Klett-Cotta.

Lewin, Kurt. 1926. Intention, Will, and Need. In *A Kurt Lewin Reader,* ed. Martin Gold. New York: American Psychological Association.

Lewin, Kurt. 1927. Gesetz und experiment in der psychologie. Reprinted in *Wissenschaftstheorie I,* ed. Alexandre Métraux as vol.1 of *Werkausgabe*, ed. Carl-Friedrich Graumann. Stuttgart: Klett-Cotta

Lewin, Kurt. 1931a. Der Übergang von aristotelischen zum galileischen Denken in Biologie und Psychologie. *Erkenntnis* 1: 429–466.

Lewin, Kurt. 1931b. The conflict between Aristotelian and Galilean modes of thought in contemporary psychology. *Journal of General Psychology* 5:141–177. Trans. D. K. Adams. Reprinted in *A Kurt Lewin Reader,* ed. Martin Gold. New York: American Psychological Association.

Lewin, Kurt. 1936. *Principles of topological psychology.* New York: McGraw-Hill Book Company.

Lewin, Kurt. 1937. Psychoanalysis and Topological Psychology. In *A Kurt Lewin Reader,* ed. Martin Gold. New York: American Psychological Association.

Lewin, Kurt. 1949. Cassirer's philosophy of science and the social sciences. In *The philosophy of Ernst Cassirer*, ed. A. Schilpp, 269–288. Evanston, Ill.: Open Court. Reprinted in *A Kurt Lewin Reader,* ed. Martin Gold. New York: American Psychological Association.

Lewin, Kurt. 1981. *Wissenschaftstheorie I,* ed. Alexandre Métraux as vol. 1 of *Werkausgabe*, ed. Carl-Friedrich Graumann. Stuttgart: Klett-Cotta.

Lewin, Kurt. 1992. Law and experiment in psychology. *Science in Context* 5: 385–416.

Lewin, Kurt. 1999. *The complete social scientist: A Kurt Lewin Reader,* ed. Martin Gold. New York: American Psychological Association.

Marrow, Alfred. 1969. *The practical theorist: The life and work of Kurt Lewin.* New York: Basic Books, Inc.

Natorp, Paul. 1912. Kant und die Marburger Schule. *Kant-Studien* 17: 193–221.

Neurath, Otto. 1930. Historische Anmerkungen. *Erkenntnis* 1: 311–314.

Neurath, Otto, et al. 1930. Chronik. *Erkenntnis* 1: 72–79.

Neurath, Otto, et al. 1931. Chronik. *Erkenntnis* 2: 310.

Padovani, Flavia. 2011. Relativizing the relativized a priori: Reichenbach's axioms of coordination divided. *Synthese* 181: 41–62.

Petzold, Joseph. 1921. *Die Stellung der Relativitätstheorie.* Leipzig: Barth.

Reichenbach, Hans. 1920/1965. *Relativitätstheorie und Erkenntnis apriori*. Berlin: Springer Verlag. Trans. and ed. Maria Reichenbach as *The theory of relativity and a priori knowledge*. Berkeley: University of California Press.

Reichenbach, Hans. 1922/1981. The present state of the discussion of relativity. In *Modern philosophy of science: Selected essays*. ed. Hans Reichenbach. London: Routledge & Kegan Paul.

Reichenbach, Hans. 1924/1969. *Axiomatik der relativistischen Raum-Zeit-Lehre*. Braunschweig: Fried. Vieweg & Sohn. Trans. M. Reichenbach as *Axiomatization of the theory of relativity*. Berkeley: University of California Press.

Reichenbach, Hans. 1928/1957. *Philosophie der Raum-Zeit-Lehre*. Berlin: Walter de Gruyter. Trans. Maria Reichenbach and John Freund as *The philosophy of space and time*. New York: Dover.

Reichenbach, Hans. 1936. Logical empiricism in Germany and the present state of its problems. *The Journal of Philosophy* 33: 141–160.

Reichenbach, Hans. 2008. *The concept of probability in the mathematical representation of reality*. Trans. and ed. Frederick Eberhardt and Clark Glymour. Chicago: Open Court Publishing Co.

Richardson, Alan. 1998. *Carnap's construction of the world: The Aufbau and the emergence of logical empiricism*. Cambridge: Cambridge University Press.

Richardson, Alan. 2006. 'The fact of science' and critique of knowledge: Exact science as problem and resource in Marburg Neo-Kantianism. In *The Kantian legacy in nineteenth-century science*, ed. Michael Friedman and Alfred Nordmann, 211–226. Cambridge, MA: MIT Press.

Ryckman, Thomas. 1992. P(oint)-C(oincident) thinking: The ironical attachment of logical empiricism to general relativity (and some lingering consequences). *Studies in History and Philosophy of Science Part A* 23: 471–497.

Ryckman, Thomas. 1994. Weyl, Reichenbach and the epistemology of geometry. *Studies in History and Philosophy of Science Part A* 25: 831–870.

Ryckman, Thomas. 2005. *The reign of relativity: Philosophy in physics, 1915–1925*. New York: Oxford University Press.

Schlick, Moritz. 1921/1979. Kritizistische oder empiristische Deutung der neuen Physik? *Kant-Studien* 26: 96–111. Translated as Critical or Empiricist Interpretation of Modern Physics? In *Moritz Schlick: Philosophical Papers*. Vol. 2, eds. H. Mulder and B. van de Velde-Schlick. Dordrecht: Reidel.

Part III
Hans Reichenbach

Chapter 5
Genidentity and Topology of Time: Kurt Lewin and Hans Reichenbach

Flavia Padovani

5.1 Introduction

Kurt Lewin and Hans Reichenbach have been central figures in the birth of logical empiricism as well as in the Berlin Group. As with many of their colleagues, their acquaintance goes back to the time of their involvement within the German youth movement known as "*Freistudentenschaft*", before World War I. At that time, they had close contact on matters related to their curriculum of studies, as we shall see in Sect. 5.2. In the early 1920s, they were among those who tackled the issue of the definition of time order. Starting from different considerations, they developed two original accounts of the topology of time, which display interesting affinities. In Sect. 5.3, I will compare these accounts and thus address some historical and theoretical questions they raise.

Before going on, let us first clarify in what sense we will use the term "topology" in this paper. Here, this term is referred not to the usual notions of topology but rather to the objective system of relations and coincidences of point-events that can be established *before* any metrical determination, thereby independent of any arbitrariness. In fact, both Reichenbach and Lewin were engaged in the attempt to derive the truth of a number of temporal propositions from the occurrence of characteristic and elementary phenomena in the world. The topological relations defined in their

F. Padovani (✉)
English and Philosophy Department, Drexel University, Philadelphia, PA, USA
e-mail: fp72@drexel.edu

N. Milkov and V. Peckhaus (eds.), *The Berlin Group and the Philosophy of Logical Empiricism*, Boston Studies in the Philosophy and History of Science 273, DOI 10.1007/978-94-007-5485-0_5,

works did not merely portray some representational artefact. On the contrary, what these two constructions intended to capture was an "ultimate fact" of nature.[1]

A common feature of these two accounts is that they both crucially make use—implicitly or explicitly—of the notion of "genidentity" (i.e., identity through/over time). The basic questions they address can be sketched as follows: (1) When are two events genidentical, that is, involving the same thing? (2) When are two genidentical events one before the other, or else simultaneous? (3) When are two non-genidentical events one before the other, or else simultaneous? The critical element of divergence is that Reichenbach aimed to illustrate the temporal order of physically *possible* events, whereas Lewin dealt with the ordering of *actual* events.

Reichenbach first mentioned the concept of "genidentity" in his *Relativitätstheorie und Erkenntnis apriori* (1920), where it was regarded as a synthetic, yet revisable a priori principle. In this book, genidentity represented a very special principle of coordination of formal structures to reality required in order to identify an object *as* the selfsame object in the passage through time. In the shift toward a conventionalist stance that Reichenbach notoriously embraced soon after 1920, coordinating principles would be turned into either coordinative (conventional) definitions or epistemological and methodological principles. However, the principle of genidentity was dropped in this passage. In both his "Bericht" (1921b) and *Axiomatik* (1924a), Reichenbach took the notion of "light signal" as the basis from which to derive the topological properties of (space-)time. By doing so, the key idea behind the principle of genidentity was dissolved into what will be later known as "the mark principle" (Sect. 5.3.2).

In *Der Begriff der Genese* (1922), Lewin used the notion of genidentity primarily as a sort of analytic tool to pursue his project of comparative sciences. In the following publication on this topic, "Die zeitliche Geneseordnung" (1923a), he addressed the question of the possibility of erecting a temporal topology only on the basis of certain most elementary features of reality. Thus, he proposed a topological account of time by taking the existential relation implied by genidentity as primitive (Sect. 5.3.1). With respect to Reichenbach's attempt, Lewin's approach turned out to involve a more fundamental level of analysis that Reichenbach should have considered given his claim that his axiomatic construction starts with the most elementary facts. In their correspondence of the early 1920s, which we shall briefly analyse in Sect. 5.3.2.2, Lewin will argue that Reichenbach's notion of "first signal" can, and actually should, be decomposed and reduced to its constituents. According to Lewin, the notion of "identity through time" as an existential relation is a primitive notion that must be presupposed even before considering any physical process, such as a light signal. The concept of signal, to Lewin's mind, already conveys the sense of *some thing* propagating through time and, therefore, requires a prior principle of temporal individuation that his formalised concept of genidentity is meant to provide.

---

[1] Reichenbach (1928/1958, 285). Cf. also Ryckman (2007, 205 ff.).

As a result of this exchange, the principle of genidentity will be reintroduced in Reichenbach's work, notably in his *Die Philosophie der Raum-Zeit-Lehre* (1928), where it will be surprisingly defined as an "empirical principle" (Sect. 5.4). In this conclusive Section, I will also briefly emphasise the importance of Lewin's influence on Reichenbach's first attempt to define the direction of time in "Die Kausalstruktur der Welt" (1925), and I will finally suggest that this principle still holds a peculiar position in Reichenbach's later work, more precisely as a constitutive principle.

5.2 Crossed Destinies: From the Youth Movements to the Erlangen Meeting, Through the Wireless Telegraphy

Like many of their German contemporaries, while they were attending university, Lewin and Reichenbach took part in the activities of a student movement known as *Freistudentenschaft*. It was essentially a libertarian, egalitarian, anti-racist, anti-authoritarian, estheticist, and non-traditionalist movement, in certain respects also anti-militarist. Oriented by the idea of the moral self-determination of the individuals and of the freedom of directing one's own future, this movement stood against any form of dogmatism, be it scholastic, religious, philosophical, political or institutional.[2] Like Lewin, Reichenbach joined the *Freistudentenschaft* around 1910, and his adhesion to the ideals of this group later developed into a prominent political commitment with socialist groups.[3]

Reichenbach's relation to Kurt Lewin dates back to at least 1911. It was likely on the occasion of the 1911 national Berlin meeting of this association that they must have first met.[4] In the summer of that year, they started a correspondence on matters related to education in German universities. In the only letter of this early exchange that was preserved,[5] Lewin warmly responded to what must have been Reichenbach's inquiry about the situation of scientific psychology (*wissenschaftliche Psychologie*) in German universities, especially in Berlin,

---

[2]For an analysis of the birth and development of the German student movement, cf. Wipf (2004). To have an idea of Reichenbach's active contribution to the activities of this group, see also Wipf (1994) and Maria Reichenbach's introductory remarks to Reichenbach (1978, Vol. I, 91–101). An overview of Lewin's involvement can be found in Ash (1995, 265 ff.).

[3]In 1918, Reichenbach drew the programme of the socialist student party and published a number of pamphlets distributed in alternative circles. Cf. Reichenbach (1978, Vol. I, 132–185).

[4]Reichenbach presented two papers that would be published as (1911a) and (1911b).

[5]In what follows, I will mainly refer to material from the Hans Reichenbach Collection (HR) available at the Archives for Scientific Philosophy (ASP) of Pittsburgh and Konstanz, except from one letter from the Moritz Schlick Collection available at the *Wiener Kreis Stichting* in Amsterdam. All the material is quoted by permission of the University of Pittsburgh and the *Wiener Kreis Stichting*. All rights are reserved.

Munich and Göttingen, i.e., the three universities where Reichenbach actually spent the following four years.[6]

Both Lewin and Reichenbach attended Ernst Cassirer's and Carl Stumpf's lectures at the University of Berlin at different times, and were variously influenced by them. As an experimental psychologist, Stumpf was a pioneer also in experimenting with tones, and in the psychology of acoustic perception—teachings that would soon reveal to be important for both of them. After a short period on the Russian front line, Reichenbach served for almost 2 years in the signal corps of the German army in Neuruppin, not far from Berlin.[7] Like Reichenbach, Lewin volunteered and served in the army, until he was injured in combat in 1917.[8] When he was transferred near Berlin, he was assigned to the construction of devices for sound measurement. Respectively, they brought their scientific and technical competences in the context of the new situation that was originating at the front, where various scientific fields proved to be of pivotal importance in solving a variety of military problems. Also psychologists contributed to this endeavour especially by providing skill tests in order to select the most suitable individuals for each specific military task.[9] It is in

[6]Lewin recommended Reichenbach to go to Munich rather than to Berlin, and to refer to the local section of the free student body, motivating his advice on the lack of interesting professors in Berlin, on the one hand, and on the opportunities that the Munich surroundings offered for skiing on the other ["Berlin kann ich Ihnen gar nicht empfehlen, wohl aber München für den Winter. Treten Sie in die Abteilung der M.[ünchener] Fr.[eien]St.[udentenschaft] ein [...] und lernen Sie skilaufen." (ASP, HR 023-13-31, Lewin to Reichenbach, 29 September 1911)]. To be sure, Lewin described the situation of the studies in psychology in Germany as not exciting at all. For example, Theodor Lipps is portrayed as an old man, almost never teaching and in general unable to deliver good lectures, despite having written a good textbook. Alexander Pfänder, student of Lipps, is presented as a good philosopher, yet with too little knowledge of scientific psychology (*von naturwissenschaftlichen Psychologie*). Georg Elias Müller appears to be one of the few having a scientific approach and as well as being truly engaged in teaching specific chapters of psychology. In Lewin's view, Hans Rupp was not particularly good as teacher, but provided the students with useful exercises, so that he seemed the only one really promoting psychology in Berlin. Curiously, the worst picture Lewin drew is that of one his own future supervisor, Carl Stumpf, described as teaching poorly and boringly ["ich kann Sie versichern, er taugt nichts und ist obendrein sprachlich langweilig"]. These discouraging words notwithstanding, besides philosophy and traditional mathematical-physical classes, Reichenbach will attend Stumpf's course *Psychologie mit Demonstrationen* and Rupp's *Experimentelle Übungen zur Psychologie* in the 1911-1912 winter semester in Berlin (ASP, HR 041-09-16). He will instead go to Munich the following academic year, and indeed become a member of the local division of the *Freie Studentenschaft*, with a specific focus on psychological research. Whether and how Lewin was influential in making Reichenbach turn at first directly to philosophy is difficult to say. But it may be a consequence of Lewin's positive appreciation of Müller if Reichenbach first (unsuccessfully) tried to have him as his supervisor in Göttingen. His doctoral thesis, published as Reichenbach (1916), was eventually supervised (albeit only formally) by Paul Hensel and the mathematician Max Noether, and was defended at the University of Erlangen in 1915.

[7]Cf. Gerner (1997, ch. 2.3.)

[8]See Marrow (1969, 10–11).

[9]Mitchell Ash illustrates this state of affairs in the following way: "The First World War saw the emergence of the technological battlefield, and it also marked a turning point in the interaction of technology and basic science. In numerous ways scientists demonstrated the usefulness of basic research by employing laboratory instruments and techniques to solve military problems, from the

the spirit of these synergies that in this period Reichenbach collaborated with Kurt Lewin and Otto Lipmann—one of the first to apply the new research tendencies in psychology to the industrial problems, and co-founder, with William Stern, of the *Zeitschrift für angewandte Psychologie*—on the development of an aptitude test for radio telegraphists.[10]

During the years 1918–1920, Lewin and Reichenbach were both in Berlin. Reichenbach was employed in the Huth radio industry,[11] while attending Einstein's lectures on statistical mechanics, special and general relativity.[12] Lewin was a member of the Institute of Psychology, around which, in contrast to the traditional Wundtian approaches, Stumpf gathered the future leading figures in this new direction in experimental and applied psychology, like Wolfgang Köhler, Kurt Koffka, and Max Wertheimer, all embracing the ideas of Gestalt psychology. At that time, Lewin was working on a lengthy monograph on the concept of "genetic series" that he unsuccessfully tried to submit as his *Habilitation* thesis to obtain the qualification for teaching philosophy at the University of Berlin.[13] The thesis would not be published until 1922, as we shall see in the next section.

After being bestowed the teaching *Habilitation* for physics with his *Relativitätstheorie und Erkenntnis apriori* (1920), Reichenbach was first appointed assistant to the physicist Erick Regener, and soon after "Privatdozent" (a sort of associate professor) at the Stuttgart Technische Hochschule. During the entire period in Stuttgart, Reichenbach kept in touch with Lewin for discussions concerning not only philosophical issues, but also one of the most important events in the early phases of logical empiricism, the famous Erlangen conference of 1923, in which both Reichenbach and Lewin participated. One of the issues discussed during that meeting was the creation of a journal for exact philosophy that was only realised with the publication of the first issue of *Erkenntnis* in 1930. Lewin was one of the most active among those who were engaged in the organisation of this

development of sound-ranging devices in physics to that of poison gas in chemistry. Psychologists in Germany participated in this process by adapting psychophysical measurement to skills testing for the selection of communications specialists, pilots, and drivers. The focus of these efforts was on the 'human factor'—on the human organism as a functioning part of a machine. Practitioners in the new field called 'psychotechnics' searched for the machine operators whose skills were best suited to the task in question—whose reactive idiosyncrasies, that is, interfered least with efficient functioning." Ash (1995, 188).

[10]This document, entitled "Entwurf zu einer Eignungsprüfung für Funkentelegraphisten" and signed by Lewin, Lippmann, and Reichenbach, was completed in 1917 (ASP, HR 024-16-02).

[11]His work focussed on particular amplifying valves. See, for instance, some of his more technical writings of the period like "Statistisches Verfahren zur Beurteilung von Verstärkerröhren" (ASP, HR 044-03-21), "Zur Theorie der Verstärkerröhren" (ASP, HR 044-03-23), and "Beschreibung des Huth'schen Verstärkers L. 43 F. 25 m Niederspannung-Verstärker" (ASP, HR 044-03-25).

[12]Cf. Reichenbach's five corresponding lecture notebooks (ASP, HR 028-01-01/05), as well as Gerner (1997, ch. 2).

[13]Carl Stumpf also supervised his *Habilitation* thesis. References to the history of this monograph, and the difficulties that Lewin had to face in order to receive that qualification with this treatise can be found in Métraux's introduction to Lewin (1983) and in Métraux (1992).

periodical, as an intense correspondence with Reichenbach demonstrates.[14] Other important topics of debate were the comparative theory of sciences, the axiomatic method in physics, and the topology of time. Each of these discussions was of clear significance to both of them, and they would lead to further exchanges of views. Despite their different approach, their topological accounts of time will have a common denominator: the use—explicit or not—of the principle of genidentity as fundamental.

5.3 The Many Faces of Genidentity

The term "genidentity" was coined by Lewin in the *Habilitation* thesis that he wrote (but did not publish) in 1920, but was officially introduced in 1922 in (what seems to be) the revised version of this thesis that was published under the title *Der Begriff der Genese in Physik, Biologie und Entwicklungsgeschichte* (1922). To be sure, the first time this term appears in a publication is in Reichenbach's 1920 book on relativity theory. In this book, Reichenbach explicitly refers to Lewin's *Die Verwandtschaftsbegriffe in Biologie und Physik und die Darstellung vollständiger Stammbäume* (1920)—where this term actually does not appear—and to another book by Lewin seemingly dealing with the order type of genetic series in various domains. Reichenbach refers to the latter book as being entitled *Der Ordnungstypus der genetischen Reihen in Physik, organismischer Biologie und Entwicklungsgeschichte*. However, this work does not appear to have ever been published, although Reichenbach even names a publisher (Borntraeger, Berlin). Presumably, this was the original title of Lewin's unpublished thesis, which Reichenbach must have read in draft while they were both in Berlin. In these early years, they were certainly very close, and most likely often discussing their respective researches. It is not a coincidence, in fact, if a few years later Reichenbach recalled that Lewin, along with Einstein, were the two persons to whom he showed the drafts of his relativity book.[15] In this work, he borrows the concept of genidentity from Lewin, but develops his account in a different fashion.

[14]From July 1923 to October 1924 they exchanged about twenty letters, almost exclusively related to the creation of their journal. See ASP, HR 016-36-07/26.

[15]In some autobiographical notes written in 1927, Reichenbach reconstructed the circumstances in which he wrote his own *Habilitation* thesis with the following words: "Im Februar (oder März) 1920 beschloss ich, meine Habilitationsschrift zu schreiben. Ich hatte in den Monaten vorher Relth. gearbeitet, auch nach Weyl; den Grund hatte ich schon in 1917–1918 in Vorlesungen bei Einstein gelegt, aus welchen meine Kenntnis der Th.[eorie] herrührt.[...] Die Schrift ist in etwa 10 Tagen niedergeschrieben. Das M[anu]s[kript] wurde dann abgetippt. Ich zeigte es Einstein u. Lewin. Durch Berliners Vermittl[un]g kam es zu Springer. Erschienen ist es im Sept. 1920, zum Naturforschertag." ASP, HR 044-06-23.

In what follows, I shall compare the meanings and roles assigned to genidentity in the works of the two authors. We will see how the differences account for two distinct topologies of time.

5.3.1 *The Concept of Genidentity in Kurt Lewin*

Lewin's first publication to tackle the problem of the order type deriving from particular sequences called "genetic" is *Die Verwandtschaftsbegriffe in Biologie und Physik und die Darstellung vollständiger Stammbäume*.[16] This is a study of some specific relationships in different sciences, namely biology and physics (where physics is interpreted in a very loose sense, including—and at times, identifying with—chemistry)[17] and represents his first treatise of comparative sciences.

The gist of Lewin's work is to identify a certain (genetic) type of order (*Ordnungstypus*) in these sciences, an order that is characteristically exemplified, for instance, by the relationship of relatedness in biology and that of affinity in chemistry. Lewin shows how these relationships can be represented through family trees. In particular, analysing biological pedigrees, he focusses on the genealogical relation, necessarily asymmetric, between ancestors and progeny, which he terms "genetic series" (*Genetische Reihe*). In general, a genetic series is an effectual relation of antecedency of the type "being-such-as-to-have-come-forth-from".[18] Interestingly, Lewin's investigation is carried out with a sort of mereological approach in which the sequences are compared with the whole of the formations they belong to.[19]

In spite of the title of this essay, the corresponding relation of existence that could obviously be thought of in physics is not considered by Lewin. In one passage, though, he highlights an essential aspect that characterises the physical sequences, that is, their continuity and extension towards infinity along their both sides, which

[16]Lewin (1920). The foreword is dated March 31, 1920.

[17]As Lewin declares in the first lines, "[i]n der Physik, worunter hier die Physik im weiteren Sinne des Wortes verstanden wird, wird der Begriff der Verwandtschaft im allgemeinen für die chemische Verwandtschaft benutzt." Lewin (1920, 5).

[18]"Die Existentialbeziehung, die in der Biologie als Verwandtschaft bezeichnet wird, kann einmal zwischen Gebilden bestehen, die *auseinander hervorgegangen* sind, z. B. zwischen Kind und Eltern oder Großeltern, oder zweitens zwischen Gebilden, die *gemeinsame Vorfahren oder Nachkommen* besitzen, ohne selbst voneinander abzustammen. Auf die Existentialbeziehung zwischen Vorfahren und Nachkommen, das existentielle Auseinanderhervorgegangensein im *Nacheinander*, soll hier nicht näher eingegangen werden." Lewin (1920, 20). The reference goes to the above-mentioned forthcoming more extended research.

[19]This approach will be cast in more formal terms in his following publication on this topic, *Der Begriff der Genese* (1922). Analyses and discussions brought up in mereological terms were quite frequent in Stumpf's circle. For an account of the theory of the whole/part relations in Stumpf and his school, cf. Smith and Mulligan (1982)

illustrates a relation of complete (*restlos*) derivation among each and every slice.[20] What this type of derivation means is better specified by Lewin only in 1922 with the expression "complete physical genidentity", as we shall see below, p. 105.

Lewin introduces and expounds on the notion of "genidentity" in *Der Begriff der Genese* (1922), where it is embedded in a much larger project of comparative sciences than the first research of 1920. More than underlining commonalities among sciences, Lewin's comparative approach also highlights the sciences' specificities. What is compared and studied are not similar objects (*Objecte*) *per se* but objects displaying an equivalence from the viewpoint of the theoretical knowledge we have of them, that is, what he defines as *"wissenschaftstheoretisch äquivalente" Objecte*.[21] The aim of this more extensive research is to show that, in spite of certain structural similarities that can be identified in such objects, they exhibit different modalities of application in the compared sciences. In 1922, this aim is pursued by investigating how the concept of genidentity functions and the way it acquires its meaning within various domains.

The concept of genidentity is again essentially characterised in mereological terms. Objects or events are temporally extended, thus their genetic series consists of a multitude of entities, representing their various phases at various times. According to Lewin, physical constructs (*Gebilde*) that have developed one from the other can be conceived as temporally distinct, so that we can define the relation of genidentity as the existential relation that holds between them.[22] Here we won't follow Lewin's complex treatment of the multiplicity of relations that can be subsumed under this same concept. To just give an idea of some different but related notions of genidentity, let us take a classical example that Reichenbach also briefly mentions

[20]"In der Physik handelt es sich, wie hier nicht weiter ausgeführt werden kann, um einen kontinuierlichen, beiderseits ins Unendliche gehenden Reihentypus. Es ist ein wesentliches Charakteristikum dieser Reihen kontinuierlich aufeinander folgender Schnitte, daß es auch zu jedem beliebig herausgegriffenen Teil eines Schnittes eine solche beiderseits unendliche Reihe von Schnitten gibt, mit denen er in der Beziehung des restlosen existentiellen Auseinanderhervorgegangenseins steht." Lewin (1920, 21).

[21]It goes without saying that the main source of inspiration for this approach is Cassirer's famous monograph *Substanzbegriff und Funktionsbegriff* (1910). As Lewin wrote in 1920: "Diese die Vergleichbarkeit begründende wissenschaftstheoretische Äquivalenz, die sich auf die ganze Stellung der Vergleichsobjekte in den betreffenden Wissenschaften stützt, kann nicht durch irgendwelche äußerliche Übereinstimmungen ersetzt oder durch äußerliche Ungleichheiten widerlegt werden. Die wissenschaftstheoretische *Äquivalenz* bestimmter Begriffsgebilde oder Einteilungsprinzipien in verschiedenen Wissenschaften bedeutet daher auch umgekehrt noch keine völlige Gleichheit der betreffenden Vergleichsobjekte oder gar eine Identität der betreffenden Wissenschaften." Lewin (1920, 3). See also Lewin (1925).

[22]"*Physikalische Gebilde, die zu verschiedenen Zeitmomente existieren, sollen also als eine Mehrheit von Gebilden aufgefaßt werden*, nicht anders als gewisse räumlich verschiedene Gebilde. [...] *Physikalische Gebilde, die auseinander hervorgegangen sind, müssen*, abgesehen von anderen möglichen Unterschieden, jedenfalls *zeitlich verschieden sein.* Wir wollen, um Verwechslungen zu vermeiden, die Beziehung, in der Gebilde stehen, die existentiell auseinander hervorgegangen sind, *Genidentität* nennen. Dieser Terminus soll nicht anderes bezeichnen als die genetische Existentialbeziehung als solche." Lewin (1922/1983, 60–62).

in his (1924c, 189). Between the egg and the hen there is a relation of biological genidentity in that they represent different stages of development of the same biological matter: they are slices of the same genetic series connecting the selfsame individual along a temporal sequence. From a physical viewpoint, though, they are not genidentical because the molecules composing them have changed. Besides, a physical genetic series can lead from the egg to the variety of other formations that can be developed from it, so it may also bring forth, say, a piece of cake.

These various ways of "cutting" reality into units reflect the different angles from which to analyse real objects or processes in temporal perspective. In 1922, Lewin introduces a different terminology to distinguish the multiplicity of genetic series, like the biological from the physical genidentity. For instance, the biological relation of genidentity between an individual and his or her descendants is defined as "avalgenidentity", whereas the relation of genidentity holding among successive temporal sections of individuals—whole organisms, but also cells—is called "individual genidentity". While the relations of genidentity used in these domains may show some affinities, they do not share the same properties when implemented in different contexts. Every science represents a specific, closed network of interrelated concepts that cannot simply be extrapolated and applied within other networks. Hence, the passage from one science to the other implies an ideal partitioning of reality.[23]

For our discussion, the relation of physical genidentity is the most salient one, as it represents a good candidate to determine temporal sequences. We have seen above that in 1920 Lewin briefly mentions an important feature of sequences in physics, that is, the relation of complete derivation among genetic slices of physical events. This relation of complete derivation is now elaborated and presented as "complete genidentity". Roughly, the idea is that physical constructs are in a relation of complete genidentity when none of their own other parts stands in a relation of genidentity with some other construct.[24] In general, the completeness (*Restlosigkeit*) of the physical, genidentical relation can be understood, Lewin explains, as typically presupposed by the activity of experimenting in physics, where it is ideally required that physical systems be in isolation from external disturbances. For cases where complete genidentity does not apply but an existential relation of partial antecedency can still be identified, Lewin speaks of "simple

[23]Lewin (1922/1983, 131 ff.). For example, the relation of complete physical genidentity is characterised by: (1) continuity in the passage between the correlated constructs; (2) independence of the relation from the direction of the series (symmetry); (3) independence from the distance between the slices along a sequence; and (4) exclusion of a partially unconnected construct simultaneously existing with one of two completely genidentical slices that is simply genidentical with the other one. Cf. Lewin (1922/1983, 89–90). Thus, complete physical genidentity has transitivity, temporal density, and continuity as properties, whereas avalgenidentity presupposes discontinuity, lack of density, and some further conditions for transitivity. See Lewin (1922/1983, 158 ff.).

[24]"Ein physikalisches Gebilde a_1 ist restlos genidentisch mit a_2, wenn (1) im Zeitpunkt 1 kein zu a_1 teilfremdes physikalisches Gebilde (Teil eines Gebilde) existiert, das mit a_2 in Genidentitätsbeziehung steht, und wenn (2) im Zeitpunkt 2 kein zu a_2 teilfremdes physikalisches Gebilde (Teil eines Gebildes) existiert, das mit a_1 in Genidentitätsbeziehung steht." Lewin (1922/1983, 82).

genidentity" (*Genidentität überhaupt*). Simple genidentity is the relation that holds between constructs that share at least one part correlated by complete genidentity. For example, consider the case of a piece of metal that is plunged into a liquid at a time t_1 and is found altered afterwards, say at a time t_2. If you then remove a part of the liquid, you will have to suppose that something of the metal that was partially dissolved in it has also been removed. Hence, the remains of the metal at the time t_2 will stand in a relation of complete genidentity with the parts of the metal they belonged to at the time t_1, but the whole formation cannot be accounted for through genidentical completeness.[25]

In general, physical genidentity does not consider qualitative differences in the properties of related objects, but mainly takes into account their temporal differences. Nonetheless, the order of genidentical formations cannot be traced back to a temporal order. The concept of genetic series is indeed more fundamental than the one of temporal order that can be drawn from it.[26] This type of order is not to be understood in terms of an external order that can be used to determine temporal relations among objects or events. As a matter of fact, it captures an internal, immediate, and most fundamental relation of the objects (or events) considered.[27] In this sense, it can be deemed a constitutive category (*constitutive Kategorie*)[28] pertaining to the existence relation.

Consequently, physical genidentity gives rise to a specific type of order, namely the existential relationship expressed in the concept of the "one-after-the-other" (*im Nacheinander*).[29] The idea behind this genetic type of order will be borrowed by Reichenbach as a model for a temporal topology, as we shall see in the next subsection. Interestingly, Lewin also addresses this question in his "Die zeitliche Geneseordnung" (1923a), where he proposes a temporal topology—still construed in mereological terms—based on the notion of a genetic series and time order of *actual* events.[30] The theory of relativity, he explains in the introductory remarks, employs light signals for the determination of the temporal relations. In particular,

---

[25]Lewin (1922/1983, 84 ff.).

[26]"[Die] Zeitverschiedenheit der Relata [bildet] eine notwendige Voraussetzung für das Vorliegen der physikalischen Genidentitätsbeziehung zwischen ihnen. Diese Verschiedenheit bezieht sich jedoch nicht auf solche mit der Zeit zusammenhängenden "Eigenschaften" wie die Geschwindigkeit, sondern lediglich auf die Verschiedenheit der Stellung des Gebildes innerhalb der Ordnung des Nacheinander. Es wird sich später zeigen, daß die Ordnung innerhalb der Reihen genidentischer Gebilde nicht auf die Zeitordnung zurückzuführen ist, sondern daß der Begriff der genetischen Reihe wahrscheinlich fundamentaler ist als der der Zeitordnung." Lewin (1922/1983, 65).

[27]"Der Ausdruck "Ordnungstypus" soll nicht bedeuten, daß es sich um subjektive, nicht gegebene, sondern gemachte Ordnungen handelt. Er wird vielmehr in einem aller Aktivität oder Passivität des Erkenntnissubjektes gegenüber völlig neutralen Sinne benutzt und besagt lediglich, daß die hier wesentlichen Verschiedenheiten der Existentialbeziehungen Verschiedenheiten der "inneren Geordnetheit" sind, und zwar weniger was den Grade als was den Typus der Ordnung anbelangt." Lewin (1922/1983, 317).

[28]Lewin (1922/1983, 69).

[29]Lewin (1922/1983, 73). Cf. also above, footnote 18.

[30]See also Lewin (1923b).

for its construction it makes use of a specific property, constancy (*Konstanz*), of a specific physical process, the propagation of light. In the Minkowski system of world lines, the point of intersection of the event series is uniquely determined. Even so, for Lewin the assignment of all those world lines to concrete objects fundamentally relies on an existential relation. The subsumption under a specific world line actually depends on the relation of historical derivation (*geschichtliche Herkunft*) of an event from another and does not depend on other features like equal measure or energy. For this relation of historical derivation, expressed by a genetic series, all other determinations (measure, velocity, volume, and in general all physical values of the members of the series) are not decisive. The essential one, according to Lewin, is the simple relation of existence that makes all other properties significant.

Thus, as a development of his previous research, Lewin focusses on the existential relation that grounds the concrete object's characteristic belonging to a specific world line. For instance, to determine the temporal succession of a real series, Lewin starts by defining the direction of the genetic series. Let a_n and a_m be distinct cuts (*Schnitte*) of a series. If a_n has derived from a_m, we can characterise a_m as "genetically earlier than" a_n, like in the first example of Fig. 5.1, where the arrow indicates the direction of the formation. He then introduces Axiom I, which states that a cut of a genetic series occurs in this series only once.[31] This means that a genetic series does not loop back on itself and that the chain cannot be closed. Obviously, a cause cannot be, at the same time, the result of its own effects. In this framework, the relation of cause to effect clearly does not cover the many meanings of genidentity. In general, we can speak of a cause only in association with a conjunction of events, not when considering one single event.[32]

Up to now, we have dealt with genidentical sequences considered separately. In order to fulfil the declared aim of the paper, namely provide a definition of temporal order, the analysis must be extended to situations where several series interact. Hence, Lewin exploits the idea that temporal determinations can obtain where genetic series share some cuts or are linked by other commonly connected series. To begin with, a series can produce separate series through a "splitting-off" (*Abspaltung*) or can be observed as their "reunion" (*Vereinigung*), as shown in Fig. 5.1 (second and third example, respectively). Besides, as we shall see, it can also split off into separate series and then reunite again.

There are cases of genetic series in which the direction of the sequence is rather obvious (e.g. from an actual egg to the hen resulting from it), but in physics the points along an existential world line do not unambiguously entail the idea of antecedency of one relatively to the other. Such instances motivate Lewin to broaden the concept of genetic series so as to embrace cases presenting a causal relation

---

[31] "*Axiom I*: Ein Schnitt einer Genesereihe kommt in dieser Reihe nicht mehrmals vor." Lewin (1923a, 66).

[32] Thus, a stone at a time b_1 is not the cause of the stone being the same at a time b_2. Cf. Lewin (1922/1983, 72).

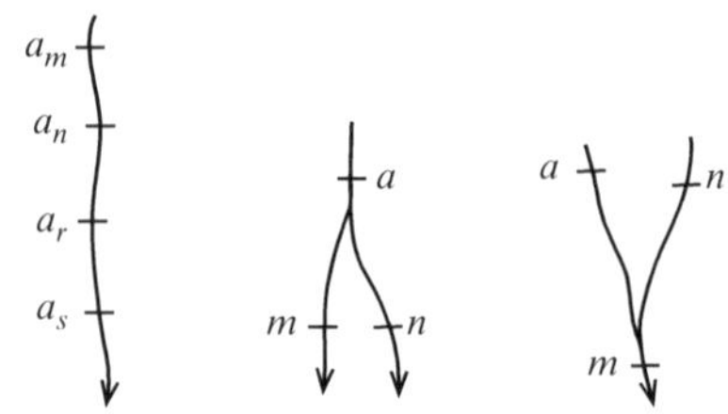

Fig. 5.1 Antecedency between multiple series cuts, splitting-off, and reunion of genetic series, Lewin (1923a, 69)

within a conjunction of events. To capture these cases, he introduces the notion of "genetic series of succession" (*Genesefolgereihe*), for which he also uses the term "causal series" (*Kausalreihe*). This type of series is characterised as follows: we have a genetic series of succession when there is a choice of cuts such that any two consecutive cuts are genidentical in the same direction, without the requirement that the other cuts be genidentical all among themselves.[33] This addition allows for the temporal comparison among series that haven't derived directly one from the other, but that are instead joined via other series. In this sense, the concept of *Genesefolgereihe* enables the univocal determination of a temporal order on the basis of the structural features of the nets realised by "interacting" series. Thus, Lewin introduces Axiom II which expresses the openness of causal chains by stating that a genetic series of succession does not lead back onto itself.[34] This clearly represents an extension of the first Axiom onto the newly defined class of series. Finally, Lewin denotes as "temporal order of the genesis" (*zeitliche Geneseordnung*) any system-time which has been determined by means of the relationship of the "one-next-to-the-other" among genetic series of succession.[35]

Besides the relation between two series that intersect or part at a certain time, particular importance is assigned to those series that play the role of "carriers" or "messengers" (*Boten*) connecting two separate series (Fig. 5.2, first example). Their function is similar to that of a clock since they enable a comparison and thereby a univocal temporal determination between separate series.

As the second (carriers V^2 and V^3) and third example (direction of the arrows resulting from the two carriers' system intersecting at x and y with respect to the direction of the genetic series B) of Fig. 5.2 show, these cases are inconsistent for they give rise to temporal relations among slices that imply the contradiction of Axiom II, the one asserting the impossibility of genidentical (therefore, temporal) loops.

[33]"Eine Reihe, in der sich eine Anzahl von Schnitten so herausgreifen läßt, daß je zwei aufeinanderfolgende Schnitte genidentisch, und zwar in derselben Richtung genidentisch sind, ohne daß sämtliche Reihenschnitte untereinander genidentisch zu sein brauchen, heiße "*Genesefolgereihe* (*G*-Folgereihe). Die Zeitbeziehung ihrer Glieder heiße: "*zeitlich früher*" (→) und "*zeitlich später*" (←)." Lewin (1923a, 67).

[34]"*Axiom II*: Eine Genesefolgereihe führt beim Fortschreiten in einer Richtung nicht in sich zurück." Lewin (1923a, 67).

[35]"Ein Zeitsystem, das lediglich durch dieses Axiom über das Nacheinander in Genesefolgereihen bestimmt wird, sei als "zeitliche Geneseordnung" bezeichnet." Lewin (1923a, 67).

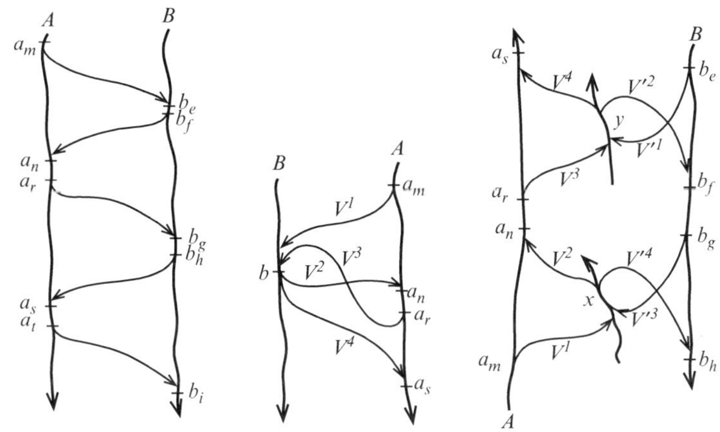

Fig. 5.2 Connection between series via *"Botenzüge"*, Lewin (1923a, 75)

Intuitively, we can take the extension to the genetic series of succession to cover cases in which there are constructs (*Gebilde*) that are properly speaking not derived from a certain sequence, but that can be temporally related to it provided that a continuous chain of genetic series makes their connection effectual. This specification allows Lewin to articulate the temporal relations between concrete individual things or events historically ordered through their own temporal paths when they are (causally) interacting among themselves. Unconnected events can thus be correlated by a common (causal) series and this is the key to define a temporal order. Lewin's strong intuition is that such temporal determinations emerge only on the occasion of conjunctions and interactions of series, so that the structural features of the resulting net make possible the introduction of an objective and univocally defined time order. Besides, this net suffices to determine the direction of the causal chain. This is an element that Reichenbach will eventually appreciate in his "Die Kausalstruktur der Welt und der Unterschied von Vergangenheit und Zukunft" (1925).

We have seen that the comparison between series requires some sort of connecting device, of which a clock is an obvious realisation. Yet, the "*Boten*" are not to be reduced to the usual concept of a clock: this concept provides the additional feature of a coordination (*Zuordnung*) with the series of real numbers, a coordination which is not at all required in Lewin's model. In a footnote to this passage, Lewin explicitly refers to Reichenbach's "Bericht" (1921b) as an example of this type of coordination, interestingly adding that it follows the same line as Weyl's.[36] In order to define a "system-time", in his (1921b) Reichenbach does indeed introduce

[36]As he writes: "Auch Weyls Definition benutzt entsprechend der Absicht, für die Metrik verwendbare Uhren zu bestimmen, Genesereihen (und zwar restlose Genidentitätsreihen), deren Schnitte spezielle physikalische *Eigenschafts*beziehungen zeigen, wie sie in der zeitlichen Geneseordnung außer Ansatz bleiben." Lewin (1923a, footnote 1 to p. 79).

a notion of "clock". However, he still considers it fulfilling a purely topological function, being an auxiliary concept devoid of metrical meaning and valid for any arbitrary metric. Reichenbach's "topological clock" is thus described as any mechanism with no specific material characterisation "that coordinates each event to a point according to the sequence of real numbers".[37] Even so, in Lewin's view such a move already entails one step further into a more complex stage, where specific properties of physical processes (like a physical system having a cyclical period) do play a role. According to the psychologist, such a step is not necessary and can indeed be dispensed with within his own construction—at least given his project: a purely topological description of time order *without* any definition of or reference to simultaneity. The character of this topological relation is such that a comparison among cuts and intervals of unconnectible series of events is excluded by definition. No temporal determinations can be given between separate series unless metrical considerations come into play. So, the concept of simultaneity is bound to not fit this construction, which would render it meaningless.[38]

The temporal metric necessarily presupposes the study of the existential relations between series, and conversely this study is supplemented by the metric in two respects: on the one hand, it is only by virtue of the metric that we can speak of temporal relations expressed by numbers; on the other, the metric allows us to consider relations of temporal lengths.

Lewin's framework embodies the very general type of structural relations that can be evinced only on the basis of the existential relation that genidentity implies. In that sense, Reichenbach's use of the concept of "first signal", as we shall see, does not represent anything more than a token case—comparable to the genealogical sequences in biology—of this most general type of relation. Lewin's construction should, therefore, be interpreted like a foundation for Reichenbach's, who still uses relations derived from characteristic properties (*Eigenschaftsbeziehungen*) of a particular physical process (a light signal). What makes the notion of the signal operative is precisely the net of which Lewin has formalised the structural properties.

Most likely, Lewin's background in psychology and hence viewpoint from which he studied the notion of genidentity granted him a more general perspective on the whole issue than Reichenbach. Conversely, the very specialised context in physics in which Reichenbach worked made him unable to see the form of primordial relations analysed by Lewin, which lay below his alleged "elementary facts". In the next subsection, we will see how Reichenbach understands his model, his objections to Lewin, and their consequent discussions.

---

[37]Reichenbach (1921b/2006, 47).

[38]"Die *Beschränktheit der zeitlichen Geneseordnung* gegenüber der gewöhnlichen Zeitordnung zeigt sich darin, daß sie die *Ordinalbeziehung nur zwischen zeitverschiedenen Ereignissen* zu bestimmen gestattet (also nur die Beziehungen "früher" und "später", aber nicht die "Gleichzeitigkeit" getrennte Ereignisse) und die *Zeitlängenbeziehung nur* zwischen Geschehnisreihen, von denen die kürzere ganz "*innerhalb derselben Zeit*" stattfindet muß wie die längere, jedoch nicht zwischen Genesereihen, die ganz zu verschiedenen Zeiten stattfinden." Lewin (1923a, 78).

5.3.2 From Genidentity to the Mark Principle

5.3.2.1 Genidentity in Reichenbach's Early Works

The term "genidentity" is introduced by Reichenbach in his *Habilitation* thesis, *Relativitätstheorie und Erkenntnis apriori* (1920), where it appears in the form of an a priori principle of cognition. The philosophical background of this monograph is particularly important, so let us just briefly recall one of its central features, the suggested "liberalised" version of Kant's synthetic a priori. In this writing, Reichenbach still defends the idea of constitutive, yet revisable a priori principles of knowledge and accordingly proposes a distinction between axioms of connection and axioms of coordination.[39] The axioms of coordination are related to the conceptual part of knowledge and are constitutive of the concept of physical object in that they determine the meaning of the axioms of connection, the laws of physics. Thus, they determine the rules of their application to reality, i.e., the rules of the connection. The specificity of genidentity as a constitutive principle is that of allowing us to indicate "how physical concepts are to be connected in sequences in order to define 'the same thing remaining identical with itself in time'."[40] As an example, Reichenbach mentions the fact that when we speak of the path of an electron, we have to consider it as the selfsame object through the passage of time. In other terms, we *assume* its identity through time, and therefore make use of the principle of genidentity as a conceptual presupposition. To be sure, among the cognitive principles, the principle of genidentity (like that of probability) plays a specific and most fundamental role. Being in fact *pre*supposed by a number of other coordinating principles—especially those related to the process of measuring—genidentity seems to involve a higher level of coordination. As such, it is a condition of the possibility for utilising other principles and, therefore, appears to have a "meta-constitutive" function.[41]

As previously mentioned, when introducing this term in his (1920) Reichenbach makes explicit reference to Lewin's *Die Verwandtschaftsbegriffe in Biologie und Physik und die Darstellung vollständiger Stammbäume* (1920), where genidentity is not considered. In the same year, Reichenbach writes a review of this essay, very much appreciative of the analytical methodology used by the author in the prospect of a positive turn in the philosophy of natural sciences.[42] Despite that title,

[39]In recent years, Reichenbach's conception of the cognitive principles has been increasingly discussed after Michael Friedman's attempt to revive his original idea. See Friedman (2001) and references therein.

[40]Reichenbach (1920/1965, 53).

[41]On these aspects, see Padovani (2011).

[42]Reichenbach (1921a). Here, he also correctly indicates the forthcoming volume *Der Begriff der Genese*. At the end of 1920, Lewin read a draft of this review before it was sent to the editor of the journal. Later, he wrote to Reichenbach in order to clarify some crucial points of his comparative approach. In the first phase of this correspondence, the point at issue was Reichenbach's (mis)understanding of the intrinsic difference among the notions of genetic series used in the various sciences—a difference which cannot be reduced to each science's degree of development, as Lewin made clear: "Es liegt mir an, und für sich sehr am Herzen, zu betonen, dass

he points out, the idea of a genetic series in physics is not examined by Lewin in this early essay. Reichenbach accordingly argues that also in physics there are certainly interesting examples of genetic series—like those representing the existence of a material thing in time, i.e., the world line of a material point—unfortunately not treated by Lewin (1920).[43] However, as we have seen in the previous section, Lewin's idea is actually that the world line of a material point is contingent, so to speak, on the genetic series, which is expressed by the existential relation embodied by the concrete events' persistence through time. Reichenbach, to the contrary and from the outset, tends to interpret the genetic series *tout court* as the world line.

In 1921, Reichenbach starts working on an axiomatisation of Einstein's theory of space and time, which he will finally achieve in his *Axiomatik* (1924a). He will write a brief report on the occasion of the famous Bad Nauheim meeting of the German Society of Natural Scientists in 1921. In this "Bericht" (1921b), Reichenbach proposes the idea of a light geometry, a geometry based on the physical properties of light and world line of a material point, independently of any relation to other material objects. One crucial feature of this project is the distinction between light and matter axioms. The first ones define the light geometry whereas the second ones "imply the identity of the developed 'light geometry' with the space-time theory of rigid rods and clocks".[44] The other fundamental trait of this report is epistemological in nature and entails a first step towards conventionalism by drawing a clear-cut distinction between empirical axioms and arbitrary coordinative definitions. Finally, synthetic a priori principles are no longer considered and are turned into conventional definitions—as a matter of fact, a conceptual shift that is quite problematic for some "meta-constitutive" principles like genidentity. In Reichenbach's view, this geometry has an important advantage over other physical geometries, as he explains in a letter to Moritz Schlick in January 1922:

> I think that especially the axiomatic analysis will be of interest to you. It provides a validation of conventionalism, but it clearly reveals those facts that also conventionalism cannot interpret. Particularly remarkable is the fact that it allowed for the complete elimination of rigid rods and clocks. I managed to define the entire metrics simply by using light signals. This is of course also a *real* definition, but one can actually show that light suffices as reality, it even enables to define rigidity.[45] [Reichenbach to Schlick, 18 January 1922 (*Schlick Collection*)]

---

der Unterschied der genetischen Reihentypen in Physik und Biologie *nicht* auf dem verschiedenen Entwicklungsgrad dieser Wissenschaften beruht [, vielmehr grade etwas ist, was in der Einstellung einer Wissenschaft konstant bleibt.] 2 Sätze in den Referat könnten nun leicht dahin missverstanden werden, dass gerade dies meine Meinung ist." ASP, HR 015-57-14, Lewin to Reichenbach, 23 December 1920. Reichenbach must have changed his review accordingly.

[43]Reichenbach (1921a, 51).

[44]Reichenbach (1921b/2006, 46).

[45]"Ich glaube, dass die axiomatische Analyse Sie besonders interessieren wird. Sie liefert natürlich eine Bestätigung des Konventionalismus, aber sie deckt auch jene Tatsachen auf, an denen auch der Konventionalismus nicht interpretieren kann. Besonders merkwürdig ist es, dass es möglich war, die starren Massstäbe u. Uhren völlig zu eliminieren. Ich konnte allein durch Benutzung von Lichtsignalen die ganze Metrik definieren. Das ist natürlich auch eine *Real*definition, aber

While in this brief report the light signal is taken as the basis of the definition of time order,[46] in the *Axiomatik* the axioms of time order will have another starting point, the following definition:

> *Definition 1.* Of two events E_1 and E_2 happening at P, the event E_2 is called later than E_1 if a signal chain can be chosen in such a way that its departure coincides with E_1 and its return with E_2. In this case, E_1 is called earlier than E_2.
>
> *Axiom I, 1.* There is no signal chain such that its departure and its return coincide at P. [Reichenbach (1924a/1969, 29)]

With the help of what will be later known as the "mark principle", Reichenbach clarifies the notion of light signal: a light signal is a physical process or event traveling from a point P to a point P' and having the property that when this event is marked at P, the mark will also be observed at P'. In this case, "the word 'signal' pinpoints this very property because it means a transmission of a sign. The word 'causal chain' is also frequently used in such instances".[47] The physical world consists of causal chains, whose structural relations can be formulated as topological and metrical axioms. Metrical axioms are statements about certain topologically specified chains (or first signals). Topological axioms are statements defining time order and concerning the possibility of connecting all space points by causal chains. The topological axioms defining time order assert that no causal chain is closed. If we possess a time order, it is indeed thanks to the causal chain. Thus, time provides the description of "an objective state of the physical world just like any other scheme of order—for instance, genealogical order." So, he concludes, "*time is the order type* [Ordnungstypus] *of causal chains*."[48] The allusion to the genealogical order and to the notion of "*Ordnungstypus*" is clearly an echo of Lewin's work (not mentioned at this point, by the way) but it is not developed here. Instead, it will be developed in his essay on the theory of motion in Newton, Leibniz and Huygens (1924b), published in the same year. In the analysis of the Leibniz-Clarke correspondence, Reichenbach takes Leibniz's reference to genealogy as an opportunity to discuss the parallelism between genealogical and causal order, eventually citing Lewin:

> In a genealogical order, every individual has a 'place', and exactly as in a spatial order the place of an individual indicates nothing but certain relations he bears to other individuals. [...] The genealogical order schematizes the structure of ancestral relations between individuals, and is not something else existing in addition to this structure. [...] According

---

es zeigt sich eben, dass das Licht als Realität genügt, es vermag sogar die Starrheit zu definieren." Reichenbach to Schlick, 18 January 1922 (*Schlick Collection*).

[46]"1. *Axioms of time order.* We first define the time order at a point. A light signal sent from a point A to an arbitrary point B (which may be moving) is reflected and returns back to A. *Definition 1.* The departure of the signal from A is called "earlier" (written $<$) than its return to A." Reichenbach (1921b/2006, 46). I shall point out that Reichenbach does not seem to be acquainted with a similar attempt to define the temporal order carried out by the mathematician Robb already in his *A Theory of Time and Space* (1914).

[47]Reichenbach (1924a/1969, 27).

[48]Reichenbach (1924a/1969, 15–16).

> to the investigations by K. Lewin, we may regard the genealogical order as the space-time order of biological evolution, in exactly the same sense as we have to regard the causal order as the space-time order of physics. [Reichenbach (1924b/1959, 54–55)]

As I pointed out before, genealogical sequences are comparable to light signals in the sense that they both represent a token case of the more primordial structure type expressing the actual, temporal (i.e., causal) order, which Lewin has constructed in his (1923a).

Probably to avoid any Kantian overtones, the notion of genidentity no longer appears in Reichenbach's *Axiomatik*. Its role is transferred onto the mark principle that he defines only after introducing the notion of light signal. Direct reference to genidentity and its fundamental significance will be made explicitly in *Die Philosophie der Raum-Zeit-Lehre* (1928). The return to the concept of genidentity will be the result of an exchange with Lewin, but will not be without consequences from a philosophical point of view, as I shall briefly highlight in the concluding Sect. 5.4. Let us first analyse their correspondence.

5.3.2.2 Reichenbach's Correspondence with Lewin

Reichenbach and Lewin were very close friends and colleagues. Most likely, Reichenbach informed Lewin about the larger manuscript on the axiomatics of time on which he was working in that period. In his first letter related to the topic of the topology of time, dated 17 March 1922,[49] Lewin addressed Reichenbach to thank him for sending his "Zeitaxiomatik",[50] and for informing him of the progress of his work about the genidentical series.[51]

The following letter, sent in September 1922, accompanied a copy of the manuscript of "Die zeitliche Geneseordnung" (1923a), which to Lewin's mind represented "a good completion" (*eine ganz gute Ergänzung*) of Reichenbach's axiomatics of time. In the diplomatic terms used by Lewin, this "completion" actually meant a clarification of the lacking elements of Reichenbach's topological foundation. Lewin deemed in fact his approach to be starting from the most fundamental level of analysis, differently than Reichenbach's. As he wrote:

> I see which time order arises merely on the basis of the genetic series in terms of genidentical series. Thereby, I would like to desist from presupposing formally the concept of 'simultaneity' in the definition of 'complete' genidentity. [...] Yet, I would like to take

[49]The Pittsburgh Archives contain many letters from Lewin to Reichenbach, but unfortunately no replies have been found to the letters we are using in this section.

[50]This manuscript represents the first draft of the *Axiomatik*, which was originally supposed to be dealing more specifically with time. Cf. ASP, HR 023-35-01.

[51]As he wrote: "Ich bin dabei, mir die Beziehungen zum Genidentitätsreihenbegriff zu überlegen. [...] Die Stellung der Zeit als Parameter der genetischen Reihen löst vielleicht gewisse Schwierigkeiten, die beim Gedanken des Raumpunktes bestehen." ASP, HR 015-57-13, Lewin to Reichenbach, 17 March 1922.

> the genetic series as a basis by emphasising its peculiarity, and not to use any specific 'relation of property or of magnitude'. The question is how far I can go simply on the ground of the *existential* relation of the genetic series. I obtained a 'general time order' of very limited use, which involves relations of time *length* (that are of course only of 'topological' nature, here). On the other hand, it is of very general nature because it does not make any assumptions about constancy, velocity or physical kinds of events, 'reference systems' or carriers, and constitutes a foundation of any factual determination of time. [...] Your axiomatisation treats just all the properties of time that I do not take into consideration.[52] [ASP, HR 015-57-12, Lewin to Reichenbach, September 1922]

This motivation notwithstanding, Reichenbach criticised Lewin's proposal in his *Axiomatik*. His disagreement is basically centred around the concept of signal and the corresponding definition of simultaneity, and it results in two footnotes, one in § 5 ("The Concepts of Real Point and Signal") and the other in § 7 ("Axioms of the Comparison of Time").

In the first footnote, Reichenbach mentioned Lewin after introducing the definition of signal as a physical process of "sign transmission" that we have seen above[53]:

> In earlier presentations I started with light signals. The validity of theorem 9 for any real events then follows from the fact that the light signal PP' arrives at P' earlier than any other signal departing simultaneously from P (limiting character of the velocity of light). K. Lewin drew my attention to the fact that all properties of signals can be formulated directly for any signal (*Zeitschr. f. Phys.*, vol. 13 [1923], p. 62). Starting from differently oriented investigations concerning genealogical sequences, Lewin (*Begriff der Genese* [Berlin: Springer, 1922]) arrived at a conception of time order similar to mine. [Reichenbach (1924a/1969, 27)]

As in 1924b, Reichenbach correctly associates the outcomes obtained by means of his notion of signal with Lewin's genealogical sequences. Yet, it is clear that Reichenbach keeps considering his own level as the most elementary, and that he holds his notion of signal to be comparable to Lewin's genetic series. However, as we have seen, in Lewin's work the genetic series provide a primordial level of

[52]"Ich sehe zu, welche Zeitordnung ergibt sich lediglich auf Grund der Genesereihen im Sinne der Genidentitätsreihen. Dabei möchte ich davon absehen, dass ich bei der Definition des Begriffs der "restlosen" Genidentität formal den Begriff der "Gleichzeitigkeit" voraussetze. [...] Sondern ich möchte die Genesereihe zugrundelegen unter Betonung ihrer Eigenheit, keine bestimmten "Eigenschaft-" oder quantitativen "Grössenbeziehungen" zu verwenden. Die Frage lautet, wie weit komme ich auf Grund lediglich der *Existential*beziehung der Genesereihen. Ich erhielt eine "allgemeine Zeitordnung" von sehr beschränkter Anwendung, was Zeit*längen*beziehungen (die hier natürlich nur "topologischer" Natur sind) betrifft. Anderseits ist sie sehr allgemeiner Natur, weil sie keine Annahmen über Konstanz, Geschwindigkeit oder physikalische Art der Ereignisse, "Bezugssysteme" oder Boten macht und bildet so eine Grundlage für jede faktische Zeitbestimmung. [...] Ihre Axiomatik geht ja gerade auf alle die Eigentümlichkeiten der Zeit, die ich ausser Acht lasse." ASP, HR 015-57-12, Lewin to Reichenbach, September 1922

[53]As he points out in the previous footnote, the fact is that there is an epistemological problem related to the concept of mark. Interestingly, here, Reichenbach writes that "for our purposes the concept of ordinary language suffices. This concept indicates already that the mark at P' is not exactly the same as the mark at P, but shares only certain fundamental features. The mark may be distorted (sounds over the telephone). We shall leave open the question of whether a reference to the concept of Gestalt is necessary." Reichenbach (1924a/1969, 27)

analysis that is meant to be more fundamental than Reichenbach's—a point that Reichenbach doesn't seem to be willing to admit at this point.[54]

Theorem 9, presented a few pages later, presupposes all previously defined topological axioms of the time order and the comparison of time, and it asserts that simultaneity can be defined from a purely topological point of view "in such a way that a continuous positive time interval is assigned to every physical process (signal, causal chain)", that is, provided that a univocal coordination be established between the order of temporal sequences and the sequence of real numbers (Theorem 2).[55] Theorem 9, according to Reichenbach, "contains the most important *topological problem of simultaneity*. Only this theorem permits the consistent application of definition 1[56] in a uniform continuous order of time."[57]

The second footnote where Lewin is mentioned by Reichenbach follows this attempt to show that his own definition of simultaneity is of a topological and not of a metrical nature. We have seen that Lewin deemed this kind of coordination to already imply a step beyond the purely topological dimension, and that it was therefore not to be regarded as so fundamental as his. However, Reichenbach contended that Lewin was mistaken in regarding simultaneity as a metrical determination.

> This is the reason—he argued—why [Lewin] misrepresents my earlier investigation,[58] which also begins with the topological problem of time; in this publication axioms I and II are topological axioms. The clock introduced by definition 2[59] has only topological qualities because of the arbitrariness of the metric. The difference between Lewin's investigations and my own lies in another direction, as was revealed in a personal discussion: whereas my axioms make assertions about all *physically possible* signals and my 'there exist' means 'except for technical difficulties they can be produced experimentally', Lewin restricts himself to assertions about *actual* signals and does not speak of possible ones. Therefore, Lewin's investigation cannot (and will not) go as far as mine, not even topologically. Above all, the concept of first signal cannot occur in the sense of a limit. Cf. Lewin's report about his studies in *Physikalische Berichte* [Lewin (1923c)]. [Reichenbach (1924a/1969, 39)]

At the beginning of 1924, Reichenbach sent the final draft of the *Axiomatik* to his colleague in Berlin. In a letter of clarification, Lewin strongly emphasised, once more, that his topology represented a supplement to Reichenbach's, and was definitely neither a substitute nor a similar attempt of limited application. This is a

---

[54]As he makes clear, "[t]hat signals exist, that we can produce them, send them to a given real point, combine them, and reflect them are elementary facts [*elementare Tatsache*]; axioms I [i.e., the Axioms of Time Order, § 6] and II [i.e, the Axioms of the Comparison of Time, § 7] contain everything concerning these facts that is necessary for the construction of the order of time." Reichenbach (1924a/1969, 28).

[55]Theorem 2: A one-to-one correspondence can be established between the real numbers and the temporally ordered events at one point. Reichenbach (1924a/1969, 30).

[56]See above, footnote 46.

[57]Reichenbach (1924a/1969, 39).

[58]Reichenbach (1921b).

[59]In the *Bericht*, Definition 2 is the first one of the second group of axioms, those regarding the comparison of time, which include the above-mentioned definition of the topological clock.

positive aspect in Lewin's standpoint that Reichenbach continued not to be willing to concede. Certainly, as Lewin remarked, the distinction between the topological and the metrical must be regarded as secondary with respect to the distinction between actual and possible. But it is precisely the latter, along with the fact that Lewin deals with the *actual* order derived by genidentical series, that rendered his approach more fundamental and not just alternative, as he clarified in the above-cited report on his own work published in the *Physikalische Berichte* in autumn 1923.[60]

There, Lewin presents the essential elements of his topology of time, namely the fact that temporal relations can obtain from actual processes and lead to a definition of temporal order if we simply consider, in mereological terms, the formations or events that follow one from the other by means of the existential sequences. The peculiarity of this definition of temporal order is that it does not depend on the specific physical nature of the genetic series connected, or whether, for instance, we are talking about light or matter transport. Once more, Lewin underlines that no consideration of speed, measure or proportion between physical values or even of simultaneity between sections of sequences is taken into account. In this sense, any change in the genetic series taken as reference does not affect the topological relations derived. So, he concludes, each metrical temporal determination (including the ones defined by the theory of relativity) should in principle presuppose his more basic approach, the only one leading to the fundamental structural net of temporal relations.[61]

Lewin's letter dated January 1924 goes along the same line and in order to avoid further misunderstandings accordingly presents a list of suggested corrections to the two footnotes where his work is mentioned in Reichenbach's axiomatisation. As far as the first one is concerned, Lewin notes that his notion of signal does not correspond to the one used by Reichenbach, especially since he does not need the additional property provided by the mark principle, already implied by the concept of genetic series.[62] Clearly, Reichenbach originally understood the genetic series as having the limited character of some existential connections that for no reason

[60]"Ich habe leider die Arbeit in den beiden Tagen noch nicht sehr eingehend studieren können und daher im wesentlichen die Anmerkungen gelesen. Ich glaube, es ist Ihnen noch nicht ganz deutlich geworden, dass ich wirklich eine gegen Ihre Axiomatik beträchtlich verschiedenes Ziel verfolgt habe. Dass der Unterschied: topologisch—metrisch sekundär ist gegenüber dem Unterschied: wirklich—möglich kommt schon in meinem Eigenbericht [Lewin (1923c)] in den Physik. "Berichten" zum Ausdruck, den ich Ihnen beilege (*bitte zurücksenden*!!). Dass meine zeitliche Geneseordnung weniger weitreichend ist als alle auf "Möglichkeiten" bezugnehmende Zeitordnungen, habe ich ja immer zu betont. Es ist also irrefürend, von "nicht erschöpfend" zu reden! Denn diese Beschränkung auf das "Wirkliche" gibt der zeit[lichen] G[enese]-Ordnung m. E. ja gerade die Fundamentalität gegenüber allen auf "Möglichkeiten" bezugnehmenden Ansätzen." ASP, HR 016-36-13, Lewin to Reichenbach, 24 January 1924.

[61]Lewin (1923c, 977).

[62]"Ad Anmk. S. 20: 'der freilich—beschreibt.': streichen! Ich sehe nicht den Sinn des Satzes, da ich gar nicht eigentlich von "Signalen" in Ihrem Sinne rede, z. B. brauche ich nicht die Erkennbarkeit der Eigenschaft des Signals $PP'P$, dass es P' erreicht hat; resp. die fehlenden Eigenschaften sind ja durch die Definition als Genesereihen gegeben, deren Axiome in meinem Buch angegeben sind.

can ground an axiomatic account of the topology of time, which up to this moment he still interprets as regarding only possible—i.e., not necessarily actual, contrary to Lewin—connections. As Reichenbach points out in the second footnote, the divergences between them relate to the kind of events they want to account for. In his letter, Lewin correspondingly spells out their specificity:

> In the footnote to p. 63 of my work, which you are most likely referring to, I did not talk about the metrical character of your work, but, rather, about the fact that you make use of property relations. This applies in any case at least to the light signals. And even now, for instance with Axiom I, 2[63] you still make assumptions about certain say quasi-continuous distributions of the physical reality (without which that axiom, as far as I can see, would not hold for a finite velocity of the signals). In principle, [these distributions] derive from the sphere of conditions of the temporal order of the genesis. [...] The earlier version appears to assign the temporal order of the genesis purposes that it does not have. Also, I have always explicitly stressed ([1923a,] p. 80 bottom) that it is possible to say more from a topological point of view with the help of a (variable or constant) metrics.[64] [ASP, HR 016-36-13, Lewin to Reichenbach, 24 January 1924]

According to Lewin, making use of light signals as the starting point of a topological account of time actually calls for more analyses concerning the specific properties of the physical process assumed as the elementary fact. This fact already entails a further level of elaboration in comparison to his own "minimalist" topological approach, let alone the introduction of a "topological clock" in order to define simultaneity. Thus, with his topological construction Lewin obtains much with a restricted number of assumptions (namely, only the assumption of genidentity as an existential relation), whereas Reichenbach has to introduce light signals supplemented by the mark principle, virtually presupposing all the considerations about genidentity explored by Lewin without using any concept of clock.

Despite what he wrote to Schlick in the missive we have seen above, Reichenbach neither provided the most basic account, nor did he accept Lewin's criticism concerning the unanalysed assumptions upon which his construction actually rests.

Dort steht übrigens auch was von den "Kennzeichen" (S. 15 u. a.), wenn auch nur in Bezug auf feste Körper." ASP, HR 016-36-13, Lewin to Reichenbach, 24 January 1924.

[63]Axiom I,2 is the axiom of connection of temporal series and it is so defined: "Axiom I, 2. For any two events E_1 and E_2 at P there exists always a signal whose departure coincides with E_1 (or E_2) and whose return coincides with E_2 (or E_1)." Reichenbach (1924a/1969, 30).

[64]Ich habe in Anmk. S. 63 meiner Arbeit, die Sie offenbar meinen, nicht von dem metrischen Charakter Ihrer Arbeit gesprochen, sondern davon, dass Sie "Eigenschaftsbeziehungen" mitbenutzen. Das traf auf die Lichtsignale zumindest auf jeden Fall zu. Und auch jetzt machen Sie z. B. mit Axiom I,2 noch Annahmen über gewisse, sagen wir quasikontinuierliche Verteilungen der physikalischen Realität (ohne die das Axiom, soviel ich sehe, bei endlicher Geschwindigkeit der Signale nicht gelten würde), die über die Voraussetzungssphäre der zeit[lichen] G[enese]-Ordnung prinzipiell hinausgehen. [...] Die alte Fassung scheint der z[eitlichen] G[enese]-O[rdnung] Absichten unterzulegen, die sie nicht hat. Auch dass man mit Hilfe einer (variabel oder konstanten) Metrik topologisch mehr aussagen kann als die z[eitliche] G[enese]-O[rdnung], habe ich selbst ausdrücklich hervorgehoben (S. 80 unten)." ASP, HR 016-36-13, Lewin to Reichenbach, 24 January 1924.

5.4 Reichenbach's Return to Genidentity

Reichenbach adopted a different perspective in his innovative work "Die Kausalstruktur der Welt" in 1925, when he eventually started formalising the causal correlations between *actual* series of events and applied his probabilistic approach to Lewin's topological model in order to define the direction of time. The idea here was to develop a topological account of the probabilistic implications that can be obtained starting from an analysis of the behaviour of interacting causal chains. Most importantly, now Reichenbach puts forward—clearly influenced by Lewin—a description of the causal processes in terms of nets. It is in fact in this paper that, for the very first time, he uses the so-called fork asymmetry account, which he will amend and improve only at the end of his life, in *The Direction of Time* (1956). In its basic form, this 1925 account relies quite consistently on Lewin's analysis of the splitting and intersecting series, which we have seen above. Following Lewin, here Reichenbach emphasises that only the relations between *actual* events belonging to different series can provide a good ground for identifying the direction of causal chains. The direction of time "can first be gained with the emergence of connecting points. In this way we are led to base the temporal order upon the characteristics of a net structure."[65]

After this "turn", the principle of genidentity will reappear in all its significance in *Die Philosophie der Raum-Zeit-Lehre* (1928), where it is regarded as a fundamental axiom, but of empirical nature. Let us recall that Reichenbach's causal chains presuppose, at the more primordial level highlighted by Lewin, some form of genidentity as a condition of their definability. In his famous book of 1928, Reichenbach refines some of the issues involved by the consideration of causal chains and the related notion of genidentity. Different states can be genidentical only if they are causally related. As he explains, "this conception agrees with our definition of causal connection, which considers the causal chain a signal, i.e. the transmission of a mark".[66] In the section devoted to the definition of time order, Reichenbach insists in particular on the necessity of considering the causal chains open, and he addresses the question whether closed causal chains could occur or even merely be imagined at all. Although this could not be excluded a priori, he points out that the uniqueness of time order as well as our familiar concept of identity through time of the individual would be lost. The properties of the causal chains fundamentally underlie our concept of individuality, and this concept, Reichenbach goes on, "originates in the fact that there are no closed causal chains" (p. 142). If the causal chains were closed, the principle of genidentity

[65]Reichenbach (1925/1978, 93). Due to space limitations, we cannot follow these issues in detail here, nor can we follow the fate of genidentity in 1956 where genidentity still plays a central role.

[66]Reichenbach (1928/1958, 271).

would be violated, and we should admit cases in which a person could meet his/her former self. However, this has never been observed and would moreover be rather difficult to accept. It is in fact our concept of individuality that *requires* that the causal chains be open. In other terms, genidentity must be assumed as a very deep principle of our physical knowledge (*ein sehr tiefes Prinzip der Naturerkenntnis*) because

> [i]t enables us to speak of a unique time order and a unique now-point. Furthermore, it makes possible the concept of the individual that remains identical during the passage of time. It is therefore the most important axiom regarding time order, and we realise to what an extent the familiar concept of time order is based on this characteristic of causality. Of course, this axiom is a result of experience [*es ist klar, daß es sich in diesem Axiom um einen Erfahrungssatz handeln kann*]. [Reichenbach (1928/1958, 142–143)]

This principle is a fundamental presupposition of our knowledge for it allows us to preserve our most important concept of individuality. Despite its fundamental role, Reichenbach would say, it cannot be considered necessary as there cannot be necessary principles in nature. Nor can we take it to be a convention. Thus, Reichenbach labels it as an "empirical principle". The justification is by exclusion and in line with his shift towards conventionalism. Since genidentity, as a principle, cannot be deemed to be conventional, nor can it be interpreted as a methodological assumption, it must be empirical. In this way, the risk of interpreting it as an otherwise inexplicable principle is seemingly avoided. Yet, this justification clearly sounds quite artificial. To be sure, the principle of genidentity really looks like an anomalous principle within this framework and to define it as a mere empirical principle would not explain why we rely on it as a principle grounding our notion of individuality, as Reichenbach emphasises. In this sense, as a (temporary) condition of possibility of our knowledge of nature, it still has the same features it used to have when it was introduced in *Relativitätstheorie und Erkenntnis apriori* (1920), namely those of a synthetic, yet revisable a priori principle—which is tantamount to reintroducing constitutive principles by the back door.

Acknowledgements Research for this paper was variously supported by a Swiss Research Fellowship from the Canton of Ticino, as well as by a grant from the Swiss National Science Foundation (PA00P1–134177). Early versions of this paper were presented at several conferences including the conference "The Berlin Group: Knowledge Probability, Interdisciplinarity" held in Paderborn in September 2009, and to the Visiting Fellows' reading group at the Center for Philosophy of Science at the University of Pittsburgh in Spring 2009. On both occasions, I have greatly benefitted from the remarks made by the audience, especially the invaluable suggestions from Chris Pincock and Ulrich Krohs. I am also grateful to Sheila Sandapen, Joe McPeak, and Jacques Catudal for comments on a more recent draft of this paper. Finally, I wish to thank both the Directors of the Archives of Scientific Philosophy in Pittsburgh and Konstanz, and of the *Wiener Kreis Stichting* in Amsterdam for their permission to quote from the Hans Reichenbach and the Moritz Schlick Collections. My special thanks goes to Brigitte Parakenings from the Konstanz Archives for her help in supplying unpublished material.

References

Ash, M.G. 1995. *Gestalt psychology in German culture, 1890–1967: Holism and the quest for objectivity*. Cambridge: Cambridge University Press.

Cassirer, E. 1910. *Substanzbegriff und Funktionsbegriff. Untersuchungen über die Grundfragen der Erkenntniskritik*. Berlin: Bruno Cassirer. Translated as *Substance and Function*. Chicago: Open Court, 1923.

Friedman, M. 2001. *Dynamics of reason*. Stanford: CSLI Publications.

Gerner, K. 1997. *Hans Reichenbach. Sein Leben und Wirken*. Osnabrück: Phoebe Autorenpress.

Lewin, K. 1920. *Die Verwandtschaftsbegriffe in Biologie und Physik und die Darstellung vollständiger Stammbäume*. Berlin: Borntraeger.

Lewin, K. 1922. *Der Begriff der Genese in Physik, Biologie und Entwicklungsgeschichte*. Berlin: Springer. (Repr. in Lewin (1983), 47–304.)

Lewin, K. 1923a. Die zeitliche Geneseordnung. *Zeitschrift für Physik* 13:62–81.

Lewin, K. 1923b. Berechtigung. *Zeitschrift für Physik* 15:64.

Lewin, K. 1923c. Kurt Lewin, Die zeitliche Geneseordnung. *Physikalische Berichte* 4:977.

Lewin, K. 1925. Über Idee und Aufgabe der vergleichenden Wissenschaftslehre. *Symposion* 1(1):61–94.

Lewin, K. 1983. Kurt Lewin: Werkausgabe. In *Wissenschaftstheorie II*, eds. Graumann, C.-F., and A. Métraux. Bern/Stuttgart: Huber–Klett-Cotta.

Marrow, A.J. 1969. *The practical theorist: The life and work of Kurt Lewin*. New York/London: Basic Books.

Métraux, A. 1992. Kurt Lewin: Philosopher-psychologist. *Science in Context* 5(2):373–384.

Padovani, F. 2011. Relativizing the relativized a priori. Reichenbach's axioms of coordination divided. *Synthese* 181(1):41–62.

Reichenbach, H. 1911a. Universität und Technische Hochschule. Ein Vergleich. *Berliner Freistudentische Blätter* 4(16):243–247.

Reichenbach, H. 1911b. Universität und Technische Hochschule. *Berliner Freistudentische Blätter* 4(20):310–312.

Reichenbach, H. 1916. *Der Begriff der Wahrscheinlichkeit für die mathematische Darstellung der Wirklichkeit*. Leipzig: Barth. Translated as *The concept of probability in the mathematical representation of reality*, eds. Eberhardt, F., and C. Glymour. Chicago–La Salle (Ill.): Open Court, 2008.

Reichenbach, H. 1920. *Relativitätstheorie und Erkenntnis apriori*. Berlin: Springer. Translated as *The theory of relativity and a priori knowledge*, ed. Reichenbach, M. Berkeley/Los Angeles: University of California Press, 1965.

Reichenbach, H. 1921a. Rezension von Kurt Lewin, Die Verwandtschaftsbegriffe in Biologie und Physik und die Darstellung vollständiger Stammbäume. *Die Naturwissenschaften* 9:51.

Reichenbach, H. 1921b. Bericht über eine Axiomatik der Einsteinschen Raum-Zeit-Lehre. *Physikalische Zeitschrift* 22:683–686. Engl. transl. in *Defending Einstein: Hans Reichenbach's writings on space, time and motion*, eds. Gimbel, S., and A. Walz, 45–55. New York: Cambridge University Press, 2006.

Reichenbach, H. 1924a. *Axiomatik der relativistischen Raum-Zeit-Lehre*. Braunschweig: Fried. Vieweg & Sohn. Translated as *The axiomatization of the theory of relativity*, ed. Reichenbach, M. Berkeley/Los Angeles: University of California Press, 1969.

Reichenbach, H. 1924b. Die Bewegungslehre bei Newton, Leibniz und Huygens. *Kant-Studien* 29:416–438. Engl. transl. in *Modern philosophy of science: Selected essays*, ed. Reichenbach, M., 46–66. London: Routledge & Kegan Paul, 1959.

Reichenbach, H. 1924c. Rezension von Kurt Lewin, Der Begriff der Genese in Physik, Biologie und Entwicklungsgeschichte. Eine Untersuchung zur vergleichenden Wissenschaftslehre. *Psychologische Forschung* 8:188–190.

Reichenbach, H. 1925. Die Kausalstruktur der Welt und der Unterschied von Vergangenheit und Zukunft. *Sitzungsberichte der Bayerische Akademie der Wissenschaft*, November: 1933–1975

Engl. transl. in *Selected writings: 1909–1953*, Vol. II, eds. Cohen, R., and M. Reichenbach, 81–119. Dordrecht/Boston: Reidel, 1978.
Reichenbach, H. 1928. *Philosophie der Raum-Zeit-Lehre*. Berlin/Leipzig: De Gruyter. Translated as *The philosophy of space and time*, eds. Reichenbach, M., and J. Freund. New York: Dover, 1958.
Reichenbach, H. 1956. *The direction of time*. Berkeley: University of California Press.
Reichenbach, H. 1978. *Selected Writings: 1909–1953*, 2 Vol., eds. Cohen, R., and M. Reichenbach. Dordrecht/Boston: Reidel.
Robb, A.A. 1914. *A theory of time and space*. Cambridge: Cambridge University Press.
Ryckman, T.A. 2007. Logical empiricism and the philosophy of physics. In *The Cambridge companion to logical empiricism*, eds. Richardson, A.W., and T. Uebel, 193–227. New York: Cambridge University Press.
Smith, B., and K. Mulligan. 1982. Pieces of a theory. In *Parts and moments. Studies in logic and formal ontology*, ed. B. Smith, 15–109. Munich: Philosophia.
Wipf, H.-U. 1994. 'Es war das Gefühl, daß die Universitätsbildung in irgend einem Punkte versagte...' – Hans Reichenbach als Freistudent 1910 bis 1916. In *Hans Reichenbach und die Berliner Gruppe*, eds. Danneberg, L., Kamlah, A., and L. Schäfer, 161–181. Braunschweig: Vieweg.
Wipf, H.-U. 2004. *Studentische Politik und Kulturreform. Geschichte der Freistudenten-Bewegung 1896–1918*. Schwalbach/TS: Wochenschau Verlag.

Chapter 6
Did Reichenbach Anticipate Quantum Mechanical Indeterminism?

Michael Stöltzner

When reflecting upon his most important achievements, Reichenbach typically mentioned his solution of Hume's problem of induction and his anticipation of the probabilistic nature of atomic physics. A case in point is his 1936 paper "Logistic Empiricism in Germany and the Present State of Its Problems", one within a series of papers written by various members of the movement of Logical Empiricism to introduce American readers to the new scientific philosophy. The paper became a source of major controversy between Reichenbach and his Vienna Circle allies because it publicly emphasized differences at a time when Otto Neurath pursued the movement's internationalization by including a large number of diverging approaches and by devising historical narratives broader than the 1929 manifesto in order to integrate the various traditions.

Reichenbach's paper started with a description of the historical background of Logical Empiricism that seems pretty uncontroversial. But he quickly came to claim authorship for the new method of analysis of science (*wissenschaftsanalytische Methode*), proposed for the first time in his *Relativitätstheorie und Erkenntnis apriori* (Reichenbach 1920c). When outlining the progress along these lines reached since, he was more than willing to share credit. For, the new method's core consisted in grasping the formerly Kantian 'reason' "only in the concrete form of scientific statements—an idea which found a more precise formulation in Rudolf Carnap's theory that philosophy must be an analysis of scientific language" (Reichenbach 1936, 142). After providing a brief of the Vienna Circle that almost exclusively centered on Carnap's *Aufbau* program and the contributions of Ludwig Wittgenstein, he came to the characteristics of the Berlin group.

> In line with their more concrete working-program, which demanded analysis of specific problems in science, they avoided all theoretical maxims like those set up by the Viennese school and embarked upon detailed work in logistics, physics, biology, and psychology. The

---

M. Stöltzner (✉)
Department of Philosophy, University of South Carolina, Columbia, SC, USA
e-mail: stoeltzn@sc.edu

N. Milkov and V. Peckhaus (eds.), *The Berlin Group and the Philosophy of Logical Empiricism*, Boston Studies in the Philosophy and History of Science 273, DOI 10.1007/978-94-007-5485-0_6,

> central problem selected for analysis was probability and induction. Their enquiries led to a new mathematical theory of probability, and to a solution of the problem of induction. For this purpose, a generalization of logic was developed in which the two truth-values of propositions "true" and "false" were replaced by a continuous scale of probability values. (Reichenbach 1936, 144)

A decisive step in developing the new method was to overcome the synthetic a priori not only for the categories of space and time, but also of causality. As did Philipp Frank (1929, 1932), Reichenbach adopted the empirical theory fathered by Hume and Mach following which causality was

> a very general *a posteriori* principle. The suspicion that there was a close connection between causality and probability received logical confirmation, in the course of which the logical priority of the concept of probability was established. The principle of causality could only be stated in the form of a proposition about the limits of "probability implications." In this context, the author suggested a possible generalization of the "causality-connection" of the world to a "probability-connection" (*Wahr-schein-lich-keits-zu-sam-men-hang*), a logical process which afterwards was realized in quantum mechanics by Heisenberg's principle of indeterminacy. (Reichenbach 1936, 146–147)

Relying upon the apparent confirmation of his own view through the development of science, Reichenbach devoted significant space (pp. 153–159) to it, surprisingly without discussing any differences with other members of the movement. For in actual fact their internal differences in matters of causality and probability were substantial and went far beyond the question whether Reichenbach was justified to proud himself of a solution to Hume's problem. As I have argued elsewhere (Stöltzner 2009), during the 1920s and 1930s there existed a triangle of disagreements between Frank (and von Mises), Schlick (and Waismann), and Reichenbach, that put each of them in opposition to the other two with respect to a core aspect of causality and probability. But my point in the present paper is a different one.

In the 1936 paper and on other occasions, Reichenbach claimed simultaneously (i) that the particular strength of the Berlin group consisted in its close adherence to the factual progress of modern science, (ii) that the concrete program he was pursuing in matters of probability and induction constituted the centerpiece of this enterprise, and (iii) that this program was ultimately confirmed by the progress of science itself. Reichenbach did not simply claim that the Berlin group was closer to the sciences because it embraced—at least within the larger setting of the "Gesellschaft für wissenschaftliche Philosophie" (Society for Scientific Philosophy)—eminent scientists, among them the gestalt psychologist Wolfgang Köhler. Even the works of members of his narrower circle, among them Walter Dubislav and Kurt Grelling, were cited only to the extent that they contributed to his own program.

The aim of the present paper is to investigate to what extent Reichenbach was justified in simultaneously asserting (i)–(iii). By investigating the history of his works on causality, probability and induction before 1936, I argue that he failed to do so. My point is that across two decades Reichenbach consistently pursued a philosophical agenda—executed with manifold twists and turns—that, in contrast to his work on relativity theory, was not prompted by the current developments of science. With the benefit of hindsight Reichenbach was nevertheless quick to argue that his philosophical insights provided a justification for the newest physics or even had anticipated them as a possibility. However, there is a major difference between the evolution of Reichenbach's agenda and the thinking of most physicists between 1916 and 1936. While statistical physics, radiation theory, fluctuation phenomena, and the investigation of atomic dynamics prompted the physical developments, they only played a marginal role in Reichenbach's growing emphasis on probability over causality. Statistical theories, for Reichenbach, enjoyed a unique status, in comparison with other scientific theories, in virtue of the statistical character of any scientific judgment—not because they described microphysical phenomena.

The thrust of the present paper is consistent with Frederick Eberhardt's recent analysis of Reichenbach's doctoral dissertation. Albeit failing to achieve a satisfactory account of objective probability, the dissertation "furnished him with a lifetime's supply of interesting problems, to which he would make influential, though rarely uncontroversial, contributions" (Eberhardt 2011, 134). Additionally, my analysis complements the recent debates about Reichenbach's abandonment of the relativized a priori in relativity theory that occurred, in correspondence with Schlick, right after the 1920 book (cf. Friedman 1994; Padovani 2011). As I will argue, transcendental arguments were surprisingly long-lived on the fields of causal ity, probability and induction. This difference makes the unequivocal distinction, in the 1936 paper, of Reichenbach's 1920 as origin of all the developments to come somewhat surprising.

Among those Vienna Circle members and associates disagreeing with Reichenbach's claim to have resolved one of philosophy's deepest enigmas and anticipated the quantum revolution, were especially those who had a science background and entertained close connections with the German-speaking scientific community, most importantly Moritz Schlick, Richard von Mises and Philipp Frank. All of them strongly objected to Reichenbach's identification of physical and logical probability, of laws of nature and inductive reasoning, even though Reichenbach in the end sided with Frank and von Mises—against Schlick—in interpreting probabilities as the limit of relative frequencies. Their disagreements on causality and probability notwithstanding, Schlick (1937) and Frank (1937) joined forces to combat the—on their account—metaphysical misinterpretations of the new quantum mechanics and to provide a philosophical justification for the Copenhagen Interpretation. Instead Reichenbach, who had played a comparably prominent role in the struggles about relativity theory, primarily continued to advance his own philosophical agenda, constantly declaring in retrospect that it had been confirmed by the new physics.

The situation described in the present paper is markedly different from the years after 1936. In his 1938 *Experience and Prediction*, Reichenbach presented his new pragmatic justification of induction to the English speaking community. It had a significant impact on the discussions within the emerging discipline of philosophy of science (see the classic Salmon 1991 and, more recently, Galavotti 2011). Reichenbach's (1944) *Philosophic Foundations of Quantum Mechanics* became highly influential for the post-war debates about the foundations of quantum mechanics and provided a detailed analysis of the theory. Thus the post-war Reichenbach can justly be considered as a philosopher of physics interacting with the comparably few physicists who, by then, were interested in foundational matters. But he had also substantially modified his philosophical approach and emphasized the problem of realism, fully in line with the shift of the overall debate away from the irreducibly probabilistic character of the quantum world to its ontology.

Let me give an overview of the development of Reichenbach's thinking on causality and probability until 1936. Already in his Ph.D. thesis written in 1915, Reichenbach expressed two ideas that would remain central to his philosophy. First, the principle of causality, to become at all applicable to the description of physical phenomena, must be supplemented with a second principle, then called the principle of lawful distribution or the principle of the continuous probability function. Second, there exists in point of principle no difference between the theory of error presupposed by any measuring science and the probabilistic theories of physics. The development of Reichenbach's conception of causality and probability, from his dissertation to his emigration in 1933, was mainly marked by a change in the epistemological status of, and a gradual shift of emphasis between, those two principles. It can be divided into three phases.

(i) From his Ph.D.-dissertation until his papers in *Die Naturwissenschaften* (Reichenbach 1920a, b), he considered both principles, causality and the principle of lawful distribution, as synthetic a priori. In contrast to the categories of space and time, causality was not historically relativized. Still, the departure of Reichenbach's alleged resolution of Hume's problem from the Kantian one, and from the neo-Kantian conception of causality, was substantial because in virtue of the second principle all physical laws, at least in their actual application, had an irreducibly probabilistic component.
(ii) In the mid-1920s, Reichenbach called causality a complex of principles, the common core of which was the inductive principle of causality. It represented a hypothesis about nature, such that physics one day could be compelled to abandon causality. Reichenbach now considered the division between both principles as merely formal and proposed a theory that was based on the concept of probable determination alone. After the advent of quantum mechanics he proudly declared that he had foreseen the demise of determinism.
(iii) After 1930, Reichenbach advocated a consistently probabilistic conception of physical theory. In order to maintain his identification of inductive inference, the theory of error, and probabilistic physics against the criticism that the former could not be translated into a statement about relative frequencies, he returned to a transcendental argument according to which the statement that probability laws do not hold was self-contradictory because it already presupposed the principle of induction. To my mind, Reichenbach in effect treated induction—in the same vein as the principle of lawful distribution more than a decade before—as an a priori condition for the possibility of experience, the only difference being that no transcendental deduction was available to justify it.

The paper proceeds in six steps. In the first section I analyze Reichenbach's PhD thesis, giving particular emphasis to those elements which remained central to his subsequent work. Among them are the above-mentioned two principles and the idea of probabilistic determination, which is at the roots of the disanalogy between geometry and probability theory. In the second section I discuss Reichenbach's 1923 attempt to distinguish causal from non-causal theories by the criterion of the governing sequences. This paper, which was published only in 1932, led to an exchange with the physicist-philosopher Erwin Schrödinger which is the subject of the third section. While Schrödinger emphasized the continuities between statistical mechanics and quantum physics, Reichenbach's project maintained the dichotomy between determinism and indeterminism and had no systematic space for fluctuation phenomena. While Schrödinger understood the application of laws as a physical measurement process, Reichenbach took it as a basic act of experience. The fourth section is dedicated to Reichenbach's probabilistic topology through which he wanted to implement probability as a logical relationship between events, not unlike his analysis of relativity theory. The fifth section investigates Reichenbach's new solution of Hume's problem according to which it was impossible to deny the existence of inductive inference because any such denial necessarily applies inductive methods, which makes it self-defeating. I take this as a pragmatic and probabilistic variant of a transcendental argument, not as the return to the Kantian reasoning of the dissertation. The sixth section investigates three core aspects of Reichenbach's 1930/1931 controversy with Schlick that show to what extent Reichenbach was unaffected by the developments of statistical mechanics. They concern the basis on which he concluded to have anticipated quantum mechanical indeterminism and his understanding of the arrow of time. In a short envoi I show that Reichenbach changed his position with respect to quantum mechanics after his move to the U.S. in 1938.

6.1 Two Synthetic Aprioris: Reichenbach's PhD Dissertation

In an autobiographical note written in 1927, Reichenbach described the results of his dissertation:

1. "I showed that the condition of equal probability can be reduced to a continuity condition—at least for a class of problems.
2. [a.] I showed that this continuity condition does not only apply to problems in probability, but is assumed for all physical claims; [b.] without it causal claims would be vacuous.
3. I attempted to base the probability claim on a claim of certainty. ...
4. I attempted to prove that the probability condition is a synthetic a priori judgment that is necessary for all knowledge." (quoted from the editors' introduction to Reichenbach 1916, 22 23)

While Reichenbach considered 1. and 2. successful, he had meanwhile given up on 3. and 4. Crediting Poincaré for 1., he judged 2. "to be the most important discovery that has been made with regards to the problem of probability since Hume." (ibid., 23) "On 4., I note that only a single publication convinced me of the impossibility of synthetic a priori judgments: my own" (ibid., 23) *Relativitätstheorie und Erkenntnis apriori* (Reichenbach 1920c) the same book which, in 1936, he claimed to be the source of the new method of analysis of science. As regards 3., he noticed that one can only conclude with probability that a probability claim is correct in virtue of the weak law of large numbers because there is no way to know after which member N an infinite series will no longer deviate from its limit for more than a given ε.

Let me start with 1. In the same vein as Poincaré, Reichenbach considered an ideal apparatus where a moving tape of black and white stripes was punctured by projectiles shot on it from above. If the number of shots increases and the width of the stripes decreases—and if some additional conditions, such as independence, are met—the ratio of hits registered on black and white stripes becomes approximately equal, whatever the distribution of hits actually is. This argument secures the existence of a continuous probability function and renders obsolete Laplace's principle of insufficient reason or the principle of indifference. This principle had directed us to assign equal probability to those elementary events we have no reason to consider as different, which introduced subjective ignorance into the foundations probability theory. Reichenbach's approach instead yielded an objective concept of probability that only required the existence of a suitable physical process. "A probability distribution only appears because the projectile requires a slightly different drop time on each occurrence. A claim is made about the shooting process, we detect that the variations [*Schwankungen*][1] of the drop time are subject to a particular law, which becomes visually apparent in the fact that an equal number of black and white stripes are hit" (Reichenbach 1916, 55). If such a probability function existed, one could even drop the requirement of independence because it only guaranteed that "the regularity [*Gesetzlichkeit*] of the variation [*Variation*] of the shooting times ... is ... *clearly visible*." (ibid., 72) While independence was an empirically verifiable condition, the existence of a probability function was not. It represented "a metaphysical principle of the understanding of nature" (ibid., 74) because it was impossible to decide "which specific regularity [*Gesetzmäßigkeit*] has to underlie the processes in nature" (ibid., 74) such that the results of probability calculus could be applied correctly.

To establish 2.a., Reichenbach carried through a similar analysis for the theory of measurement errors. Since there are no closed systems, "every process in nature is in principle subject to an infinity of external influences" (ibid., 90) that are all encompassed within the measured value of a quantity. Presupposing, in the sense of Gauß's approach, that the observable measurement error is composed from a series of elementary errors, Reichenbach formulated three basic principles: "(1)

[1] I translate here "*Schwankungen*" as 'variations' while I translate them as 'fluctuations' if they are intended as a physical process in its own right, which Reichenbach—as the present paper argues—does not assume.

The frequency of the elementary error is determined by a probability function. (2) These [functions] are combined according to the theory of composite probability. (3) Many mutually independent elementary errors of the same order of magnitude must work together." (ibid., 94) It is entirely an empirical matter to decide whether (3.) is fulfilled. The same holds true for (2.), given that there exist probability functions for both the elementary errors and the composite error. This assumption (1.) transcends finite experience because it comprises an infinite number of influences. More generally, while the special form of any probability function is determined by experience, its existence, i.e. that the frequencies ultimately approach a Riemann integrable function, is not. The situation, to Reichenbach's mind, is analogous to the law of causality, whose special content is empirical while its universal validity is not. "Nevertheless, the lawfulness [*Gesetzlichkeit*] postulated here is very different from the causal one. After all it describes a law about processes that are *not* causally related. Anywhere that we are unable to make any progress on events in nature with the help of the causal law, the principle of the probability function applies" (ibid., 104).

Reichenbach now invokes a transcendental argument to prove that the principle of the probability function is synthetic a priori, i.e., that it represents a condition for the possibility of scientific knowledge. For the synthetic a priori "principle of lawful connection of all events, which causality brings about, is insufficient for the mathematical representation of reality. A further principle has to be added, which connects the events—one could say—orthogonally [or laterally: "*in der Querrichtung*"]; this is the principle of lawful distribution" (ibid., 126). Reichenbach obviously believed that 4. was the only way to guarantee 2.b. A natural law may claim validity in reality only if it "represents the real processes to numerical approximation. ... The possibility of physical knowledge has thereby been traced back to the assertability of numerical approximations" (ibid., 112). Interestingly, Reichenbach phrases this fact in terms of coordination, a concept which became most influential through his subsequent work on the theory of relativity. "The actual task of physics is the coordination of a class, which is initially only given as a system of mathematical theorems, with objects of empirical intuition; this includes the numerical determination of constants" (ibid., 116).

The central role of constants was based in their double role in the process of science. On the one hand, their specification and measurement "does not mean anything other than progress along the path whose endpoint is the individual case" (ibid., 114). On the other hand, they are measured through experiments involving other laws. "This is the general approach of physics: to resolve constants into functions, to find more general laws that contain the previous law as a special case" (ibid., 114). From a contemporary perspective, it is important to understand that Reichenbach is not primarily talking about the few fundamental constants of nature, i.e. the velocity of light or Planck's constant, and the fact that these characterize a certain fundamental theory, i.e. special relativity or quantum theory, such that their resolution into other functions or laws would bring about a reduction of the respective theory to a universal theory. Rather does he intend all numerical values in whatever laws of nature, fundamental or phenomenological, and argues that only his second principle guarantees that their numerical values are distributed according to a law that permits us to assert "approximate equality" (ibid., 126) among real objects. Since this law contains in principle an infinite number of functions and constants, "a claim is made about the size of influences whose numerical values and laws one does not know" (ibid., 122).

While, as we shall see below, Reichenbach after 1920 abandons 4. and no longer considers the existence of the probability function, or the lawful distribution of constants, as a synthetic a priori, he maintains the idea that, on pain of rendering all causal claims vacuous, there must be a condition for the possibility of the constants' measurable distribution. Following 2a, this distribution embraces both measurement errors and physical fluctuation phenomena, such that, to my mind, there remains no criterion to distinguish statistical laws about fluctuations from measurement errors other than to resolve all constants into functions. However this does not necessarily amount to a return of nomological determinism, but corresponds to the maximum of determination. Since, in virtue of the infinite complexity, even the best determination invokes probabilities, "it would be a mistake to believe that probability will become less important for the mathematical representation of the world as physical knowledge grows" (ibid., 144). And in a later paper that summarizes his thesis, he writes that probabilistic laws are not "escape routes sought out by the physicist when he lacks a more precise knowledge of the connections involved" (Reichenbach 1920b, 153/326).[2]

Let me come to 3., which Reichenbach would abandon as well. Within the framework of Kant's transcendental philosophy, 3. at bottom follows from 4. "We have deduced the existence of a probability function in the same sense as Kant uses the word deduction for transcendental philosophy. Ultimately the necessity of such a law must be intuited,[3] and hence we refer to it as a synthetic a priori judgment ... [B]y showing that such a principle is a necessary condition for all possible knowledge, its validity for experience has been *proven*. Consequently, the assertions probability theory makes are true with certainty for the objects of reality[;] ... it is certain that with an increasing number of instances we get an approximation" (Reichenbach 1916, 130) to a probability distribution. It would be ridiculous to test this metaphysical principle empirically. Equally: "It would be mistaken to believe that the principle of distribution is void of content merely because [for a given set of functions] a particular N at which the desired approximation [smaller than a given ε] occurs cannot be specified." (ibid., 136) Since the existence of such an N was proven by transcendental analysis, "this is a rational expectation ... based on an objective law of events" (ibid., 138).

---

[2] I am citing both the German original and the English translation, if available. At places I have however modified the English translation to restore a terminological relationship existing in the German original. Translations from German originals, where no translation was available, are mine.

[3] Eberhardt and Glymour here translate "*einsehen*" as "recognized". I prefer "intuited" because it better captures Reichenbach's argument; cf. the next paragraph.

In establishing 4. and 3. by transcendental deduction, Reichenbach relied on Kant's conception of pure intuition. "In the end we are dealing with an insight, an unmediated recognition [*Erkenntnis*] that these claims are valid, and this spontaneous act is ultimately not further analyzable and cannot be justified" (ibid., 106). This feature of the principle of lawful distribution brings probability calculus closer to geometry, even though it "does not enjoy such undisputed respect" (ibid., 104). For only the discovery of the principle of the probability function "completed the parallel to geometrical theorems" (ibid., 140) that already existed on the level of the axiomatized calculus. Still there remains a major difference. Geometric theorems describe relations between ideal objects which are "imaginable and it is not ruled out that a real object might at some point assume exactly the same shape [,] ... even though we do not know them in every detail. In the case of the theorems of probability, the situation is different. There is no event that constitutes the ideal case of the probability calculus" (ibid., 142) because the events will never form a continuous function. From this Reichenbach concludes that it "makes no sense to consider the concept of coincidental [*zufällig*] but causally unrelated events, since the only concepts that are imaginable are those whose content ... [after due specification] can be experienced. We can only speak of coincidental events as a limiting case" (ibid., 142). Randomness accordingly becomes a limiting concept such as infinity in geometry.

Eberhardt rightly argues that the main philosophical reference of Reichenbach's dissertation was the *Spielraum*-interpretation[4] of the philosopher-psychologist Johannes von Kries (1886). "Reichenbach sought to develop Kries' account of probability in such a way that the principle of insufficient reason became redundant and probability claims could be couched in a purely objective fashion." (Eberhardt 2011, 127) For Reichenbach held, as many contemporaries, that Kries had not fully succeeded in avoiding subjective elements. In the 1910s and 1920s, von Kries' interpretation was still a viable competitor to the increasingly popular relative frequency interpretation–or Fechner's (1897) theory of collectives that was given a mathematically rigorous formulation by Richard von Mises (1912, 1919). Already in his book, von Kries had applied his interpretation to the kinetic theory of gases and won Boltzmann's endorsement.[5] Almost three decades later, in a long paper in the scientific weekly *Die Naturwissenschaften*, he justified a version of the ergodic hypothesis that linked the micro- and the macro-level of statistical mechanics (Cf. von Kries 1919, 19–21).

Von Kries understood probability as a feature of the physical world without abandoning the Kantian idea of a universe governed by strict laws of nature.

[4]In Reichenbach 1916, Eberhardt and Glymour translate "*Spielraum*" as "event space", rather than the commonly used "range". This is certainly a good choice from a contemporary perspective, but hides the psychological aspects of the concept and the fact that it represents the "free play" left by the laws of nature – a view which Reichenbach explicitly criticizes. For this reason I leave the German expression untranslated.

[5]Made almost in passing in a 1886 address, cf. (Boltzmann 1905, 37/22).

His approach was based on the distinction between nomological and ontological regularities (*Gesetzmäßigkeiten*).[6] While the former express the laws of nature, the latter correspond to purely factual conditions, for instance, "which bodies are present at all, which velocities they have at a given moment, hence what is described as 'initial conditions' in the mathematical treatment" (von Kries 1919, 5). Because of ontological regularities, facts and events can be assigned a probability of their occurrence. In order to define this probability, one has first to list which states of the system are at all possible within the framework of the nomological regularities. This *Spielraum* must now be subdivided in such a way as to compare different parts of it and define probability as the ratio of the *Spielraum* bringing about the event to the entire *Spielraum*. The major difficulty is to find a practical and objective numerical measure that avoids Laplace's principle of insufficient reason. Von Kries believed to achieve this objectivation by demanding that the *Spielraum* be indifferent, original, and comparable. *Indifference* means that the *Spielraum* is subdivided into a set of exclusive and exhaustive equal alternatives; a *Spielraum* is *original* "if the sizes of the alternatives remain stable or stationary when taking into account the prehistory of the individual alternatives; or, to put it differently, if a *Spielraum* cannot be derived from another, more fundamental one;" it "is *comparable*, if there is a unique, objective, non-arbitrary and compelling method of subdividing it" (Heidelberger 2001, 179).

In his dissertation, Reichenbach shows that all three conditions can be replaced by the existence of a continuous probability function, which can always be partitioned in the appropriate way. Thus as long as one can imagine a suitable physical construction of the probability distribution, the remaining subjective traits of von Kries' approach are purged. Interestingly, Reichenbach did not discuss the relative merits of this approach as compared to the relative frequency interpretation. Even though he used frequency ratios in the definition of probability values, he apparently did not yet consider the limit of relative frequencies a suitable concept for the foundations of probability.

Reichenbach criticized the Kriesian distinction between nomological and ontological determinations in an important respect. Von Kries defined a process as objectively possible if it is logically possible and nomologically possible, i.e. does not contradict the empirical laws. Reichenbach however considered this distinction as moot. "The impossibility of the existence of a process can in general be traced back to a contradiction between the process and mathematical laws" (Reichenbach 1916, 116). The crucial point is determination. "A process that is only nomologically determined is still indeterminate and therefore cannot yet be called a process because only fully determined processes are possible" (ibid., 118). It becomes real only once it can be empirically perceived, which requires that the mathematical laws are supplemented by specifying the constants. But Reichenbach rejects to parallel this problem with the distinction between differential equations

[6]The German *Gesetzmäßigkeit* or *Gesetzlichkeit* stand between a strict law (*Gesetz*) and any nomologically weaker kind of statistically established regularity (*Regel* or *Regelmäßigkeit*).

and initial conditions, which von Kries had used to explicate his own distinction within the context of statistical mechanics, especially in the face of Loschmidt's reversibility objection. For the Kriesian conception is only valid for finite systems because in this case "one can view every real process as an instance of a class and this class represents all possible processes" (ibid., 120). There is a finite number of differential equations that already specify the class of possible values for the constants. But this is not true in general.

Reichenbach follows Kant's rejection of the idea that something has to be added to the possible to make it real.[7] Each empirical reality requires its own genuine constitution process. For "an instance [an element of a class] is not real because the constants have assumed particular values; and it is not possible just because the values have been left undetermined. Lack of determination [*Unbestimmtheit*] is not the criterion for possibility, and determination is not the criterion of reality. The meaning of these fundamental categories can only be grasped as plain insight, as pure intuition" (ibid., 120). Thus the point is not simply to maximize determination. "It is easy to make a whole class of processes real,[8] for example if one lets a gas pass continuously through all states of compression. In contrast, if only one instance becomes real, then this needs special justification on the given circumstances" (ibid., 120). It is of course true, that in this case we have not a differential equation but a state equation, which few philosophers would take as an instance of causality. But Reichenbach's general point seems to be that both nomological determinations, i.e. the principle of causality, and probabilistic determinations, i.e. the continuous probability function, have to be combined to determinate reality, not necessarily in the sense of unique determination, but possibly also by reference to a class that contains more than one instance.

As vague as these passages in Reichenbach's dissertation may sound, to my mind, they foreshadow the route which his philosophical agenda would take in subsequent years. First, in contrast to Schlick, uniqueness is not a satisfactory criterion of determination. Second, probability functions are an essential element at the lowest level of determination because they contain—as he writes in 1920—the "irrational remainder of the determinants" (Reichenbach 1920b, 148/315) not expressed in any additional laws. The principle of lawful distribution ensures that this largely unknown infinite class remained small enough. Third, by tightly connecting the theory of error and physical probabilities, Reichenbach effectively collapses the micro–macro distinction that had been characteristic for the debates about statistical mechanics. No wonder that such examples would play only a little role in his subsequent works, even after he openly endorsed the relative frequency interpretation in 1930. Fourth, Reichenbach retains at least the general thrust of the Kriesian distinction between nomological and ontological determinations when claiming that the probability function "connects the events

[7]The respective passage from Kant's *Critique of Pure Reason* (A 231, B 284) appears in a footnote on p. 122.

[8]The German original reads "*wirklich*" (p. 120), but the editors' translation "possible" (p. 121).

laterally" (ibid., 152/324). But rather than following von Kries' Kantian approach, Reichenbach models the distinction in the sense of a Minkowskian picture of special relativity, which identifies the causal order with the temporal direction and an orthogonal hypersurface. Also Schlick (1920) saw the latter as the domain of von Kries' ontological determinations. The only difference was that in Reichenbach's account there is no well-defined orthogonality relation, such that probability cannot be simply contained within the ontological determinations.

6.2 Years of Transition: The Mid 1920s

In a paper written in 1923 but published only in 1932, Reichenbach partly changed his mind. Although he considered the a priori conception of causality as irrefutable, because one could still claim the existence of causal laws that have not been found to date, the principle of causality was not positively required for the existence of natural laws. The application of the other principle, which was now called probabilistic (or inductive) inference [*Wahrscheinlichkeitsschluß*], only required that causality was not a priori excluded.

At the beginning of the paper, Reichenbach called causality a complex of principles and provided an avowedly non-exhaustive list of its elements. It contained Schlick's (1920) assertion that space-time coordinates must not figure in the laws themselves, the principle of action by contact, and the temporal order of events. But all three were only partial claims that supplemented the more general inductive principle of causality. This "says that by means of a functional relationship unobserved events can be predicted from observed ones, no matter whether the observed events lie in the future, or in the past, or happen at different space points simultaneously with the act of observation" (Reichenbach 1932a, 34/347). As he made clear in a rather similar list in his entry for the prestigious *Handbuch der Physik* (Handbook of Physics) that was written in the mid-1920s, causality was not exhausted by the concept of a function, as Mach (1883) had held, because it represented "a functional connection of a very specific character" (Reichenbach 1929, 59/193). Presupposing Laplacian determinism as a Kantian category, on the other hand, represented an unwarranted extrapolation beyond the implication from causes to effects. Reichenbach, having abandoned the synthetic a priori, in the mid-1920s positioned himself between the Kantian and the empiricist tradition: the principle of causality could be empirically false but the principle of probabilistic or inductive inference remained a condition for the possibility of scientific knowledge.

Reichenbach's inductive principle of causality operated as such: Starting from a presumed law $F_r(p_1, \ldots, p_r)$ we find further relevant causes $p_{r+1}, \ldots, p_{r+s}$ that lead to a modified function $F'_{r+s}(p_1, \ldots, p_r, p_{r+1}, \ldots, p_{r+s})$. This new governing function is the simplest function that, without being ad hoc, approximates the additional parameters in the least squares. Iterating this procedure with new classes of observed points M', M'', ... we obtain either the infinite governing sequence (I) $F_r, F'_{r+s}, F'_{r+s}, F'_{r+s}, \ldots$ or (II) $F_r, F'_{r+s}, F''_{r+s+t}, \ldots, F^{(i)}_{r+s+t+\ldots+w}, \ldots$.

In case (I) we have found a causal law, whereas in case (II) the connection between the observations is random. Both cases "characterize an objective state of affairs" (Reichenbach 1932a, 43/354), a conclusion for which the requirement of inductive simplicity is crucial. For only then, "the subclasses M, M′, M″ furnish different governing functions from that furnished by the total class $M^{(i)}$." (ibid., 45/355) Otherwise, (I) could trivially be obtained by an arbitrarily complex function. Apart from finding the governing function F'_{r+s}, inductive simplicity implies that the intermediate values between two observed values, that is, future measurements, are described by F'_{r+s}. This assumption of smoothness shows that inductive simplicity had taken the place of the principle of the continuous probability function. To sum up: "Either no continuous causal laws exist or they can be obtained by the requirement of simplicity" (ibid., 51/361).

Other than descriptive simplicity, which guided the choice by convention of a geometry in relativity theory, inductive simplicity represented a hypothesis about nature. In virtue of this difference, Reichenbach also rejected the conventionalist conception of causality because "the principle of causality constitutes a restrictive statement about the behaviour of physical phenomena, and may therefore encounter contradictions" (ibid., 59/367). How can inductive simplicity and, accordingly, the principle of causality be justified? Evidently, the sequences (I) and (II) are infinite, while further observations yield only finitely many data points. Thus we only know with probability whether causality holds or not in a given case. But the empiricist argument that our experiences (probabilistically) confirm whether causality holds, to Reichenbach's mind, misses the point. For, each single case contains the problem of induction in its entirety. "Whether causality holds in a specific instance can ultimately be decided only by investigating that instance. If causality holds in other cases, the probability that causality holds in the specific case under consideration merely *increases*" (ibid., 60/367). But it never actually reaches unity, that is, certainty. "It is therefore not impossible that physics will some day be confronted by phenomena that compel it to abandon causality." Mentioning quantum theory, Reichenbach concluded that "[i]n principle, it is possible to determine on the basis of experience whether causality holds." (both ibid., 63/370) Little wonder that, when publishing the paper with a decade of delay, Reichenbach (cf. 1932a, 32) proudly announced that quantum mechanics had meanwhile led to a breakthrough of his conception by providing a physical theory of type (II).

6.3 The Debate with Schrödinger

In the appendix to his 1923 paper, Reichenbach published a letter Erwin Schrödinger had sent him in response to the manuscript on January 25th, 1924. Schrödinger added a copy of his then likewise unpublished 1922 Zurich inaugural address (1929a) that elaborated his own conception of probabilistic physics (see Stöltzner 2012b). Schrödinger understands "the *deep* problem of causality" in the Humean sense as the question

> why under *completely* identical conditions [*Umständen*] do we always expect a *completely* identical outcome? ... We will ultimately be able to accept an influence however *strong* on the outcome of conditions changed however *small*, but will never be willing to admit the *smallest* change in outcome for *really* unchanged conditions. I call this the riddle of inductive inference. I do not believe that it is solvable for us in its proper sense. If one ponders about it for a longer time, an awfully distressing feeling arises, ... a kind of intellectual rotary vertigo [*Drehschwindel*], because one constantly believes to have understood the matter, but then realizes that one is moving in circles that are becoming narrower and narrower. (Schrödinger 1932, 65)

All attempts to solve the riddle of induction resulted in tautologies. Reichenbach, to Schrödinger's mind, had merely "locked this up in probabilistic inference" (ibid., 67), which itself was still poorly understood. Thus: "In actual fact one does not get beyond the fact *that* we constantly infer inductively, ... *that* all our living is based upon it." (ibid., 66)

Schrödinger criticized Reichenbach's criterion of the governing sequences as "useless" (ibid., 68) for the distinction between causal and non-causal laws. First, if one disregards fluctuation phenomena, the observed values will oscillate wildly and ultimately not only fill a curve y(x), but a two-dimensional strip around some y_0 and x_0. "The observer will then have to admit that from a certain density of the observations on he will not gain anything for the determination of the *form* of the function (including the number and character of the parameters appearing in it) from stacking up further observations, but only for the precision of the parameters. *Hence the sequence necessarily reaches type I after a finite number of steps*." (ibid., 68) Thus Reichenbach's criterion is self-refuting. Second, let us disregard observation errors but allow fluctuations, "which, to be sure, occur in principle for every physical quantity. Then our observer discovers the sequence type II in its pure form, whereupon he may convince himself, by controlling his measurement instruments in other ways, that errors of observation are *not* present. Will he accordingly deny any kind of causality for the phenomenon under investigation? He delves further into the issue, sees it through, and recognizes that his observation points are grouped around a certain function *as if* they were errors of observation, and he derives an entirely exact *law* (e.g., of radioactive decay $a = a_0 e^{-\lambda t}$) not just for the average curve but also for those fluctuations, which is now a genuinely statistical law." (ibid., 68) In short, if one takes into account the finite precision of any observation, and accordingly measurement errors, one arrives at type I, while if one includes actual physical fluctuations, which are always present, one arrives at type II.

Thus "the factual approach of natural scientists seems to confirm the apriori philosophers: causality is an undetachable element of our mode of comprehension. ... One may nevertheless decide to retain your sequence criterion [*Reihenkriterium*]. But then the experience available today decides with high probability *against* causality, probably in *all cases*. It is of great value to me that through careful and unbiased analysis you have been led to this conclusion, which was certainly not your intention" (ibid., 69). Reichenbach, to Schrödinger's mind, had unintentionally subscribed to the indeterminist view that he himself was advocating since his Vienna days and which he traced back to his teacher, the physical chemist Franz Serafin Exner. This tradition—which I have called "Vienna Indeterminism" (cf. Stöltzner 2012b)—was based (i) on the assumption of a liberal Machian notion of causality—as compared to a Kantian category—; (ii) an empiricist shift of the burden of proof on the determinist's shoulders—who had to provide a satisfactory theory of microphenomena before claiming victory over the assumption that random microscopic processes reproduced the observed macroscopic regularities in the limit; and (iii) the relative frequency interpretation of probability. The latter opened the large space between the macroscopic and the microscopic as a domain of transition in which fluctuations, and the statistical laws governing them, were dwelling because, e.g., the number of molecules impacting Brownian particles or of decaying atoms was still too low to reach the macroscopic limit.

Schrödinger contemplated how an Exnerian picture of nature, "as it will probably be the case in a few decades," (Schrödinger 1932, 69) could look like. It would not contradict the determinism of classical physics that we observe on the macroscopic level, but using the law of large numbers derive it from possibly random atomic processes at the microscopic level. Its basic elements would exhibit a certain persistence presumably based on the conservation laws. But the riddle of inductive inference would just be shifted to the atomic level.

> There one will have to assume—or so I imagine—laws of the kind that *sharply* defined conditions are related to a whole continuum of possible outcomes—or perhaps, with certain restrictions of stability [*Beständigkeitseinschränkungen*], to *all* possible outcomes. The "riddle" [of induction] will have retreated to [the position] that by repeated preparation of sharply defined initial conditions the distribution of the outcomes over this continuum will be a specific one, e.g., a uniform one.—To be sure, one cannot know whether this idea (which is obviously modeled after a game of chance) will turn out to be useful. However, at some point an axiom will slip in that is no less enigmatic than causality. For problems do not resolve themselves. (Ibid., 70)

This was quite a successful prediction about how quantum mechanical results would look like just two years later, even though the indeterminacy inherent in Heisenberg's uncertainty relations would usually be interpreted in the sense that the antecendent of the causal principle was violated, that is, that for non-commuting observables one could not realize sharply defined conditions.

Discussing the details of Schrödinger's physical outlook was not the main point in Reichenbach's (1932b) "Concluding Remarks". Rather did he defend his sequence criterion as appropriate for classical physics and emphasized that only on the basis of quantum mechanics could one conclude that performing more and more precise observations necessarily yields a sequence of type II. Without hesitation he recommended his own conception for a precise formulation of the uncertainty relations. However, his sequence criterion did not presuppose that any adaptation of the form of the function to the observations would necessarily increase the precision. Hence,

> "once we reach the domain where fluctuations already play a significant role, according to the conception of classical physics the sequence type I emerges; by way of the sketched procedure one would ultimately obtain as many parameters ... as one needs for the determination of the motion of all single molecules and then accordingly capture the deterministic law of the fluctuation phenomena, that is, the causal lawfulness [*Gesetzlichkeit*] of all single molecular collisions" (Reichenbach 1932b, 71).

The mere fact that one had only a statistical law did not distinguish both sequence types.

> "The emergence of statistical laws can derive from two possibilities: either it is *practically* impossible to determine the remaining [*bleibende*] function, that is, to make the probability of the single case large (classical interpretation of the kinetic theory of gas), or it is in principle impossible because the governing function is of type II (interpretation of quantum mechanics)" (Reichenbach 1932a, 63).

Reichenbach also rejected Schrödinger's contention that depending on how we treat fluctuations we necessarily obtain sequence type I or II. For one, "it represents a property of nature" (Reichenbach 1932b, 71) whether a sequence is of type I because the influence of the errors is limited. Moreover, Schödinger's reconstruction of the factual approach of natural scientists was of little relevance for epistemology, which intended "the *rational reconstruction* of the knowledge procedure" (ibid., 72). "The path of physical discovery can be entirely different from its later justification in the face of the empirical material" (ibid., 72). But this early distinction between context of discovery and context of justification—a thesis which would later be considered as one of his signature contributions—does not resolve the problem that Reichenbach's epistemological distinction was rooted in the objective constitution of the world and still required a verdict on the smallness of the "irrational remainder". This outcome of the discussion and the strict distinction between classical physics and quantum physics appears to me somewhat surprising, given that Reichenbach, in 1932, defended a philosophical outlook in which objective probability represented a more basic concept than causality. But he wanted probability to take hold at a much more local, if not individual, level than did the physicists and required a stronger determination of the events than mere membership in a statistical collective, which was required for the relative frequency interpretation to hold.

Schrödinger's conception of probabilistic physics was markedly different and emphasized the close relationship between fluctuation physics and quantum mechanics. No wonder that Reichenbach's concluding remarks (1932b) did not appraise the remarkable continuity between what Schrödinger wrote in the early 1920s (Schrödinger 1924, 1929) and the quantum mechanics as of 1932. Not that Reichenbach was unfamiliar with the Viennese tradition Schrödinger had come from. After all, Reichenbach had reviewed Exner's *Lectures on the Physical Foundations of Natural Science* (1919), where the basic ideas of Vienna Indeterminism were broadly outlined. The reviewer commended the "unbiased attitude of the natural scientist [Exner] who dislikes metaphysical speculations and who is conscious of the inductive character of all regularities discovered, even of the most general ones. .... Of particular importance seems to me that Exner unequivocally advocates the objective meaning of the probabilistic laws in which he rightly conceives a very general regularity of nature." (Reichenbach 1921, 415) Also the first footnote of *Philosophic Foundations of Quantum Mechanics* cites Exner as "perhaps the first" (Reichenbach 1944, 1) to have criticized the assumption of strict causality.

The signature scientific achievement of the Viennese tradition, beyond its philosophical impact on Schrödinger and Frank, was to consider fluctuations as a quantity in its own rights that could figure in statistical laws of nature—rather than indicate deeper, hitherto undiscovered laws. This required, or so I have argued elsewhere (Stöltzner 2012a), an open-minded empiricist indeterminism that did not require a deterministic foundation for laws of nature. The main physical problem however was to distinguish fluctuations from equally statistically distributed measurement errors. Since Schrödinger held that there was no difference in principle between laws which had a deterministic foundation and those which had not, this required a physical theory of measurement rather than Reichenbach's epistemological analysis of measurement errors. Let me elaborate on Schrödinger's stance at the example of radioactivity mentioned in his letter.

In 1905, Egon von Schweidler, then Exner's assistant, had shown that the phenomenological law of radioactive decay, the Rutherford-Soddy law, was only valid for a large number of decaying atoms, while for a small number of atoms the decay constant λ exhibited fluctuations (Von Schweidler 1906). Independently of Albert Einstein, Exner's former assistant Marian von Smoluchowski (1906) derived the formula for the position fluctuations of a Brownian particle. While most physicists quickly accepted Brownian motion as an experimental proof of atomism, Schweidler fluctuations, until the advent of quantum mechanics, were typically considered as a convenient phenomenological regularity still to be explained by the hitherto unknown laws governing the decay of the single atom. The Viennese thought differently and accepted them as a statistical law. Schrödinger undertook a detailed statistical analysis of the measurement of radioactive fluctuations and developed a theory of the preferred measuring device, the electroscope, treating the motion of its pointer as a Brownian process, or as he put it, "a Smoluchowski motion" (Schrödinger 1919, 184). Thus he used a phenomenon with a deterministic foundation as the theory of measurement of a genuinely probabilistic phenomenon without such a foundation. The important point is that Schrödinger did not distinguish between indeterminacy in principle and indeterminacy in practice, but took both kinds of fluctuations as physical phenomena. For the Vienna Indeterminist, the burden of proof was with the determinist. "The *possibility* [that deterministic causality] may be in reality the case must be admitted, but this duplication of natural law so closely resembles the animistic duplication of natural *objects*, that I cannot regard it as at all tenable" (Schrödinger 1929a, 11/145). While in his Zurich speech, and even more after the Bohr-Kramers-Slater theory (Schrödinger 1924), he had been quite optimistic that a breakthrough of the Exnerian picture was near, his 1929 inaugural address as a member of the Prussian Academy of Sciences gave the problem a different twist.

> In my opinion this question [whether causality is present] does not involve a decision as to what the real constitution of nature is, but rather as to whether the one or the other predisposition of mind be the more purposive and convenient one with which to approach nature. Henri Poincaré has illustrated that we are free to apply Euclidean or any kind of non-Euclidean geometry we like to real space. ... The same probably applies to the postulate of rigid causality. One can hardly imagine empirical facts which ultimately decide on whether the natural phenomena are in reality absolutely determined or partially indetermined, but at best on whether the one or the other conception permits a simpler survey of what is observed. Even this question will probably take a long time to decide. (1929b, 732/xvii f.)

Schrödinger's discovery of quantum mechanics and in particular his proof of the equivalence between wave mechanics and matrix mechanics substantially changed the nature of the alternative. There was, on the one hand, a deterministic differential equation the application or interpretation of which permitted only statistical predictions. There was, on the other hand, an abstract and openly indeterministic theory which nonetheless integrated the whole conceptual apparatus of classical mechanics in a quantized form. Schrödinger's equivalence proof corresponded to the systematic classification of all possible geometries achieved at the end of the nineteenth century which had nurtured Poincaré's conventionalism. For conventionalist choice required a precise formal characterization of the alternatives.

Schrödinger's line of reasoning went against Reichenbach's contention that probability was special and could not fully be compared to geometry. Perfecting instead the analogy between probability and geometry, Schrödinger (1934) went as far as to argue that geometry was not applicable on the atomic level because it required concepts, such as a particle trajectory, that were well-defined only within the limits set by Heisenberg's uncertainty principle. This blocked even the possible existence of ideal objects in geometry which Reichenbach (1916) had cited as geometry's major difference to probability theory.

Despite his criticism of the Copenhagen interpretation as introducing concepts with limited validity and his contemplating a conventionalist resolution of the problem of causality, Schrödinger continued to emphasize the continuity between statistical mechanics, fluctuation physics, and quantum mechanics. Reichenbach did not approach quantum physics from the perspective of statistical mechanics, which for most physicists had been the entrance into the physics of atomic phenomena since Planck had used Boltzmann's methods in the derivation of his quantum theory of radiation. Instead, Reichenbach applied his ideas about probability in a setting that was, it appears, partly motivated by considering the issue of causal order in the sense of special relativity and contemplated how a probabilistic order of events would look like.

6.4 Probabilistic Topology

Reichenbach's 1925 paper on the causal structure of the world was ambitious. Entirely dispensing with the hypothesis of strict causality, he proposed a conception based on "the concept of probable determination alone." (Reichenbach 1925, 136/83) This "accomplishes everything that is achievable by physics and … furthermore possesses the capacity to solve the problem of the difference between past and future, a problem to which the strict causal hypothesis has no solution" (ibid., 133/81). Although he maintained his earlier convictions that physics rests upon both "the principle of causal connection and the principle of probable distribution" (ibid., 135/82), and that one can in principle separate the causal connection between the determining factors and the probabilistic distribution of the remaining factors, he now considered the latter division as purely formal. It "can be replaced by the single assumption that a connection of a probabilistic nature exists between cause and effect" (ibid., 138/84). This connection was now anchored on the level of logic, which yielded an exclusively probabilistic determination.

Reichenbach replaced the causal connection of events 'A causes B' by '*A* implies *B* with probability', or $A \ni B$, which he understood as a primitive concept and provided a list of laws fulfilled by it, "which claims neither to be exhaustive nor to represent a table of independent axioms" (ibid., 146/91). While logical implication ($\rightarrow$) connects propositions, probability implication ($\ni$) connects events. The most striking formal novelty was that $(A \ni B) \rightarrow (A \ni \neg B)$. One thus obtained a topology of probability implications, while the probability measure remained unspecified.

This topology, Reichenbach claimed, was sufficient to define a temporal order of events. "*If probability implication is valid in only one direction* [i.e. $(B \ni A) \wedge \neg(A \ni B)$], *then the antecedent* [*B*] *is the temporarily later event*" (ibid., 150/94). The main difference was that "[n]othing short of the totality of all causes is required for inferences into the future, but inference about the past can be made on the basis of a partial action [of causes]" (ibid., 151/96). The future was thus objectively undetermined. Reichenbach also provided a detailed analysis of various inferential scenarios between three or more causes in the form of causal forks. This approach in its mature, and more rigorous, form outlined in the posthumous *The Direction of Time* (1956) became quite influential on the philosophical debates about causality in the 1960s and 1970s.

In the handbook entry, Reichenbach (1929) also discussed the relationship between causality and the special theory of relativity on the basis of his method of mark transmission (cf. Reichenbach 1924). A mark represents a small variation in an event. If we attach a mark to the cause *A*, this mark will also be observable in the effect *B*, but not vice versa. This asymmetry is "the distinctive characteristic of the causal relation ... [and] can, in turn, be used in defining the sequence of time" (1929, 53/186). Accordingly, the "objective significance of time consists in its formulating the type of order of causal chains. It is, then, a physical theory of a very general nature, but not in any way the product of a special human faculty" (ibid., 57/190)—as Kant had assumed. And, referring to his 1925 definition, he argued that the microscopic events in nature could be subjected to temporal order. Boltzmann's contention that irreversibility and the direction of time emerge only as statistical features at the macroscopic level, while atomic collisions remain reversible as in Newtonian mechanics, to Reichenbach's mind, was too closely connected to the false ideal of Laplacian determinism. Boltzmann's reasoning by way of the probability of initial states, he continued, did not get around the reversibility objection. Consequently he criticized Schlick's (1925a) claim "that every indication of temporal direction must conform to the Boltzmann scheme" (Reichenbach 1929, 62/196). But Schrödinger's (1929a) above-discussed stance demonstrates that the distinction between a micro-level and a macro-level, which stood behind Boltzmann's conception of time, does not necessarily require determinism. As long as it reproduced macroscopic observations in the limit, randomness at the micro-level was equally acceptable; it even had the advantage of a more unified probabilistic world picture.

Other than in 1925, Reichenbach's handbook entry anchored probability implication on the level of perception rather than space-time events. These perceptions were coordinated to the things—and for the positivist this coordination amounted to identity—or to the concepts denoting the things—a difference which the realist Reichenbach considered as crucial. He now criticized Schlick's (1925b) claim that the uniqueness of this coordination represented the only feasible definition of truth. For this characterization, first, "pertains solely to the ultimate goal of knowledge" (Reichenbach 1929, 28/154) and, second, "does not offer any means whereby the truth of a given physical proposition can be tested" (ibid., 29/155). The only way to solve the second problem, Reichenbach held, was to analyze our observations and propositions by means of probability implications. "We will no longer be able to speak strictly of the truth of a proposition, but only of its degree of probability" (ibid., 29/155). And he called a proposition correct if it was highly probable. The first problem led Reichenbach to regard truth so conceived as a property of a coherent system of scientific knowledge. A similar kind of holism was advocated by most Austrian members of the Vienna Circle, chief among them Neurath.

6.5 Inductive Inference as an a Priori

The 'First Meeting on the Epistemology of the Exact Sciences' that was held in Prague in 1929 and in which the Vienna Circle first went public with its famous manifesto, figured a large section on 'Probability and Causality'. The papers and the extensive discussion that followed them were printed in *Erkenntnis* (Zilsel et al. 1930). There existed a great variety of opinions (Cf. Stöltzner 2009). Waismann (1930)—in full agreement with Schlick—advocated the Kriesian interpretation as regards both physical probability and the logical probability of judgment, but insisted that both concepts were distinct and that the so-called application problem was meaningless. Reichenbach (1930) advocated, more explicitly than ever before, the relative frequency interpretation and conjectured that every assertion of probability could be translated in an assertion of frequency. This prompted a discussion with von Mises in which the analogy between geometry and probability was a stake. To von Mises, Newtonian mechanics and relativistic geometry, on the one hand, and (classical and quantum) probabilistic physics, on the other, stood on a par. While in the former cases, the symbols of the theory were uniquely coordinated to individual experiences, in the latter cases they were coordinated to mass phenomena. Frank and von Mises followed Schlick in considering unique coordination as the only meaningful criterion of truth applicable in science. Reichenbach vehemently disagreed.

> For in the coordination of physical bodies to a mathematical theory, the concept of approximation appears, and this contains the concept of probability. ... In the case of geometry, it is true, one is allowed to separate the problem of coordination from the mathematical theory because the problem of coordination does not contain any *geometrical* concept, but in the theory of probability the concept constituted by this theory enters into the problem of coordination. (Zilsel et al. 1930, 275)

The association between finite observations and an infinite collective in the frequentist account, to Reichenbach's mind, was based on probability (or inductive) inference. Reichenbach's problem was not that a statistical collective contained infinitely many elements; this was a common criticism against frequentism shared, among others, by Schlick and Waismann. Rather did a statistical collective not qualify as an ideal object in the same sense as geometric objects because there was no difference between the errors necessarily arising in coordination and the probabilistic object theory. Thus his endorsement of the relative frequency interpretation had not changed a core part of Reichenbach's dissertation: the emphasis of probabilistic determination on the event level and the disanalogy between geometry and probability. But the endorsement made him vulnerable to von Mises's criticism that the association between the formal calculus and assertions of probability "was not translatable into a frequency statement" (Zilsel et al. 1930, 282).

To avoid this criticism, Reichenbach shifted the problem to the most basic level. "Probability logic cannot be squeezed into the Procrustes bed of strict logic" which leads to the "catastrophe of undecidability" (Reichenbach 1930, 170) about whether a law of nature is actually confirmed or not. It embraces strict logic as a limit in the same vein as truth arises as the limit of high probability. Probability logic itself can only be justified by "the fact that we cannot think differently." For: "The statement that probability laws do not hold is equivalent to predicting that, in repeated sequences, the regularity implied by the principle of induction does not hold—and this statement is empirically meaningful only if it can be decided inductively, i.e. if the principle of induction holds. The statement that probability laws do not hold is thus self-contradictory and makes no sense" (ibid., 187/343; translation readjusted to original). Since Reichenbach did not presuppose strict logic to hold, this contradiction did not amount to an indirect proof of the principle of induction. Rather, it dissolved Hume's problem, as Reichenbach proudly announced.

But, to my mind, this ambitious claim is unwarranted. Reichenbach in effect treated induction—in the same vein as the principle of lawful distribution a decade before—as an a priori condition for the possibility of experience, the only difference being that there was no longer a transcendental deduction available to justify it. Still one might wonder, whether it was not at bottom a transcendental-pragmatic argument, to hold that rejecting induction was self-defeating. Since Reichenbach granted that the principle of causality could be empirically inadequate, it becomes clear that the two principles had finally changed rank. While initially the second principle—be it lawful distribution or probabilistic inference—had only represented an indispensable complement to causality, it had now assumed the lead.

Given his repeated claims to have anticipated important epistemological characteristics of quantum mechanics, it is quite surprising that in those years Reichenbach did not embark on a more detailed discussion of it and only criticized two interpretative claims of Heisenberg's. First, the 'positivistic' maxim to omit unobservable quantities from the theory "must be correctly reformulated as the stipulation that *dispensable* quantities should be eliminated." Yet this was, to Reichenbach's lights, a simple consequence of probability inference. Second, Heisenberg's elucidation of the uncertainty relation as a disturbance effect, that is, that "the *influence of the instruments of observation* cannot be ignored, ... is not viable" (both Reichenbach 1929, 78/215). For, as he argued a year later, "separation in object of observation and means of observation is an idealization that is to a certain extent fulfilled for certain macroscopic phenomena, but it cannot be regarded as a necessary presupposition of the exact sciences in the sense of the principle of causality" (Reichenbach 1930, 180–181/338 translation modified). The crucial point was rather that one could not push the probability to predict certain combinations of parameters arbitrarily close to unity.

6.6 Statistical Mechanics and the Direction of Time

Reichenbach's (1931) paper to a large extent was a criticism of Schlick's (1931) new theory of causality that took the fulfillment of predictions as the one and only criterion of causality. Without aspiring at a complete account of this controversy, its origins, and its consequences for the inner dynamics of the movement of Logical Empiricism,[9] let me mention three issues at stake that are of relevance for the present paper. They show that Reichenbach only superficially connected statistical mechanics and quantum theory, in contrast to many physicists of the day.

First, Schlick had argued that "the new contribution made by present physics to the causality problem does not consist in contesting the validity of the principle of causality as such ... nor in the recognition of a purely probabilistic validity of natural laws having replaced belief in their absolute validity. ... The novelty, rather, consists in the discovery, never previously anticipated, that a limit of principle is set to the exactness of prediction by the laws of nature themselves" (1931, 153/191). Reichenbach disagreed and cited a footnote from his 1925 paper where he discussed the question as to whether by a more and more precise determination of the participating factors, the probability of an event can approach 1 without previously reaching a limit <1. "Justifiable as such an assumption [of a limit <1]—which would be confirmed if quantum theory abandoned all attempts at causal explanation and contended itself with probability jumps of electrons—may appear, we shall not discuss it here, and everything that follows is also compatible with a probability capable of approaching 1 without limit" (Reichenbach 1931, 716/332).[10] In the handbook entry of the late 1920s, Reichenbach mentioned Exner's contention that all causal laws might share the fate of the second law of thermodynamics and become only statistical, but not without mentioning also Planck's opposition against such a view and the latter's claim that statistical laws only have provisional validity (cf. 1929, 71/224). Back then he had also argued, more cautiously, that Heisenberg's uncertainty relations were "an entirely new kind of restriction to our knowledge of

[9]Among them Schlick's very negative evaluation of Reichenbach's work for the Prussian Ministry of Science, cf. Stadler (2011).

[10]The footnote originally appears in Reichenbach (1925, 139/118).

nature, the existence of which was never before suspected" (Reichenbach 1929, 78/216). In short, Reichenbach had not made the bold predictions he would later ascribe to himself. It is true, Schlick at the same time had written that "only in the utmost case of emergency will the scientist or philosopher decide to postulate purely statistical micro-laws ... . The principle of causality would be abandoned ... and hence the possibility of exhaustive knowledge would have to be renounced" (Schlick 1925a, 461/61). And when the emergency occurred in 1926, Schlick needed five years to work out his new theory of causality. Reichenbach was no doubt better prepared for the historical turn of events in atomic physics than Schlick, but he had not anticipated the indeterminism of quantum mechanics.

Second, Reichenbach emphasizes the probabilistic character of the second law, but when talking about the kinetic theory of gases in 1931 he reads the history in different fashion, contradicting the passage just quoted from the handbook article. "No doubt some scientists expressed even then the thought that the first law [energy conservation], too, and perhaps all causal laws, would meet with the same fate, but others, and probably the majority of the physicists, could not make up their minds to recognize the concept of probability as enjoying a status equal to the concept of causality and assign to the probability concept a merely provisional role in the theory of gases as well." And they used the theory of error to consider statistical laws as makeshift laws "called for by human imperfection" without recognizing that they had thus applied the same concept of probability. "Boltzmann himself was evidently satisfied with this conception, as is shown by his manifold attempts to establish his statistical principle as a consequence of the basic laws of mechanics; to be sure, his and all successive attempts in this direction failed" (all Reichenbach 1931, 714/328). It is quite interesting that this time he added a footnote that explicitly denied Exner's priority for the statistical character of energy conservation, even though this idea was present in Exner's *Lectures* and Boltzmann's writings, a fact which Schrödinger tirelessly stressed. Exner's book appeared "at a time when the idea of this possibility had already been the common property of physicists for decades" (ibid., 714/342).

Third, Boltzmann's arguably deterministic foundations of statistical mechanics had another consequence. Reichenbach claims to have shown in his *Handbook* entry that "if determinism holds, it is impossible to define an asymmetry in temporal direction on the Boltzmannian concept" (1931, 718–719/336). His argument had two parts. (i) If we assume that the microscopic processes are reversible, does Boltzmann's inference about the irreversibility of macroscopic processes hold? (ii) "What justification is there for establishing the thesis that the elementary event is reversible?" (Reichenbach 1929, 62/197). In order carry home the first point (i), Reichenbach took up the reversibility objection—which he ascribed to Gibbs instead of Josef Loschmidt who had actually raised it against his Vienna colleague Boltzmann—according to which a sudden inversion of the velocities of all particles also changes the sign of the entropy. Following an argument of the Göttingen physicist-philosopher Paul Hertz, Reichenbach argued that Boltzmann's rejoinder, that the states for which entropy decreases are highly improbable as compared to those in agreement with the second law, was circular for a closed system because in distinguishing an initial state one had already used the direction of time. Only for an open system, in which the origin of the initial state "may be traced back to outside causes" (ibid., 63/198), could one infer the temporal order for the sequence of states. Reichenbach also rejected to assign a cosmological direction of time. For if the Universe contains finitely many reversible elementary subprocesses, there would not be a homogeneous flow of time. The second point (ii) emerged from the dilemma to reject either the finiteness assumption or the reversibility of the elementary processes. Reichenbach took the second route because the reversibility "arose essentially through the influence of the mechanical view of the world, which can only be maintained in conjunction with the idea of determinism" (ibid., 64/199). Instead he advocated his own approach of substituting causality with probability and defining the direction of time on the level of elementary processes by means of his probabilistic topology. In this way, he argued, one could establish irreversibility and avoided the problematic idea of mixing processes and—or so I read a remark that Reichenbach made in passing later—the ergodic hypothesis, which he preferred to turn from a metric assumption into a topological one (cf. ibid., 69/204–205). Interestingly, Reichenbach did not follow von Mises's (1922, 1930) argument that a consistently probabilistic approach would overcome "the notorious ergodic hypothesis" (1922, 29) by dispensing with the necessity of any connecting link between microscopic and macroscopic physics altogether.

All three points listed here show that Reichenbach mainly advocated his own program and did not really discuss in depth the problems that were plaguing the—at least until 1926—most important statistical theory of the day. This is not to say that Reichenbach was wrong to consider Boltzmann's line of reasoning as inconclusive. Moreover, concepts like mixing, entropy and the relationship between the second law and the direction of time—let alone on the cosmological scale—were anything but clear. In fact, they are still keeping mathematical physicists and philosophers of science busy. (cf. Uffink 2007) My point is just that he did not involve himself into the details of the physical discussions, but pursued a genuinely philosophical agenda. Let me conclude this paper with an envoi that shows that this would change in the 1940s.

6.7 Envoi

Having emigrated to the U.S., Reichenbach published extensively on the philosophy of physics. Especially his 1944 book *Philosophic Foundations of Quantum Mechanics* inspired the subsequent debates on the subject. He now considered the "quantum mechanical criticism of causality … as the logical continuation of a line of development which began with the introduction of statistical laws into physics" (Reichenbach 1944, 3), but emphasized the peculiarities of the quantum world. The central claim of the book was that causal anomalies were unavoidable if one insisted that interphenomena, i.e. the states between the phenomena actually observed, possess definite values. Reichenbach's definition of a normal system was based on the idea that neither the laws of nature nor the states depend upon their being observed—while in 1930/1931 he had rejected precisely this kind of separation between object system and measurement apparatus. He thus had accepted Schrödinger's contention that measurement represented a physical process, not an induction that could be modeled by a suitable probability machine.

The major innovation of the book was Reichenbach's three-valued semantics for quantum mechanical statements. He was dissatisfied with the Copenhagen criterion for physically meaningful statement because this restriction was of a meta-linguistic kind, and physics could not get by without any description of interphenomena. He shared this dissatisfaction with Schrödinger, who however insisted that the interphenomena could not be described by trajectories and other classical concepts (cf. Stöltzner 2012b).

In his posthumous *The Direction of Time* (1956), Reichenbach modified his causal theory of time because it became clear to him that the mark method was not free of temporal concepts. At the very end of the book, Reichenbach was worrying whether his idea of basing time on the microscopic order was faulted by Richard P. Feynman's contention that a positron corresponded to an electron going backward in time. In this way, a definite causal chain would exist merely locally and causal loops could not be excluded. Luckily, positrons are short-lived and "the vast majority of particles thus conform to the rules of ordered time" (Reichenbach 1956, 268). Still, this was a statistical argument applied to the most elementary level of determination.

References

Boltzmann, Ludwig. 1905. *Populäre Schriften*. Leipzig: J.A. Barth; Part. Trans. as *Theoretical physics and philosophical problems*, ed. Brian McGuinness. Dordrecht: Reidel, 1974.

Eberhardt, Frederick. 2011. Reliability via synthetic a priori: Reichenbach's doctoral thesis on probability. *Synthese* 181: 125–136.

Exner, Franz S. 1919. *Vorlesungen über die physikalischen Grundlagen der Naturwissenschaften*. Leipzig/Wien: Franz Deuticke.

Fechner, Gustav T. 1897. In *Kollektivmaßlehre*, Leipzig: Engelmann.

Frank, Philipp. 1929. Was bedeuten die gegenwärtigen physikalischen Theorien für die allgemeine Erkenntnislehre? *Die Naturwissenschaften* 17: 971–7 & 987–94; English Trans. in Frank 1961. 96–125.

Frank, Philipp. 1932. *Das Kausalgesetz und seine Grenzen*. Wien: Springer; English Trans. as *The law of causality and Its limits*. Dordrecht: Kluwer, 1998.

Frank, Philipp. 1937. Philosophische Deutungen und Mißdeutungen der Quantentheorie. *Erkenntnis* 6:303–317; English Trans. in Frank 1961, 158–170.

Friedman, Michael. 1994. Geometry, convention, and the relativized a priori: Reichenbach, Schlick, and Carnap. In *Logic, language, and the structure of scientific theories*, ed. Wesley Salmon, and Gereon Wolters. 21–34. Pittsburgh, PA–Konstanz: University of Pittsburgh Press–Universitätsverlag Konstanz.

Galavotti, Maria Carla. 2011. On Hans Reichenbach's inductivism. *Synthese* 181: 95–111.

Heidelberger, Michael. 2001. Origins of the logical theory of probability: von Kries, Wittgenstein, Waismann. *International Studies in the Philosophy of Science* 15: 177–188.

Mach, Ernst. 1883. *Die Mechanik in ihrer Entwickelung. Historisch-kritisch dargestellt.* Leipzig: Brockhaus; English Trans. as *The science of mechanics*, La Salle, IL: Open Court 1989.
Padovani, Flavia. 2011. Relativizing the relativized a priori: Reichenbach's axioms of coordination divided. *Synthese* 181: 41–62.
Reichenbach, Hans. 1916. *The concept of probability in the mathematical represenation of reality.* Trans. and ed. Frederick Eberhardt and Clark Glymour. Chicago: Open Court, 2008; orig. as Der Begriff der Wahrscheinlichkeit für die mathematischer Darstellung der Wirklichkeit. *Zeitschrift für Philosophie und philosophische Kritik* 161:210–239 & 162:9–112, 223–253.
Reichenbach, Hans. 1920a. Die physikalischen Voraussetzungen der Wahrscheinlichkeitsrechnung. *Die Naturwissenschaften* 8:46–55; Nachtrag, 349; English Trans. in Reichenbach 1978. vol. II. 293–311.
Reichenbach, Hans. 1920b. Philosophische Kritik der Wahrscheinlichkeitsrechnung. *Die Naturwissenschaften* 8:146–153; English Trans. in Reichenbach 1978. vol. II. 312–327.
Reichenbach, Hans. 1920c. *Relativitätstheorie und Erkenntnis apriori.* Berlin: Springer; English Trans. as *The theory of relativity and a priori knowledge*, with an introduction by Maria Reichenbach, Berkeley: University of California Press 1965.
Reichenbach, Hans. 1921. Review of Exner, Franz, *Vorlesungen über die physikalischen Grundlagen der Naturwissenschaften. Die Naturwissenschaften* 9: 414–415.
Reichenbach, Hans. 1924. *Axiomatik der relativistischen Raum-Zeit-Lehre.* Braunschweig: Vieweg.
Reichenbach, Hans. 1925. Die Kausalstruktur der Welt und der Unterschied von Vergangenheit und Zukunft. *Sitzungsberichte der Bayerischen Akademie der Wissenschaften, mathematisch-naturwissenschaftliche Abteilung.* 133–175; English Trans. in Reichenbach 1978. vol. II. 81–119.
Reichenbach, Hans. 1929. Ziele und Wege der physikalischen Erkenntnis. In *Handbuch der Physik.* vol. 4. Berlin: Springer. 1–80; English Trans. in Reichenbach 1978. vol. II. 120–225.
Reichenbach, Hans. 1930. Kausalität und Wahrscheinlichkeit. *Erkenntnis* 1:158–188; part. Trans. in Reichenbach 1978. vol. II. 333–344.
Reichenbach, Hans. 1931. Das Kausalproblem in der Physik. *Die Naturwissenschaften* 19: 713–722; English Trans. in Reichenbach 1978. vol. I. 326–42.
Reichenbach, Hans. 1932a. Die Kausalbehauptung und die Möglichkeit ihrer empirischen Nachprüfung. *Erkenntnis* 3:32–64; English Trans. in Reichenbach 197. vol. II. 345–371.
Reichenbach, Hans. 1932b. Schlußbemerkung. *Erkenntnis* 3: 70–71.
Reichenbach, Hans. 1936. Logistic empiricism in Germany and the present state of its problems. *Journal of Philosophy* 33: 141–160.
Reichenbach, Hans. 1938. *Experience and prediction.* Chicago: University of Chicago Press.
Reichenbach, Hans. 1944. *Philosophic foundations of quantum mechanics.* Berkeley: University of California Press.
Reichenbach, Hans. 1956. *The direction of time.* Berkeley: University of California Press.
Salmon, Wesley C. 1991. Hans Reichenbach's vindication of induction. *Erkenntnis* 35: 99–122.
Schlick, Moritz. 1920. Naturphilosophische Betrachtungen über das Kausalprinzip. *Die Naturwissenschaften* 8:461–474, English Trans. in Schlick 1979. vol. I. 295–321.
Schlick, Moritz. 1925a. Naturphilosophie. In *Lehrbuch der Philosophie: Die Philosophie in ihren Einzelgebieten*, ed. Max Dessoir, 397–492. Berlin: Ullstein. English Trans. in Schlick 1979. vol. II. 1–90.
Schlick, Moritz. 1925b. *Allgemeine Erkenntnislehre*, 2nd ed. Berlin: Springer.
Schlick, Moritz. 1931. Die Kausalität in der gegenwärtigen Physik. *Die Naturwissenschaften* 19:145–162; English Trans. in Schlick 1979. vol. II. 176–209.
Schlick, Moritz. 1937. Quantentheorie und Erkennbarkeit der Natur. *Erkenntnis* 6:317–326; English Trans. in Schlick 1979. vol II. 482–490.
Schrödinger, Erwin. 1919. Wahrscheinlichkeitstheoretische Studien betreffend Schweidler'sche Schwankungen, besonders die Theorie der Meßanordnung. *Sitzungsberichte der Österreichischen Akademie der Wissenschaften, Mathematisch-naturwissenschaftliche Klasse*, Abt. IIa, 128, 177–237.

Schrödinger, Erwin. 1924. Bohrs neue Strahlungshypothese und der Energiesatz. *Die Naturwissenschaften* 12: 720–724.
Schrödinger, Erwin. 1929a. Was ist ein Naturgesetz? *Die Naturwissenschaften* 17:9–11; English Trans. by James Murphy and W.H. Johnston as *Science and the human temperament*, 133–147. New York: W. W. Norton & Co.
Schrödinger, Erwin. 1929b. Aus der Antrittsrede des neu in die Akademie eintretenden Herrn Schrödinger. *Die Naturwissenschaften* 17: 732; English Trans. in the Introd. to *Science and the human temperament*, xiii–xviii.
Schrödinger, Erwin. 1932. Anmerkungen zum Kausalproblem. *Erkenntnis* 3: 65–70.
Schrödinger, Erwin. 1934. Über die Unanwendbarkeit der Geometrie im Kleinen. *Die Naturwissenschaften* 22: 518–520.
Stadler, Friedrich. 2011. The road to *Experience and Prediction* from within: Hans Reichenbach's scientific correspondence from Berlin to Istanbul. *Synthese* 181: 137–155.
Stöltzner, Michael. 2009. The logical empiricists. In *Oxford handbook of causation*, ed. Helen Beebee, Christopher Hitchcock, and Peter Menzies, 108–127. Oxford: Oxford University Press.
Stöltzner, Michael. 2012a. Zur Genese der Schweidlerschen Schwankungen und der Brownschen Molekularbewegung. In *Kernforschung in Österreich: Wandlungen eines interdisziplinären Forschungsfeldes 1900–1978*, ed. Silke Fengler and Carola Sachse, 309–340. Wien: Böhlau.
Stöltzner, Michael. 2012b. Erwin Schrödinger – Vienna indeterminist. In *Probabilities, laws, and structures*, ed. Marcel Weber et al., 481–495. Dordrecht: Springer.
Uffink, Jos. 2007. Compendium to the foundations of classical statistical physics. In *Handbook for the philosophy of physics*, ed. Jeremy Butterfield and John Earman, 924–1074. Amsterdam: Elsevier.
Von Kries, Johannes. 1886. *Prinzipien der Wahrscheinlichkeitsrechnung*. Freiburg i.B: Mohr.
Von Kries, Johannes. 1919. Über Wahrscheinlichkeitsrechnung und ihre Anwendung in der Physik. *Die Naturwissenschaften* 7:2–7 & 17–23
Von Mises, Richard. 1912. Über die Grundbegriffe der Kollektivmaßlehre. *Jahresbericht der Deutschen Mathematiker-Vereinigung* 21: 9–20.
Von Mises, Richard. 1919. Fundamentalsätze der Wahrscheinlichkeitsrechnung. *Mathematische Zeitschrift* 5: 52–99 & 100.
Von Mises, Richard. 1922. Über die gegenwärtige Krise der Mechanik. *Die Naturwissenschaften* 10: 25–29.
Von Mises, Richard. 1930. Über kausale und statistische Gesetzmäßigkeit in der Physik. *Die Naturwissenschaften* 18:145–53 and In *Erkenntnis* 1:189–210.
Von Schweidler, Egon. 1906. Ueber Schwankungen der radioaktiven Umwandlung. In *Premier Congrès International pour l'étude de la Radiologie et de l'Ionisation, tenue à Liège du 12 au 14 Septembre 1905. Comptes Rendus Section de Physique – Langue allemande*, Paris: H. Dunod & E. Pinat, 1–3.
Von Smoluchowski, Marian. 1906. Zur kinetischen Theorie der Brownschen Molekularbewegung und der Suspensionen. *Annalen der Physik* 326: 756–780.
Waismann, Friedrich. 1930. Logische Analyse des Wahrscheinlichkeitsbegriffs. *Erkenntnis* 1: 228–248.
Zilsel, Edgar, et al. 1930. Diskussion über Wahrscheinlichkeit. *Erkenntnis* 1: 260–285.

Hans Reichenbach, on March 13, 1947, in New York City (by Fred Stein)

Chapter 7
Everybody Has the Right to Do What He Wants: Hans Reichenbach's Volitionism and Its Historical Roots

Andreas Kamlah

7.1 Introduction[1]

When reading Reichenbach, one notices a frequently recurring and puzzling emphasis upon freedom of choice within every possible context: in the philosophy of science, in epistemology, in ethics, and last but not least in the theory of the freedom of the will. We read again and again that *everybody has the right to do what he wants*. This statement has not much to do with epistemology properly speaking, it is rather an unmistakable sign of Reichenbach's *anti-authoritarian ideology*. In his ethics, however, this "anarchist principle" is being modified and clarified to such an extent that (in Sect. 7.5) it can be shown to act as a bridge between his scientific, and political, and social world views.

However, this "anarchistic principle," is modified for ethics in a way that sheds light upon it. Thus natural science becomes here a model for ethics, in a manner analogous to Kant's famous expression about the starred sky and the moral law. Just as the order of nature is, for Kant, a model for the moral law; the conventionality of language becomes, for Reichenbach, a symbol for the absolute—quasi-anarchistic—freedom of human action.

My discussion in the present chapter is not only part of Hans Reichenbach's biography, but about the movement of logical empiricism in general, and about the way how the interpretation of science reflects a certain conception of man.[2] We start with a stock taking in philosophy of science.

---

[1] I am indepted to Wendy Wilutzky and Lothar Ern for checking my English grammar and style.

[2] For a Biography of Hans Reichenbach see (Gerner 1997).

A. Kamlah (✉)
Institute of Philosophy, University of Osnabruck, Osnabruck, Germany
e-mail: andreas.kamlah@uni-osnabrueck.de

N. Milkov and V. Peckhaus (eds.), *The Berlin Group and the Philosophy of Logical Empiricism*, Boston Studies in the Philosophy and History of Science 273, DOI 10.1007/978-94-007-5485-0_7,

7.2 Taking Stock of Reichenbach's Philosophy of Science

7.2.1 Coordinative Definitions

In his first book *Relativitätsthorie und Erkenntnis a priori* (1920; *Theory of Relativity and a priori Knowledge* 1965), Reichenbach replaced Kant's synthetic a priori principles with "coordinating principles" meant to do the same job, but which can be chosen freely.

In this book classical physics is characterised by the following list of principles:

> "relativity of uniformly moving coordinates," "irreversible causality," "action of contact," "approximate ideal," "normal induction," and "absolute time". (ibid. 15; *GW* vol. 3, 207)

The combination of these principles together with the empirical data, however, does not lead to a consistent description of the world. The special theory of relativity and, soon after it, the general theory of relativity, describe the world with different sets of principles. One year after his first book (about 1921) Reichenbach set out to replace these principles with "coordinative definitions". Herein, he argued that the merits of Einstein's theory of relativity was its replacement of alleged "facts" by "definitions". In *Philosophie der Raum-Zeit-Lehre* (1928; *Philosophy of Space and Time* 1953) Reichenbach's conventionalist operationism is perspicuous. He states:

> The philosophical significance of the theory of relativity consists in the fact that is has demonstrated the necessity for metrical coordinative definitions in several places where empirical relations had previously been assumed. (1928, 26; *GW* vol. 2, 34; 1958a, 15)

Paradigmatic of this achievement is Einstein's definition of simultaneity in his famous paper of 1905, "Zur Elektrodynamik bewegter Körper:"

> [...] it is not possible without further assumption to compare, in respect of time, an event at *A* with an event at *B*. We have so far defined only an "*A*-time" and a "*B*-time". We have not defined a common "time" for *A* and *B*, for the latter cannot be defined at all unless we establish *by definition* that the "time" required by light to travel from *A* to *B* equals the "time" it requires to travel from *B* to *A*. (Einstein 1905, § 2)

This definition implied enormous support for Reichenbach's conventionalist operationism. For, if we want to define a concept, we can do this in more than one way. And, every convention can be replaced with another one. Reichenbach does not restrict conventionalism to simultaneity. For him it is valid for all physical concepts, especially for length.

As he argues:

> The problem does not concern a matter of *cognition* but of *definition*. There is no way of knowing whether a measuring rod retains its length when it is transported to another place; a statement of this kind can only be introduced by a definition. (1928, 25; *GW* vol. 2, 33; 1958a, 16)

It is however a fact that, given such a definition, two measuring rods which are equally long at one place in space—and are equal in this respect—are so at other places as well.

> This consideration can only mean that the factual relations may be used for the simple definition of congruence where any rigid measuring rod establishes the congruence. If the factual relations did not hold, a special definition of the unit of length would have to be given for every space point. Not only at Paris, but also at every other place a rod having the length of a "meter" would have to be displayed, and all these arbitrarily chosen rods would be called equal in length by definition. The requirement of uniformity would be satisfied by carrying around a measuring rod selected at random for the purpose of making copies and displaying these as the unit. [. . .]
>
> Such a definition would complicate all measurements, but *epistemologically it is equivalent* to the ordinary definition, which calls the [rigid] rods equal in length. In this statement we make use of the fact that the definition of a unit at only one space point does not render general measurements possible. For the general case the definition of the unit has to be given in advance as a function of the place (and also of the time). *It is again a matter of fact that our world admits of a simple definition of congruence because of the factual relations holding for the behaviour of rigid rods; but this fact does not deprive the simple definition of its definitional character.* (1928, 26; *GW* vol. 2, 34; 1958a, 17; [the first emphasis in the quotation is my own—A. K.])

Reichenbach used the fact that we have the freedom to choose among several methods of measuring length to show that Euclidian geometry can always be defended provided the measuring method is defined appropriately. However, this does not help the adherents of the synthetic a priori, since most other geometries can be defended in the same way, for example the hyperbolic geometry in which the sum of angles in the triangle is less than 180°. By this argument Reichenbach has reduced to the absurd the efforts of some Philosophers of his time who wanted to save Euclidean Geometry as a priori valid. This was certainly an interesting result of his analysis.

But can one always replace a definition with a different but equivalent one? Let us imagine an alternative definition of length which can replace the usual one such that there does not arise any loss of information. Let us consider a method of measuring length the units of which are 2 m from South to North, and 1 m from East to West with the remaining distances to be determined according to the Pythagorean theorem. Everything in physics which can be expressed in terms of the standard definition, can also be formulated with the aid of the new one. We thus have translated the physical sentences into a new language. Does this mean that the deviant definition is "epistemologically equivalent" to the usual one as Reichenbach says?

This measuring method cannot be applied if the direction of the meridian cannot be determined; that is, if one has no compass. Since if we use a measuring rod we have always to take into account the angle between the rod and the meridian. There are cases in which a certain definition, though it leads to a logically equivalent theory, cannot be applied.

Reichenbach is certainly right when he says that the translation of true sentences into another physical language does not lead to a wrong picture of nature. But "epistemological equivalence" demands more than that. First, it requires the aforementioned equivalence in the applicability of measuring methods. Second, it requires inductive equivalence. That is, it must be possible to detect the inductive characteristics of theoretical descriptions, which make it more or less simple. A theory, in which all directions in space are nomologically equivalent with respect to the natural laws, is simpler than one in which one direction in space is preferred. If we use an alternative language in which the unit of length depends on the spatial direction, the isotropy is hidden, and it seems as if the theory is lacking this important inductive characteristic, which impairs the epistemological equivalence with the description in the standard language. Consequently, such a procedure is in no way "epistemologically equivalent" to the ordinary method, since it fails in cases where the standard method works very well.

7.2.2 *Relativity*

Until now we have not yet talked about the *principle of special relativity* which is the central point in Einstein's theory of 1905, but not in Reichenbach's philosophy of space-time. We will see that for Reichenbach this principle simply does not seem to exist at all. That is strange, since Reichenbach claims to tell us the epistemology of just this for him nonexistent theory.

Einstein defines a class of coordinate systems, the *inertial systems*, in such a way that all physical equations are exactly the same in all of them. For a certain set of coordinate systems which can be transformed into each other, by the well known Lorentz transformations, the sentences of physics are written down with the same sequence of signs, regardless of which coordinate system this is done for. In other words, the physical laws are invariant under those changes of coordinate systems which belong to the Lorentz group.

However Reichenbach does not talk about invariance. Is it possible—what is really hard to believe—that Reichenbach did not understand Einstein in the essential point of his special Relativity, the *Lorentz invariance of all natural laws*? To discuss this question I have first to explain in a nutshell what *Special Relativity* means. I start with *Galilean invariance*.

We can describe space time by using a special kind of Cartesian coordinate systems x, y, z, t, the so called inertial systems, for which I want to use letters C, C' etc. A physically possible process p in one inertial system C can have a physically possible counterpart p' which in C' has the same description as p in C. This is the invariance of physics with respect to the transformations of the inertial systems into each other. Before 1905 physicists believed that these inertial systems can be transformed into each other by shifting, turning them into other directions, and by the special Galilean transformation

$$x', y', z', t' = x + vt, y, z, t,$$

where v is the velocity of the system C' relative to C, and that the physical laws are invariant under these transformations. Galileo illustrated what later became known as *Galilean invariance* by his famous thought experiment of physical processes in the cabin of a ship, where everything runs in the same way independent of the ship's velocity.

Einstein discovered that the Galilean transformations have to be replaced with the Lorentz transformations, i.e. with shifts, turns, and with the special Lorentz transformation

$$x', y', z', t' = \beta\,(x + vt)\,, y, z, \beta\left(t + vx/c^2\right) \text{ with } \beta = \left(1 + v^2/c^2\right)^{-1/2}.$$

Thus he had discovered the Lorentz invariance of physics. Later special relativity was replaced with general relativity. But that is another story which I do not want to discuss here.

The Lorentz invariance of physics is the content of Einstein's principle of special relativity. Einstein defines it as follows:

> If, relative to C, C' is a uniformly moving co-ordinate system devoid of rotation, then natural phenomena run their course with respect to C' according to exactly the same general laws as with respect to C. This statement is called the *principle of relativity* (in the restricted sense). (Einstein 1920, section 5)

This definition of special relativity may have its flaws. It can at least benevolently be interpreted in the way in which I have characterized the Lorentz invariance in the preceding lines. This, then, is a property of all physical theories, which certainly has experimental implications. It is the empirical content of the principle of special relativity.

Let us now confront Einstein's principle to that of Reichenbach. He writes:

> The physical core of the theory, however, consists of the hypothesis that natural measuring instruments [in the German original text: "natürliche Messkörper"] follow coordinative definitions different [behave in a way which is different] from those assumed in the classical theory. This statement is, of course, empirical. On its truth depends only the *physical theory of relativity*. However, the *philosophical theory of relativity*, i.e., the discovery of the definitional character of the metric in all its details holds independently of experience. (1958a, 177; 1928, 206–207; *GW* vol. 2, 214)

In his admirable axiomatic system of special relativity (1924; engl. transl. 1969) he can derive the Lorentz transformation for rods and light signals. But I think that he never grasped that these transformations are conceived to be norms *for any physical law*, and that special relativity—if true—affects all parts of physics. Reichenbach seems simply to ignore this fact in his analysis. Certainly he had become aware of the fact that Einstein's theory has physical implications. But for him, these concerned only the mentioned measuring bodies and processes. What counted for him was only the free choice of definitions: *everybody has the right to do what he wants.*

This limitation to light rays, clocks, and measuring rods is characteristic for Reichenbach's way of thinking, and one has to admit that his axioms which use just those concepts are fascinating. In the discussion after his talk about his new axioms of relativistic space-time at the German Congress of Physics in Jena 1921 (Reichenbach 1921), someone in the audience remarked: "To these axioms the principle of relativity has to be added." To which Reichenbach answered: "That was not the problem to be solved."[3] It seems that Reichenbach considered the principle to be epistemologically unimportant. He simply did not know that the physical invariance principles play an important role in epistemology (see Kamlah 2002, chapters 11–13).

In his *Theory of Relativity and A Priori Knowledge* (1920) Reichenbach still mentions the two principles of special and general relativity. Sometime later, they seem to have lost their significance for him. What remained was his obsession of the freedom to choose ones concepts in physics: *Everyone has the right to do what he wants.*

7.2.3 Volitional Bifurcations

Reichenbach wrote *Experience and Prediction* (1938) during his stay in Istanbul, a book in which he again and again uses the expression "volitional decision". With regard to this concept, one might ask just what else can a decision be, if not a directive for the will? Or, what would it be if it was not "volitional"?

Thus, it seems that the adjective "volitional" is a clearly superfluous, even ideological, addition. Another term which appears in this book is "volitional bifurcation". With regard to this notion Reichenbach states:

> The examples chosen from the theory of space and time previously mentioned are likewise to be ranked among conventions. There are decisions of another character which do not lead to equivalent conceptions but to divergent systems; they may be called *volitional bifurcations*. (1938, 10; *GW* vol. 4, 5)

Reichenbach introduces the concept of volitional bifurcation in a discussion of the difference between positivism and realism. He thinks that distinguishing between these two viewpoints should be understood to be a matter of deciding between different languages. In his own words:

> With the reflections of the preceding section our inquiry about the difference of the positivistic and the realistic conception of the world has taken another turn; this difference has been formulated as the difference of two languages. [...] The conception of the difference in question as a difference of language corresponds also to our idea that the question of meaning is a matter of decision and not of truth-character. (1938, 145; *GW* vol. 4, 92)

In short, he thinks that we can choose between an "egocentric" (positivist) language and a "realistic language". Positivism (including solipsism) and realism are, for Reichenbach, not two different theses with empirical content, but rather two different ways to encounter the world between which we may decide.

The former language, however, is much poorer in its expressiveness than the latter. This seems to be clear, since the solipsist has in his language no personal

---

[3]See Kamlah (1979), Comments to *GW* vol. 3, 466.

pronouns. The words "I, you, he, she, we, you, they" are for him devoid of meaning. He is also lacking the concepts of love and hate, responsibility and thankfulness, and many others. Thus the language of realism offers us much further reaching possibilities than that of positivism. And the decision to accept one of both languages is not one made for one of two equivalent alternatives. And even if this may be conceded, for Reichenbach this choice between the two is basically free.

There are surely many objections to be made against Reichenbach's analysis. But that is presently not our subject. We are here rather interested in studying the role of his volitionism.

7.2.4 *Induction*

In the winter of 1933 Reichenbach must have had the idea that our whole corpus of empirical knowledge rests upon a single decision—he calls it a "volitional bifurcation"—namely the decision to accept or to reject the rule of induction.

The principle or rule of induction says that future events of a certain kind will happen nearly as frequently as they do now that means i.e. in a sample of the events hitherto observed.

Let a sample of n events be given; m events from the sample may have the property A, the other ones $\neg A$. $h^n = m/n$ is the relative frequency of A in the sample. We than have:

> For any further prolongation of the series as far as s events ($s > n$), the relative frequency will remain within a small interval around h^n; i.e., we assume the relation
>
> $$h^n - \in \leq h^s \leq h^n + \in$$

where $\in$ is a small number (1938, 340; *GW* vol. 4, 213).

If we decide to accept this postulate, we may have a chance to gain knowledge in our world and to survive in it. In other words, our survival depends on the favourable result of a wager which we make against the world. We are free to make such a wager; and at the point of that "volitional bifurcation" we choose one of two possible paths. As Reichenbach argues:

> The inductive inference is the only method of which we know that it leads to the aim if the aim can be reached; this is the reason why we must use it, if we want to reach the aim. The problem of the inductive inference finds its solution by means of the argument that it is not necessary for the application of this inference to know a *positive* condition to hold, but that the application is already justified if a negative condition is *not* known to hold.
>
> We are often confronted by similar situations in daily life. We want to reach a certain aim and we know of a necessary step, which we shall have to take in order to attain this aim, but we do not know whether this step is sufficient. He who wants to reach the aim will have to take the step, even if it is uncertain whether he will reach his aim in this way. The businessman who keeps his store well stocked so that he can sell something when a costumer comes in, the unemployed who makes an application with reference to an advertisement in the paper, although he does not know whether he will receive answer,

> the ship-wrecked man who climbs a cliff, although he does not know whether a rescue-ship will spot him—all these persons find themselves in an analogous situation; they satisfy the *necessary* conditions of reaching an aim without knowing whether the *sufficient* conditions are satisfied. (1933b, 423)

With regard to these examples, I think that everybody would apply the usual procedure of induction even if he does not know if he has any chance that his expectations are justified.

7.2.5 Result of the Preceding Subsections

I want to emphasize once more that the adjectives "volitional" and "arbitrary" which Reichenbach likes so much are absolutely redundant. Certainly every decision is volitional. Otherwise it is not a decision at all. And, the same holds for the word "arbitrary". If a decision is not arbitrary in some respect, it is not a decision but a giving way under external pressure.

Therefore, the terms "volitional" and "arbitrary" do not mean anything in this context but represent what Carnap has called "accompanying ideas" (begleitende Vorstellungen) which add nothing to the factual content (sachlicher Gehalt) of statements (Carnap 1928). These terms are purely ideological and reveal Reichenbach's extreme liberalism and decisionism.

We encounter those "volitional decisions" everywhere in Reichenbach's epistemology. I have mentioned three kinds of them: coordinative definitions, bifurcations, and the wager to accept the rule of induction. He compares these volitional decisions with the choice to do science:

> What is the purpose of scientific enquiry? That is, logically speaking, a question not of truth character but a volitional decision, and the decision determined by the answer to this question belongs to the bifurcation type. If anyone tells us that he studies science for his pleasure and to fill his hours of leisure, we cannot raise the objection that this reasoning is "a false statement"—it is no statement at all but a decision, and *everybody has the right to do what he wants* (my emphasis—A. K.). (1938, 10; *GW* vol. 4, 5)

Reichenbach puts the mentioned three kinds of decisions on the same level as the choice to pursue a certain hobby. For a hobby it is certainly essential that it is a freely chosen activity. And, as long as the interests of others are not impaired, there is nothing objectionable about it. But are the aforementioned decisions really of the same kind? Or do they have to prove to be successful in a consistent description of the world?

Reichenbach was probably aware of these doubtful questions and their implications, but he seemed to forget about them from time to time. At those moments he would unequivocally proclaim the "right to do what one wants". However, this idea, in its more radical interpretations, becomes untenable within the domain of ethics. And indeed, as we will see in Sect. 7.5, Reichenbach favoured a rather mitigated version of his principle in that field.

7.3 The Influence of the *Jugend* Movement on Reichenbach

7.3.1 Introduction

The history of philosophy is often seen as a mere record of the discourse of a number of eminent philosophers that has been going on for some 2,500 years. In a way, the philosophers themselves are not altogether innocent of this rather one-sided picture, for they have a tendency to immerse themselves exclusively into the works and thoughts of other philosophers in their writings. This way, influences on philosophy coming from the outside world go largely unnoticed. This is a pity, for, surely, philosophers, like everybody else, are children of their times and, as such, subject to changes in society. Therefore, a modern historiography of philosophy must not ignore the socio-cultural environment of its protagonists. As a matter of fact, a history of philosophy that leaves out the political, economic, and scientific developments of the time—let alone the trivia of everyday life like pop culture, the media, and the movies—will give but a distorted picture of its subject.

The point here is that these unofficial sources can be very important and a modern history of philosophy should make every effort to incorporate them. I would even go so far as to say that influences coming from the society at large are more important than many a work by erstwhile philosophers, and that acknowledging them will greatly enhance progress in modern philosophy. This way, epistemology will, at last, become a true mirror, always reflecting the latest state of social and cultural development.

With this in mind, I want to have a look at some of the socio-cultural influences which have been important for the development of Reichenbach's philosophy of science. In particular, I want to focus attention upon three social movements that took place in Germany, during the early 1910s:

(i) the *Wandervogel* ("Birds of Passage");
(ii) the *Landschulheim* movement;
(iii) the *Freistudenten* ("Free Students").

In particular, his commitment for the *Freistudenten* was decisive for his philosophy throughout his whole life.

7.3.2 The Wandervogel Movement[4]

The *Wandervogel* was the first incarnation of what later became the *Jugendbewegung* (youth movement). In 1896 some grammar school students in Steglitz (nowadays a part of Berlin) set out on their first hiking tour. They wanted to escape

[4] See the memoir of Carl Landauer (*SW* vol. 1, 25–30). This text contains nearly everything which is important for section 3 of our paper. See also Blüher (1912–1914) and Laqueur (1962).

the big city and freely roam in the woods, fields, and meadows. The first two verses of one of their songs characterizes how they saw themselves[5]:

[1. verse:] From grey cities walls we roam through woods and fields.
Who stays may rot. We travel into the world.
[2. verse:] The woods are our love, the sky is our tent
Whether bright or dull. We travel into the world.

They also wanted to escape the authoritarian education from their parents and teachers. These groups soon developed certain habits at their *Fahrten* (today *fahren* means to travel by means of a vehicle, originally it meant also "to hike"). They slept in the hay in farmers' barns and even under the open sky, and they cooked their meals over open fires. They sang songs that came from various sources: some from soldiers, hiking kraftsmen, some from sailors, and some were just ordinary folk songs. There were also old songs from the sixteenth century and, of course, there were those they composed themselves.

Their instruments of choice were the lute and the guitar. Their *Fahrten* could last an entire summer vacation, and range over some hundred miles. Their attitude was one of general escape: from the constraints of an industrialized bourgeois society as well as from a repressive school system. The *Wandervogel* was certainly not an educational institution conceived by educationists as were the Boy Scouts. Rather, it was a grass roots movement that sprang up among and was run by the teenagers themselves.

Within a few years the *Wandervogel* spread out all over Germany. Due to much disagreement among its leaders, it split up into many different associations that, together, formed a mighty movement, the *Jugendbewegung*. After some years the *Wandervogel* wanted more than just to hike. They developed a new consciousness and a new culture: a *Jugendkultur*. A new life style was created. Many groups renounced smoking and drinking alcohol. Many wore new kinds of clothes, and cultivated folk dancing.

---

[5]The song, however, with the text by Hans Riedel and Hermann Löns was composed by Robert Götz much later in 1920. So it is not really an authentic source about the *Wandervogel*. But it reflects well what the teenagers of the *Wandervogel* felt. The original Text is:

Aus grauer Städte Mauern	Der Wald ist unsre Liebe,	Ein Heil dem deutschen Walde,	Die Sommervögel ziehen
ziehn wir durch Wald und Feld,	der Himmel unser Zelt.	zu dem wir uns gesellt.	schon über Wald und Feld.
wer bleibt, der mag versauern,	Ob heiter oder trübe,	Hell klingt's durch Berg und Halde:	Da heißt es Abschied nehmen,
wir fahren in die Welt	wir fahren in die Welt.	wir fahren in die Welt.	wir fahren in die Welt.
Halli, hallo, wir fahren,	Halli, hallo, wir fahren,	Halli, hallo, wir fahren,	Halli, hallo wir fahren,
wir fahren in die Welt.	wir fahren in die Welt.	wir fahren in die Welt.	wir fahren in die Welt.

Hans Reichenbach seems to have been part of the *Wandervogel* community, and we will see in the following subsections that this remained important for his succeeding years as university student.[6]

7.3.3 The *Landschulheim* Movement

The second movement which influenced Reichenbach was the *Landschulheimbewegung* (cf. Nohl 1933). In the same year, when boys from Steglitz started their first hiking tours, Hermann Lietz founded his first *Landschulheim* in Ilsenburg near the Harz Mountains. Lietz wanted to offer a broad education to young people, and not merely academic instruction, as was done in the public grammar schools (*Gymnasien*). In some way the English boarding schools were a model for his project. But the goal of his education was not to form the perfect English gentleman. Hermann Lietz felt that education had to take the entire human being into account and, not just his brain. This is why every student had to learn a craft. Furthermore, the schools founded by Lietz—and some of them still exist—are located in the countryside, for he believed that the unspoiled atmosphere of the country was more conducive to his educational objectives than a city environment.

In 1900, Gustav Wyneken, who had studied theology, became one of the teachers at the *Landschulheim* in Ilsenburg. He worked there and at anothers of Lietz' schools, for a total of 6 years. But by the end, Wyneken refused to go along with Lietz' concept of education which was based on his firm belief in the natural authority of the educator towards his pupils. Wyneken, however, had become convinced that children and teenagers are naturally curious and that they want to learn and to discover human culture their own way rather than take over the beliefs of the older generation. They want to follow rules that they, themselves, feel to be justified. They want to deal with literature, music and art that they, themselves, feel to be convincing and honest. And, they want to learn about the things that they, themselves, feel to be relevant. The role of the educator, therefore, is to encourage and support his pupils' spontaneous initiatives. He must incite rather than stifle his students' natural urge for activity. On this point, Wyneken wrote:

> The acknowledgment of the youth's right to a self-determined lifestyle and to the feeling of their valuable and irreplaceable originality is what sets the modern educator apart from the reactionary, prevailing and feigned. This attribute does not yet make up the entire pedagogical talent, but is its necessary foundation. Considered from this perspective, the educator is no longer an educator, not a "soulsmith" or "personcreator," but a leader, indeed a leader chosen by the youth itself. Only he, who naturally attracts them and whom they

[6] Carl Landauer, a former friend and a member of the inner circle of *Freistudenten*, writes in his memory of Hans Reichenbach (*SW* vol. 1, 26): "Hans, I think, had been in the *Wandervogel* while in highschool." Hans Ulrich Wipf writes: "Hans Reichenbach is considered an eminent exponent of the generation of students which was shaped by the *Wandervogel*" (Wipf 1994, 167).

> follow, can be an educator in the new sense, not however he, who has no respect for the willpower which lies in the youth's nature, who only discerns the imperfect brain-states and wishes to alleviate this shortcoming. [...]
>
> The method of teaching is to be understood as an agreement between teacher and students to reach a certain goal through joint effort. It does not suffice to only endorse the now generally accepted right to ask questions: what is more, he shares a responsibility for the progress and success of the tuition, in other words, it is his duty to take part in the tuition's successful development using his best endeavors. And he will be the best teacher, who evokes such participation. (Wyneken 1914, 39)

It seems that the spirit of Wyneken's theory of education was the same as that of the *Jugendbewegung* and the *Wandervogel.*

In 1906, Wyneken founded his own boarding school, the *Freie Schulgemeinde* Wickersdorf (near Meiningen in Thuringia). There he tried to put into practice his own ideas about education. The *Freie Schulgemeinde* was governed by a committee of pupils who were elected by a general assembly. The official policy of this school was that the teachers, and even Wyneken himself, could not dictate to the pupils what they had to do. But, in reality, the personality of Wyneken was strong enough to persuade the committee to follow his suggestions. And in most cases it did.

One can hardly imagine that such a model of school administration would work under an average headmaster. However, Wyneken was a charismatic leader who could inspire young people. As a result, he managed to run his school more as a consultant than as a director. In 1910, however, he got into trouble with the government of the duchy of Sachsen–Meiningen (one of the eight tiny states which were later united to form the state Thuringia). The reason for his difficulties was Wyneken's concept of religious education. For, though trained as a theologian, Wyneken later became a free thinker who considered religion to be merely a cultural phenomenon—though a very important one. Such ideas were inacceptable to the government, and he was told that he either had to leave the school or else it would be closed. Wyneken decided to leave, and during the following years he traveled around in Germany giving talks on education. In these years he became well known to the *Freistudenten* at different universities where many students in his audience were training to become school teachers. It is this context that Hans Reichenbach, who was one of the leaders of the *Freistudentische Bewegung*, made Wyneken's acquaintance and became strongly influenced by him.

Before World War I, Wyneken became the theoretician of the *Jugendbewegung* which culminated shortly before the War in the festival on the Hoher Meißner. At this time, there had already existed a powerful air of congeniality between the *Wandervogel* and Wynecken's *Freie Schulgemeinde* Wickersdorf, but soon the ties between the two movements were to become even closer. In 1913, the *Wandervogel*, the *Freie Schulgemeinde* and many other groups of the *Jugendbewegung* met on the Hoher Meißner, about 50 km to the south of Göttingen, at a festival of German youth. The Hoher Meißner is a 700 m high mountain whose flat top provides space for large groups of people to congregate. It is situated not very far from the geographical centre of Germany, and it is known as the mythical place where "Frau

Holle" lives.[7] The story goes that that every time this mythical creature makes her feather bed, downy feathers will fall down to the earth in the form of snow.

More than 2,000 teenagers and young people gathered at the festival. Reichenbach went to the Hoher Meißner with a delegation of the *Freistudenten*, as well as Rudolf Carnap, who was there as a member of the *Sera–Kreis* from Jena, where he was studying.[8] At that time, however, the two philosophers did not know of each other.

The Meißner Festival was organized as an alternative to the celebration of the Centennial of the Battle of Leipzig of 1813, in which Napoleon and the French Army were defeated. On the occasion of the centennial celebrations a colossal memorial, the *Völkerschlachtdenkmal*, was to be inaugurated, and one could safely expect that every conservative and military group, especially the *Korporationen*, would come together in an orgy of nationalist fervour. At the Meißner Festival, many speeches abounding with idealism were made. Wyneken gave the main address. At the festival, too, the *Freideutsche Jugend* was founded as a parent organization of the many youth associations (*Bünde*), which were present at the festival, and one agreed at the so called *Meißnerformel* which stated that:

> The *Freideutsche Jugend* wants to shape its own life by self-determination, on its own responsibility and with inner truthfulness. It jointly defends this inner liberty under all circumstances. *Freideutsche Jugendtage* are held for exchange of ideas. All common meetings of the *Freideutsche Jugend* are free of alcohol and nicotine.[9]

One year later the Great War started, and most of the leaders of the *Wandervogel* movement, and of other groups which had emerged from it, were conscripted. Many of those fell in its bloody battles that followed.

7.3.4 Reichenbach's Involvement in the Movement of German "Freistudenten"

Hans Reichenbach's early involvement with the rather loosely organized *Freistudenten* ("free students") has determined the style of his philosophical thought for his entire life. The *Freistudenten* or *Finken* ("finches") were those students who were not members of the *Korporationen*.In earlier centuries almost every German student

[7]A character of Grimm's fairy tale, known in English culture as "Mother Holle," or "Mother Hulda".

[8]For Reichenbach see 1913e, for Carnap see (Dahms 2004, 70). Carnap was a member of the Sera-Kreis in Jena, which, like many other groups, supported the initiative of having a meeting of all groups of the *Jugendbewegung* at Hoher Meißner. Carnap, however, writes in his autobiography that he met Reichenbach for the first time in Erlangen in 1923; see Carnap (1963, 14).

[9]From Erich Weniger (1980): 1–8, quotation 3: "The *Meißnerfest* is the unforgettable peak [highlight; *Höhepunkt*] of the movement."

belonged to a *Korporation*. However, starting at the second half of nineteenth-century, less and less individuals had the money to pay for the sabres, uniforms and large quantities of beer which was drunk during their meetings. This trend continued, so that by the end of the century the *Finken* made up 50% of all German students. Of course, the *Finken*, like all students, did not want to spend all their time studying. So they founded an informal organization that was meant to provide sporting events, parties, evening lectures, and discussions on subjects of general interest.

The *Freistudenten*, however, did not want to become just another *Korporation*. But a minimum of organizational structure was indispensable. So, at many universities, they held general assemblies to which everybody had access and where they elected their leaders who were meant to represent them in front of the authorities and also to the *Freistudenten* at other universities.

While most members of the *Korporationen* adhered to a rather conservative ideology, one would find among the *Freistudenten* individuals of a more liberal or even socialist persuasion. To these students the rituals and antiquated forms of behaviour which were cultivated in the *Korporationen* did not make sense. They were especially repelled by their medieval concept of honour. Thus the *Freistudenten* generally came to be associated with a more modern attitude towards life and politics.

Among the leading circles of the *Freistudenten* the spirit of the *Wandervogel* prevailed, since most leaders had been members of that movement in their youth. So had Hans Reichenbach, it seems.[10] While many of the *Freistudenten* had been in or were influenced by the *Wandervogel* movement, there existed also the *akademische Freischar*, a group which tried to carry over the life and activities of the *Wandervogel* into the universities. The *akademische Freischar* shared with the *Freistudenten* their opposition against the *Korporationen*, and therefore they were a natural ally of them. But they were not *Freistudenten* themselves.

For Reichenbach, being a *Freistudent* meant more than only the absence of membership in a *Korporation*. He actually developed a kind of ideology of the *Freistudenten*, which contained many ideas of the *Wandervogel* and the *Landschulheim* movements. In an essay for a student journal he wrote:

The desired end of the Free Students can be summarized as follows:

> The supreme moral ideal is exemplified in the person who determines his own values freely and independently of others and who, as a member of society, demands this autonomy for all members and of all members. [...]
>
> The individual may give his life whatever form he finds to be of value and may set for himself particular goals, as, for instance, to follow the profession of an artist or a mathematician, but to demand that others pursue the very same goals is to overrate one's own particular gifts to the exclusion of others, is to be both petty and pedantic. [...] The individual may do whatever he considers to be right. Indeed, he ought to do it; in general, we consider as immoral nothing but an inconsistency between goal and action. To force a person to commit an act that he himself does not consider right is to compel him to

---

[10]See Landauer (1978), Wipf (1994), and Linse (1974).

> be immoral. That is why we reject every authoritarian morality that wants to replace the autonomy of the individual with principles of action set forth by some external authority or other. (1913, 109)

We encounter here the nucleus of the ideology of the *Jugendbewegung*, which was the urge for autonomy. And it is this urge that shaped Reichenbach's philosophical endeavour for the rest of his life. The idea of autonomy, which is most clearly spelled out here, later appears in his epistemology in the guise of conventionalism, and more directly within his conceptions of education and ethics (cf. Kamlah 1977, 480–483).

7.4 The Montessori School

After World War I, the spirit of *Jugendbewegung* continued to be influential in the pedagogical movement *(pädagogische Bewegung)* which split up into many different projects of education each with their specific theories (cf. Nohl 1933). One of these were the Montessori schools, that based education on the principle of voluntary cooperation. They were initiated by the Italian Maria Montessori.

At this time, Reichenbach was married and had two children that he sent to a Montessori school in Berlin–Dahlem. In the early 1930s he published an article, in the journal *Die neue Erziehung*, which seems to refer to the Montessori school in Berlin. In it Reichenbach describes the interplay between the principle of voluntary cooperation and group pressure that he knew from Wyneken's *Freie Schulgemeinde*. In a way, this article can be seen as a missing link between Reichenbach's early talks to the *Freistudenten* and his chapter on ethics in *The Rise of Scientific Philosophy* about 40 years later. Here are some quotes from it[11]:

> It is not at all true that children avoid work, that for them learning is inherently disagreeable. This is only the case when you lead them along enforced paths (*erzwungene Wege*), not, however, when they are allowed to learn on their own accord. (1931, 94)

> You see that in the Montessori-school too, of course, there is pressure (*Zwang*): But it is not the pressure of an external authority, but a pressure, which exists *within the endeavor* itself (*in der Sache*).

> Even the superior, the master, the department head etc. are not educators of the same kind as the teachers, because they are not concerned with the subjective achievements but instead only with the objective product of their subordinates' labor; their wishes and requests are therefore simply a component of the situational constraints (*Situationszwang*), are rated as facts, such as, for instance, the necessity to speak Spanish when establishing commercial correspondence with South America. (ibid., 96)

> Such situational pressure must also be imputed to the pressure within social groups, which asserts itself substantially throughout life. It is precisely this pressure though, which is so fervently at work in the Montessori-school. One is surprised how in this seemingly individualistic youth, for which classes are disbanded into free workplaces,

[11]Engl. translation of excerpts from quotations in Kamlah (1994).

> an overwhelming and coherent sense of community, can come about. [...] In occasional collective events, for instance in the preliminary discussion of an excursion, one can observe such a sense of community in a positive form. The child, who cannot integrate into this team spirit, is continuously drilled by the invisible social pressure until he has found his place among the others. (ibid., 97)

In other words, one simply has to let school children—and people in general—do what they like. One should not force them to cooperate with others, for they will eventually do this voluntarily—forced by their own interests as it were—and thus find their way. Behind all this one might detect, once again, Reichenbach's one and only rule of ethics: "Everybody has the right to do what he wants". But in the next section it will be shown that Reichenbach's ethical principles are a bit more sophisticated than this. Yet, his ethical non-cognitivism is already perspicuous here. And it has not changed much. For, many years later, when he dealt with ethics properly speaking in *The Rise of Scientific Philosophy* he formulated essentially the same non-cognitivism.

7.5 Ethics

According to Reichenbach (1951), there is no such thing as a science of ethics. Of course, we can, like a sociologist, study the behaviour of people and examine whether they follow general rules. But the mere description of human behaviour will not reveal the maxims and norms behind such behaviour. Neither a teacher, nor a policeman is entitled to dictate ethical rules to anyone. Everybody has to decide for himself which norms he will accept. Reichenbach compares this freedom of choice with that one has when selecting a hobby.[12]

But even there, like in all ethical decisions, the choice might not be an easy one. If I take up the hobby of, let's say, killing people, I will get into trouble with my fellow citizens. For, they have their own interests, among which is the widespread desire not to get killed. Thus, were I to take up such a hobby, I would sooner or later end up in jail. Therefore, I will have to find a way to get along with my fellow citizens.

Reichenbach firmly believes that this is what most men want anyway: to live in peace with their neighbour, and that they will think twice before selecting hobbies like murder or terrorism. Nevertheless, he stresses the fact that we are free in our decisions. Neither natural laws nor law codices like the Ten Commandments can dictate us what we must do.

This, again, sounds like a clear endorsement of the maxim that "Everybody has the right to do what he wants". However, when directing himself to a fictitious critic, Reichenbach writes:

> You see that the volitional interpretation of moral directives does not lead to the consequence that the speaker should allow everybody the right to follow his own decisions;

---

[12]See Section 3 on *Freistudenten*.

> that is it does not lead to anarchism. If I set up certain volitional aims and demand that they be followed by all persons, you can counter my argument only by setting up another imperative, for instance the anarchist imperative "everybody has the right to do what he wants". You cannot prove, however, that my system of volitional ethics is inconsistent, that logic compels me to allow everybody the right to do what he wants. (1951, 294; *GW* vol. 1, 409)

However, he did not go so far as to proclaim the "anarchist imperative". Rather, he held that:

> We may differ in many respects, perhaps about the question of whether the state should own the means of production, or whether a world government should be set up that controls the atomic bomb. But we can discuss such problems if we both agree about a democratic principle which I oppose to your anarchist principle:
>
> Everybody is entitled to set up his own moral imperatives and to demand that everyone follow these imperatives.
>
> This democratic principle supplies the precise formulation and of my appeal to everybody to trust his own volitions, which you regarded as contradictory to my claim that everybody may set up imperatives for other persons. (1951, 295; *GW* vol. 1, 410–11)

It is not easy to understand this "democratic principle". For, we normally understand the idea of being "entitled to demand something" as follows: If I have the right to demand *A*, this implies an obligation for other people to obey my command. They have to execute *A*. But that is not what Reichenbach means. Rather, on his account, I can only try to convince my fellow men to accept that *A* is desirable or to get them in any way to follow my order. To try to understand this strange principle better, let us compare it with Kant's categorical imperative:

> Act only according to that maxim whereby you can at the same time will that is should become a universal law.[13]

Reichenbach's principle can be rephrased, using some of Kant's terms, in the following way:

> Everybody is entitled to set up maxims and to demand that they should become universal laws and that all people act according to them.

All the examples for such maxims that Reichenbach gives are universal. For instance the following:

> The imperative that if there is more than one room to each person in a house, the surplus rooms should be opened to persons who have no room of their own. (1951, 295; *GW* vol. 1, 411)

Obviously, both, Kant and Reichenbach presuppose that moral rules or laws should be universal. The main difference between Kant's and Reichenbach's principles regards the distinction between a duty and a right. Kant demands that the individual obey those rules which he himself wants other people to follow. Whereas Reichenbach demands that the individual try to make other people obey his own

---

[13]Kant (1785), 17, Engl. transl. Kant 1993, 30.

rules. This distinction reminds one of the disagreement the teenagers who founded the *Wandervogel* once had with their teachers. The students wanted the notorious "thou shalt" to be replaced with "you may". Reichenbach makes it quite clear that the imperatives or maxims which different people propose can vary widely. Kant, it seems, did not realize this difficulty which can lead to different interpretations of the categorical imperative.

Let us assume that people actually succeed in obeying a codex of rules or laws, either voluntarily or under compulsion. How can we be sure that this will not result in a totalitarian society? Reichenbach seems to believe that men's nature is essentially good. His conception of human nature was a very optimistic one, in spite of the Nazi induced terrors at play in the decade before he wrote *The Rise,* and the communist rule in many countries which still existed at the time. Reichenbach's optimism that his volitional principle will work, can be illustrated by the following passage:

> Whoever wants to study ethics, therefore, should not go to the philosopher; he should go where moral issues are fought out. He should live in the community of a group, where life is made vivid by competing volitions, be it the group of a political party, or of a trade union, or of a professional organization, or of a ski club or a group formed by common study in a class room. There he will experience what it means to set his volition against that of other persons and what it means to adjust oneself to a group will. If ethics is the pursuit of volitions, it is also the conditioning of volitions through a group environment. (1951, 297; *GW* vol. 1, 412–13)

Beyond that, he seems to adhere to a kind of eudemonism, even though this is never stated very clearly:

> The exponent of individualism is short-sighted when he overlooks the volitional satisfaction which accrues from belonging to a group. Whether we regard the conditioning of volitions through the group as a useful or a dangerous process depends on whether we support or oppose the group; but we must admit that there exists such a group influence. (1951, 297; *GW* vol. 1, 413)

What Reichenbach did not see, was that to examine the rules, which best govern human society was exactly what is commonly called "ethics". Today his position seems strange to us, since discussions about morals, medical and environmental ethics are ubiquitous. Therefore we have to find an explanation for Reichenbach's puzzling conception.

7.6 The Freedom of the Will

7.6.1 Reichenbach's Discussion with Schlick and the Vienna Circle

If we remind ourselves of the importance that free choice had for Reichenbach, we should not be surprised that he was a libertarian. For somehow he was convinced that determinism contradicts the freedom of the will.

One does not necessarily have to be a libertarian if one shares Reichenbach's conventionalist attitudes in natural sciences. But this position goes well with his ethical conventionalism by which Reichenbach emphasized that the human will was free. Compatibilism, on the other hand, is very similar to determinism of the will. Many determinists have pointed out that punishment and reward are still useful instruments in human social life even if man's actions are completely determined by his past and his environment. The threat of punishment does influence the behaviour of human beings. They most likely will not commit a crime if they are afraid of its prosecution and eventual punishment.

Thus compatibilists and determinists are frequently put into the same category by libertarians. Indeed, Kant ridiculed compatibilists by calling their freedom of the will the "freedom of a roasting jack" (*Bratenwender*). The same argument was used by Reichenbach when he referred to Spinoza, who, according to him, made a distinction between internal and external causes, and called an action determined by internal causes "free".

Because of his rejection of compatibilism, Reichenbach stood in opposition to the Vienna Circle, who, like Hume, held that we are free in our actions insofar as we have made the experience that we can do what we want to do.[14]

It is true that Reichenbach had, for a while, hoped that the recently discovered indeterminism in physics might give the debate on free will a new turn. But, being a philosopher, he could not just take over the position of physicist Pascual Jordan whose arguments were rather weak anyway.[15] His friends in the Vienna Circle would have criticised him for that, especially Moritz Schlick who defended, as Hume once did, the thesis that there was no contradiction between determinism in nature and the freedom of action. Over the years, Reichenbach made several attempts to prove the freedom of action and the freedom of the will. In an interesting and deep (but partly confused) paper "The Causal Structure of the World and the Difference between Past and Future" (1925), Reichenbach claimed that he could derive the freedom of the will from the temporal asymmetry of the physical processes:

> If determinism is correct, then we cannot in any way justify undertaking an action for tomorrow but not for yesterday. No doubt it is true that it is not even possible for us to give up our *intention* to act tomorrow and our belief in freedom—we surely cannot. The point is that, given determinism, our behaviour would be senseless, for then tomorrow would be already past in the same sense that yesterday is.[16]

For the physical determinist, there cannot be a divide between past and future. That is, there is no "now". Indeed, the future is determined in the same way as the past. This thesis was met with strong opposition by Moritz Schlick, who could not accept Reichenbach's speculations. On March 20th, 1926, Schlick wrote Reichenbach that he could not follow his thoughts. Reichenbach answered:

---

[14]Hume (1748), section 8. Schlick (1930), chapter 7.

[15]Jordan (1932), Reichenbach (1935); cf. Kamlah (2008).

[16]1925a; *SW* vol. 2, 86–87.

> With respect to the connection with determinism, I still believe that the compatibility of freedom of the will with strict causality is an untenable position. [17]

Schlick criticized Reichenbach on that point publicly. After having quoted the above passage from Reichenbach's "Kausalstruktur der Welt," he continues:

> It seems to me that exactly the contrary is the case: Our actions and resolutions (*Vorsätze*) make sense only insofar as future is determined by them.[18]

After that disagreement on time and free will, the relationship between Schlick and Reichenbach deteriorated considerably.

It is true that no kind of freedom of the will could exist, if past and future did not differ from each other. For, we act in order to give a hitherto indeterminate future some definite shape. The cognitive basis of our actions is our knowledge of the past. And the time structure of the cosmos is a necessary condition for the possibility that one can act at all—that there can be any kind of action in the world. But it is no sufficient condition for freedom of action or even for freedom of the will.

In spite of many justified objections from Schlick, Reichenbach remained convinced that there was an intimate connection between the time structure of the world and the freedom of action. For him, it was quite clear that the solution of the problem of freedom of the will could neither be as simple as physicists like Pascual Jordan believed (he tried to explain the free will by quantum mechanics), nor could determinism be true.

7.6.2 Reichenbach's Logical Reconstruction of the Freedom of Action[19]

Only in the last years of his life did Reichenbach attempt, once more, to derive the freedom of action and the freedom of the will. He wrote two manuscripts, which were merged into a single article and published posthumously by his wife Maria Reichenbach. They show how Reichenbach tried to approach the problem from the phenomenological side, listing real life situations in which we consider the will to be either free or unfree. In these two manuscripts Reichenbach tries to give a state of the art treatment of the free will problem, using a newly created formalism for conditionals.

Reichenbach distinguishes the *freedom of action* from the *freedom of the will.* The first is more easily defined, and therefore we shall reconstruct it and shall deal with the second only in passing.

---

[17]Reichenbach's letter to Schlick from 20.03.1926 [HR-016-18-12].

[18]Cf. Schlick (1931, 162).

[19] 1959a; *SW* vol. 1, 431–473.

Reichenbach uses for the formulation of freedom of action a special kind of causal conditional. Let $\diamond_L A$ and $\Box_L A$ denote the necessity and the possibility of A due to the laws of nature. Then Reichenbach's conditional $A \rightarrow B$ will be:

$$A \rightarrow B := \diamond_L A \wedge \diamond_L \neg B \wedge \Box_L (A \supset B)$$

$A \rightarrow B$ is defined in such a way that absurd cases like

If I pray to Saint Mary for improvement of my intelligence, then $2 \times 2 = 4$.

or

$$\text{If } 2 \times 2 = 5, \text{I have birthday today.}$$

will not count as valid necessary implications. In both cases we will not say that B is true because of A. B is true anyway in the first case and in the second B does not depend on A. One excludes these cases from the conditional by the inserting into the definition the clause

$$\diamond_L A \wedge \diamond_L \neg B.$$

After having defined $A \rightarrow B$, we can now write down the freedom of action. We first introduce some relations:

$V_{p,t}(B) :=$ at time t, person p wants to do B; $U_t :=$ the state of the world at time t;

$H_{t^1} :=$ the action H at time t^1

As a preliminary result we obtain for the freedom of action:

> The action of doing H at time t_1 after having at time t_0 decided to do this is free
>
> if and only if
>
> $(U_{t^\circ} \wedge V_{p,t^\circ}(H_{t^1}) \rightarrow H_{t^1}))$ and $(U_{t^\circ} \wedge V_{p,t^\circ}(\neg H_{t^1}) \rightarrow \neg H_{t^1}))$

That means that

> person p is free to do H exactly when
> the volition of person p at time t° of H at time t^1 necessarily causes H at time t^1
> and the same holds for $\neg H$ instead of H.

Reichenbach does not only demand for the freedom of action, that the volition of an action would imply it necessarily, but also that the will to prevent an action would imply necessarily its not taking place. This presupposes that not only the volition of H is possible, but the same also for the volition of $\neg H$, the contrary. If determinism is true, both cannot be possible at the same time. But what, then, is determinism?

Laplace has illustrated determinism via his famous thought experiment of a perfect intelligence, frequently called "Laplace's demon:"

> We ought then to regard the present state of the universe as the effect of its anterior state and as the cause of the one which is to follow. Given for one instant an intelligence which could comprehend all the forces by which nature is animated and the respective situation of the beings who compose it—an intelligence sufficiently vast to submit these data to analysis—it would embrace in the same formula the movements of the greatest bodies of the universe and those of the lightest atom; for it, nothing would be uncertain and the future, as the past, would be present to its eyes. (Laplace 1814, 4)

If Laplace's demon can predict everything that will happen and how it will happen, then we have to say that determinism is true. All events in the world are "determined" by physical laws and by the state of the world in the past. But how does Laplace's demon get the data which he needs for his prediction? Reichenbach thinks that he will not succeed because he has the laws of physics against him. Even a demon, with his overwhelming intelligence, calculation power, and nearly infinite memory is, due to the laws of physics, unable to acquire the information he needs (cf. also Reichenbach 1932).

To come to a better understanding of Reichenbach's argument, let us look at just one example: Imagine for instance the attempt to predict the trace of an outburst of matter on the surface of the sun on a photographical plate. The appearance of this trace is doubtless a physical event. But the light from the sun is not quicker than the velocity of light in general, and therefore the information about an event on the surface of the sun, which happens just now, cannot yet have arrived. The sun is about eight light minutes away. Therefore we cannot predict what will happen on the plate 8 min later. Also the Laplacian demon cannot predict what will happen on the plate before it really happens.

This was but one restriction of acquisition of data about the world. If one now defines, like Reichenbach did, determinism as the possibility for the demon, to predict all future events of physical systems from experimental data, determinism is just not true.

Also the freedom of action will now be defined in an unusual way. U_t, the state of the world at time t, is for Reichenbach to be read as

$U_t :=$ the state of the world at time *t as far as it can be known.*

It is clear that by this interpretation of the circumstances U_t we obtain other results than the usual ones. We can call Reichenbach's concept of determinism "predictive determinism" and his concept of freedom of the action "predictive freedom of action".

We get now the following final result for the freedom of action (*SW* vol. 1, 457–460):

1. At time t^{-1} it cannot be predicted from the known circumstances U_{-t} if a person p wants at time $t^{o} > t^{-1}$ to do H or $\neg H$, and
2. It can be predicted that the volition of person p at time t^{o} of H at time $t^{1} > t^{o}$ would cause H at time t^{1} and the same holds for $\neg H$ instead of H.

Thus Reichenbach who today would have called himself a libertarian was according to the common terminology a compatibilist, for whom, even in a deterministic universe, human actions and decisions are free in many situations. It is not our job here to go into further detail of the extended discussion on the question of whether or not the human will is free.

I want only to add here that Reichenbach already defined the freedom of will of *A* properly speaking in the same way as 20 years later did Harry Frankfurt (1971). Both philosophers define the freedom of the will as a special kind of freedom of action *H* where the action *H* is again the volition to do an action *H'*. For Reichenbach and Frankfurt as well the freedom of the will is the ability to retain an intention or resolution for a longer stretch of time (1958; *SW* vol. 1, 463–469). For this aim we have simply to replace *H* in the definition of freedom of action with a second volition V_{p,t^o} (H_{t2}). The person instead of wanting to do something wants to want to do something at a later time. This freedom can also be called strength of the will. Thus Reichenbach was ahead of the other logical empiricists of his time.

7.7 Summary

In this paper I have tried to draw a line from the *Jugendbewegung* to Reichenbach's conventionalism, his ethics and finally to his theory of free action and free will. Everybody has to find his own principles and to try to defend them. What I have reported here is only a small fraction of a connection, which was prevalent in the first half of the twentieth century. The norms which were valid in nineteenth century in art, science, and society were broken down. Men reacted differently to this fact.

There is no heaven of ideas from which principles of behaviour are obtained. Other philosophers have complained this loss of orientation. For them we are condemned to be free. Some people enjoyed the freedom which they had gained, for others chaos had been erupted. Reichenbach belonged to the first kind together with many scientists and artists. I should also h1ave studied mental developments of other logical empiricists like Carnap, Schlick and the Vienna circle, of modern artists and composers. But that would have gone beyond the limits of an article based on a talk at a philosophical workshop.

References

Hans Reichenbach's works are quoted
according to his "Collected Works" as published in German and in English:

GW. Gesammelte Werke, in 9 Bänden. Maria Reichenbach and Andreas Kamlah (eds.). Braunschweig/Wiesbaden: Vieweg, 1977–1999 (vol. 1–7 have been published).

SW. Selected Writings: 1909–1953, M. Reichenbach and R. S. Cohen (eds.). 2 vols. Dordrecht/Boston/London: Reidel. 1978.

Blüher, Hans. 1978. *Wandervogel 1–3. Geschichte einer Jugendbewegung*, 5th ed. Frankfurt: Dipa-Verlag.

Carnap, Rudolf. 1928. *Scheinprobleme in der Philosophie*. Berlin-Schlachtensee: Weltkreis-Verlag; English Trans. In 1967 *The logical structure of the world and pseudoproblems in philosophy*. La Salle Illinois: Open Court.

Carnap, Rudolf. 1963. Autobiography. In *The philosophy of Rudolf Carnap*, ed. Paul Arthur Schilpp. La Salle Illinois/London: Open Court/Cambridge University Press.

Dahms, Joachim. 2004. Die Emigration des Wiener Kreises. In *Vertriebene Vernunft,* ed. Stadler Friedrich, 66–121, vol 1, 2nd ed., Münster: Lit.

Einstein, Albert. 1905. Zur Elektrodynamik bewegter Körper. *Annalen der Physik* 17: 891–921; English Trans.

Einstein, Albert. 1920. *Relativity. The special and the general theory*. New York: H. Holt and Company.

Frankfurt, Harry. 1971. Freedom of the will and the concept of a person. *The Journal of Philosophy* 68: 5–20.

Gerner, Karin. 1997. *Hans Reichenbach, sein Leben und sein Wirken*. Osnabrück: Phöbe Autoren-press.

Hume, David. 1748. *Enquiry concerning human understanding*. London.

Jordan, Pascual. 1932. Die Quantenmechanik und die Grundlagen der Biologie und Psychologie. *Die Naturwissenschaften* 20: 815–821.

Kamlah, Andreas. 1977, 1979. Erläuterungen, Bemerkungen und Verweise. In Hans Reichenbach. *GW* vols. 1, 2.

Kamlah, Andreas. 1994. Hinweise des Nachlasses von Hans Reichenbach auf sein Menschenbild, auf Motive und Quellen seiner Philosophie. In Danneberg et al. (eds.), 183–200.

Kamlah, Andreas. 2002. *Der Griff der Sprache nach der Natur*. Paderborn: Mentis.

Kamlah, Andreas. 2008. La reception de la mécanique quantique par Reichenbach et le Cercle de Vienne. In *Mathématiques et expérience. L'empirisme logique à l'épreuve,* eds. Jacques Bouveresse and Pierre Wagner. Paris: Odile Jacob.

Kant, Immanuel. 1785. *Grundlegung zur Metaphysik der Sitten*. Riga: J.F. Hartknoch.

Landauer, Carl. 1978. Memories of Hans Reichenbach. University Student: Carl Landauer, In *SW* vol. 1: 25–31.

Laplace, Pierre Simon. 1814. *Essai philosophique sur les probabilités*. Paris; English. Trans. F. W. Truscott and F. L. Emory. *A philosophical essay on probabilities*. New York: Dover 2007 [1902].

Laqueur, Walter. 1962. *Die deutsche Jugendbewegung. Eine historische Studie*. Köln: Verlag Wissenschaft und Politik.

Linse, Hans-Ulrich. 1974. Hochschulrevolution. *Archiv für Sozialgeschichte* 14: 1–114.

Nohl, Herman. 1933. *Die Pädagogische Bewegung und ihre Theorie*. Frankfurt/M: Schulte–Bulmke.

Reichenbach, Hans. 1913. The free student idea. Its unified contents. *SW* 1: 108–123.

Reichenbach, Hans. 1920. *Relativitätstheorie und Erkenntnis apriori*. Berlin: Springer; English Trans. and ed. by Maria Reichenbach: 1965 *The theory of relativity and a priori knowledge*, Berkeley: University of California Press.

Reichenbach, Hans. 1921. Bericht über eine Axiomatik der Einsteinschen Raum-Zeit-Lehre. *Physikalische Zeitschrift* 22: 683–687.

Reichenbach, Hans. 1924. *Axiomatik der relativistischen Raum-Zeit-Lehre*. Braunschweig: Vieweg; Reprinted in *GW* vol. 3; English Trans. and ed. by Maria Reichenbach: 1969. *Axiomatization of the theory of relativity*. Berkeley: University of California Press.

Reichenbach, Hans. 1925a. Die Kausalstruktur der Welt und der Unterschied von Vergangenheit und Zukunft. *Sitzungsbericht der Bayerischen Akademie der Wissenschaften. Mathematisch-Naturwissenschaftliche Abteilung*. 133–175; English Trans.: The causal structure of the world and the difference between past and future. In *SW* vol. 2: 81–119.

Reichenbach, Hans. 1928. *Philosophie der Raum-Zeit-Lehre*. Berlin: de Gruyter; reprinted in *GW* vol. 2; English Trans. 1953: *The philosophy of space and time*. Trans. Maria Reichenbach and J. Freund. New York: Dover.

Reichenbach, Hans. 1931. Montessori-Erziehung. Erziehung zur Gegenwart. *Die Neue Erziehung* 8: 91–99.

Reichenbach, Hans. 1932. Die Kausalbehauptung und die Möglichkeit ihrer empirischen Nachprüfung. *Erkenntnis* 3: 32–64.

Reichenbach, Hans. 1933. Die logischen Grundlagen des Wahrscheinlichkeitsbegriffs. *Erkenntnis* 3: 401–425. *GW* vol. 5; 401–425.

Reichenbach, Hans. 1935. Metaphysik bei Jordan? *Erkenntnis* 5: 178–179.

Reichenbach, Hans. 1938. *Experience and prediction. Analysis of the foundations and the structure of knowledge*. Chicago: University of Chicago Press; German Trans. in: *GW* vol. 4.

Reichenbach, Hans. 1951. *The rise of scientific philosophy*. Berkeley: University of California Press; German Trans.: *Der Aufstieg der wissenschaftlichen Philosophie*. In: *GW* vol. 1.

Reichenbach, Hans. 1958a. The freedom of the will. In *Modern philosophy of science, selected essays* ed. Maria Reichenbach, and in: *SW* vol. 1, 431–473.

Schlick, Moritz. 1930. *Fragen der Ethik*. Wien: Springer.

Schlick, Moritz. 1931. Die Kausalität in der gegenwärtigen Physik. *Die Naturwissenschaften* 19: 145–162; Reprint in: *Gesammelte Aufsätze 1926–1936*. Hildesheim: Olms, 1969, 41–82.

Weniger, Erich. 1980. Die Jugendbewegung und ihre kulturelle Auswirkung. In *Die deutsche Jugendmusikbewegung in Dokumenten ihrer Zeit von den Anfängen bis 1933*, ed. Wilhelm Scholz et al., 1–9. Wolfenbüttel and Zürich: Möseler Verlag.

Wipf, Hans-Ulrich. 1994. 'Es war das Gefühl, dass die Universitätsbildung in irgend einem Punkte versagte... '. Hans Reichenbach als Freistudent 1910 bis 1916. In eds. Danneberg et al. 161–181.

Wyneken, Gustav. 1914. *Schule und Jugendkultur*. Jena: Diederichs.

Part IV
Walter Dubislav

Chapter 8
Dubislav and Classical Monadic Quantificational Logic

Christian Thiel

Walter Dubislav was born in Berlin-Friedenau in 1895, and ended his life under dramatic circumstances in Prague in 1937. Since 1927 he had belonged to the Berlin Group, i.e. the Society for Scientific (or: Empirical) Philosophy. His interests and contributions were broad, as shown by his work on various topics such as the writings of Bolzano, the Friesian School, natural philosophy and the philosophy of mathematics. The best known of Dubislav's work is *Die Definition*, first published in 1926 (under the title *Über die Definition*), then in a second edition in 1927, and finally in a completely revised and enlarged third edition by the publishing house of Felix Meiner, as the first supplement ("Beiheft") of the journal *Erkenntnis* edited by Reichenbach and Carnap in 1931. On the occasion of the 50th anniversary of its publication, Meiner published a reprint (Dubislav 1981) with a new preface by Wilhelm K. Essler.

The present paper deals with Dubislav's interesting, but contentually and technically problematic contribution to the philosophy of mathematics, or more precisely, to mathematical logic and metalogic. When Dubislav turned his eye to this field, Gödel had not yet written his dissertation on the completeness of the calculus of quantificational logic, nor his paper on the incompleteness of "*Principia Mathematica* and related systems", Church had not yet proved the undecidability of classical quantificational logic, and no "Hilbert-Bernays" was at hand. This said, beyond the *Principia Mathematica*, there were promising investigations on proof theory by Hilbert, Ackermann, Behmann, von Neumann (von Neumann 1927) and some others. And, the theoretical survey provided by the first two of those scholars (entitled *Grundzüge der theoretischen Logik*) would stimulate research by Gödel, Carnap and other great logicians of the 1930s. Yet, according to Hilbert and Ackermann, the "main problem of mathematical logic" at the time was the decision

C. Thiel (✉)
Institute of Philosophy, University of Erlangen–Nürnberg, Erlangen, Germany
e-mail: Theil-Erlangen@t-online.de

N. Milkov and V. Peckhaus (eds.), *The Berlin Group and the Philosophy of Logical Empiricism*, Boston Studies in the Philosophy and History of Science 273, DOI 10.1007/978-94-007-5485-0_8,

problem, i.e. the quest for a procedure "which permits us to decide for any given formula, by finitely many operations, whether it is universally valid (or satisfiable, respectively)" (Hilbert and Ackermann 1928, 77 and 73, respectively). With regard to the calculi of classical propositional logic and classical monadic quantificational logic (treating only of one-place propositional functions) the problem had already been solved by this time. However, for classical quantificational logic, such a resolution remained elusive, even for classes of formulas with a particular logical structure. Thus, it was no wonder that Dubislav also turned his attention to this question.

On his approach, the method of the so-called "quasi truth-tables" is of great importance. In short, they are an extension of the familiar truth-tables of classical propositional logic. Operations with truth-values had been known, at least in principle, to Peirce, MacColl, Frege and others, but they were explicitly propagated at the beginning of the twentieth century, in large part, by Wittgenstein's *Tractatus* (Wittgenstein 1921). In 1921, Emil L. Post supplemented the two truth-values "true" and "false" (T and F, + and –, 1 and 0 or vice versa) with further values, not all of which could now be interpreted to strictly pertain to the truth as such—hence the prefix "quasi".

Dubislav makes use of his method (inspired by Post) in two papers, published in 1928 and 1929. In the second of these, he explicitly mentions Wittgenstein and Post as his sources. As an aside, it is worth noting that in neither of the two papers are the terms "decidable", "decidability" and "decision problem" ever deployed. For, these works serve a different set of concerns that is made clear by their titles. Specifically, the paper from 1928 is called "Zur kalkülmäßigen Charakterisierung der Definitionen". Herein, Dubislav investigates the explicit definitions, eliminable in the sense of Pascal, and goes on to describe them as rules for the replacement of a definiendum by its definiens in all, or only in some places, of an expression containing them. Examples that had already been given by Peano in 1901 make it evident that special precautions have to be taken. For, while Dubislav shows that Peano's criterion is necessary, he also demonstrates that it is not sufficient. In order to rectify this situation, he introduces a property abbreviated by the letter "E". This property is meant to hold of a well-formed expression of the calculus if and only if its valuation, by means of the usual truth-tables or of certain quasi truth-tables, finally yields a column of exclusively designated values (i.e., "true", + or 1 or 0, respectively). This property is passed on, in propositional logic and in monadic quantificational logic, by every application of the rules of substitution and of the rule of detachment ("modus ponens"). Dubislav finds the criterion for the correctness of explicit definitions in the condition that the latter, if formulated as "additional substitution rules", also transmit the property E.

By 1928 he had already mentioned another and even more elementary application of this property. For, due to the fact that the truth table of negation replaces a designated value with a non-designated one, and given that Dubislav's quasi truth-tables, being conservative extensions of the truth-tables, have the same effect, the negation of an expression whose valuation leads to a column with exclusively designated values can never yield a column of the same kind. Therefore, since all expressions derivable in the calculus get designated values, an expression and its negation cannot both be derivable. This is to say, that the calculus—in our case the calculus of propositional logic and that of classical monadic quantificational logic, respectively—is consistent.

It is this point that is the subject of the second early paper of Dubislav's mentioned above, which was published under the title, "Elementarer Nachweis der Widerspruchslosigkeit des Logik-Kalküls" in the *Journal für die reine und angewandte Mathematik*, a journal highly esteemed as *Crelles Journal*, albeit not specializing in logic. Again, in this paper there is no mention of decidability. Rather, Dubislav first explains the method of evaluation for the validity of formulas composed by propositional connectives: if "plus" (+) designates the value "true", such a formula is valid if its evaluation by the truth-tables yields a plus-column. On account of the heredity property, mentioned previously, and of the fact that all axioms of the classical propositional calculus (taken by Dubislav from Hilbert and Ackermann) yield a plus-column, we may infer the consistency of the calculus. Dubislav extends this result (quoting Post 1921 as, at least formally, a predecessor, although with different tables) to formulas of classical monadic quantificational logic. In the same manner as before, Dubislav also establishes the consistency "for the calculus operating with 'all' and 'some'" (Dubislav 1929, 110). Here, it is worth noting that he actually only treated the monadic case, but he did so without drawing attention to this restriction. That is, the axioms of propositional logic are simply supplemented by an axiom for the universal quantifier and one for the existential quantifier, while the general calculus of quantificational logic is not even mentioned by name.

In the third edition of *Die Definition* (Dubislav 1931) we are met with a closely related presentation. The subject makes it necessary to include the purely calculatory criterion for correct explicit definitions established in "Zur kalkülmäßigen Charakterisierung der Definitionen" (Dubislav 1928), and the consistency proof emerges, as it were, *en passant*. This time it is somewhat clearer why we need three-valued tables, and why this is sufficient. But once more, the argument is given only for the monadic calculus, and its validity for the general calculus is merely asserted. Of course, for a one-place predicate and a one-place complex propositional function it seems simply evident that they have to be "always true", or "always false", or "sometimes true and sometimes false". Thus, the three cases yield the three values, and reflection on their content justifies the structure of the three-valued tables employed by Dubislav.

Tables of this kind are completely absent in a survey of the philosophy of mathematics in Germany published as "Les recherches sur la philosophie des mathématiques en Allemagne" (Dubislav 1931–32), (for the most part a French translation of extracts from Dubislav's book *Die Philosophie der Mathematik in der Gegenwart* announced for 1932). In the book as well as in its partial translation, Dubislav refers the reader to Gödel's papers on the completeness of quantificational logic and on the undecidability of *Principia Mathematica* (Gödel 1930, 1931, respectively), and mentions Löwenheim's discovery of the decidability of classical monadic quantificational logic as well as Behmann's decision procedure of 1922 (Behmann 1922), to which he adds another one developed by himself (curiously quoting his paper in *Crelles Journal* of 1929 where the tables served quite another purpose). The chapter entitled "Der wissenschafts-theoretische Problemkreis" appears *nolens volens* under the title "Les problèmes épistémologiques". Incidentally, the French translation, which is at times somewhat freely done, but obviously competently, comes courtesy of Emmanuel Levinas.

The more comprehensive German monograph (Dubislav 1932) appeared under the title *Die Philosophie der Mathematik in der Gegenwart*, as Heft 13 of the series *Philosophische Forschungsberichte* of the publishing house Junker und Dünnhaupt in Berlin. Dubislav presents his quasi truth-valuation in Chap. 4 "Das Entscheidungsproblem und das Vollständigkeitsproblem" (op. cit., 22–27). It is here that, for the first time, he also offers his own appraisal of the procedure. On page 24, he explicitly calls it a decision procedure, and after recalling the classical truth-table method, announces its extension to the functional calculus (i.e. to quantificational logic): "to begin with ['*zunächst*'], to the functional calculus in which the fundamental logical connectives join only formulas with one and the same variable" (loc. cit., 25). As a point of application and for the purpose of illustration, he examines the syllogistic inference from "all men are mortal beings" and "Caius is a man" to "Caius is a mortal being". The schema of evaluation is presented on page 27 (with "M" for "Mensch", "S" for "sterblich" and "c" for "Caius"):

$M\hat{z}$,	$St\hat{z}$	$Mx \supset Stx$	$(x)\{Mx \supset Stx\}$	Mc	Stc	$[(x)\{Mx \supset Stx\} \cdot Mc] \supset Stc$	
+	+	+	+	+	+	+	
+	—	—	—	+	—	+	
—	+	+	+	—	+	+	
—	—	+	+	—	—	+	
*	+	+	+	+; —	+	+	(bei allen durch die Mehrdeutigkeit entstehend. Kombinationen)
+	*	*	—	+	+; —	+	"
*	—	*	—	+; —	—	+	"
—	*	+	+	—	+; —	+	"
*	*	+; *	+; —	+; —	+; —	+	(bei allen durch die Mehrdeutigkeit entstehend. Kombinationen, die auf Grund der vorgängigen Wertungen nicht von selbst ausfallen).

The method by which the lines and columns are calculated will be explained later. For the moment, it is sufficient to realize that the last line shows a horizontal combination of plus signs at the ambiguous entries, and that the rightmost plus-column (the result of the evaluation) is flanked by two not very perspicuous marginal notes stating that in the last five lines the combinations caused by ambiguities either disappear automatically, or yield a plus.

Dubislav was reproached for this irritating opacity, in reviews of *Die Definition* (1931) and of the volume on the philosophy of mathematics. The consistency paper had been announced briefly but without complaints by Fraenkel in the *Jahrbuch über die Fortschritte der Mathematik* of 1929. But in the 1931/1932 edition of *Zentralblatt für Mathematik und ihre Grenzgebiete* a member of the Hilbert circle in Göttingen, Arnold Schmidt, reviewed the third edition of Dubislav's *Die Definition* and criticized some of its central points severely. In his words:

> The consistency proof given for the restricted functional calculus, using three-figure value tables [...] is not correct; to render it correct, one would not only have to modify and formally extend the tables given on page 84/85, but also to supply the (necessarily) ambiguous places of the tables with a special instruction for the distinction of values. (Schmidt 1932, 1)

What is more, he asserted that the validity of the procedure claimed by Dubislav for all formalizable disciplines, is lacking. Since even for the hitherto known consistency proofs for the elementary theory of numbers the method of valuation is insufficient. In Schmidt's opinion:

> The author overlooks the circumstance that his consistency proof for the functional calculus does not cover every axiom of a discipline that can be expressed by the symbols of this calculus. The *sufficiency* of the criterion becomes evident—with the proviso that the necessary consistency proofs are available at all—only if one may be sure (as in the case of the propositional calculus) the property E does not pertain to a formula which is expressible by the symbols of the corresponding calculus but is not (and ought not be) provable in it. But this is a condition far from being a matter of course, a condition the fulfilment of which (or its proof) will in many a case turn out to be a rather difficult task. E.g., the Hilbert-Ackermann consistency proof for the functional calculus mentioned by the reviewer does not fulfil the condition. (loc. cit., 2)

In the 1931/1932 edition of *Zentralblatt* we also find a review of Dubislav's *Die Philosophie der Mathematik in der Gegenwart.* Again, the reviewer is Arnold Schmidt, and, as such, it is not surprising that Dubislav's use of the procedure of 1931, for the restricted functional calculus, is the subject of criticism. So too is the continuing lack of support for the claim "that the criterion offered (the table of plus values) *suffices* for provability; the proof of the *necessity* is open to the objection [made by Schmidt in the review just quoted] that we miss particular instructions for distinguishing values at ambiguous places" (Schmidt 1933, 145). Schmidt also criticizes the opacity of this work, that has already been discussed herein. As he notes:

> [...] in an example (27) we are told of 'combinations which disappear by themselves thanks to valuations performed before'. But combinations of this kind do not exist in the procedure described by the author; rather, he refers to *contentual* considerations, whereas the aim of the procedure is just to make the decision independent from contentual considerations. (Ibid.)

On a more positive note, Schmidt tentatively presents a conceivable amendment to Dubislav's procedure, but not without immediately pointing out that even this augmentation would not be sufficient to remedy the situation in all possible cases.

With regard to Dubislav's book on the philosophy of mathematics, I will only mention the review published by Heinrich Scholz in the *Jahresbericht der Deutschen Mathematiker-Vereinigung* of 1934 (Scholz 1934). Scholz speaks of how Dubislav proposes "interesting tables of evaluation for the solution of the decision problem in the elementary predicate calculus". However, he closes his review "with a few critical remarks" (loc. cit., 89). The second of these reads:

> A stringent proof of the efficiency of the valuation tables on page 25 has not been given; neither for the rules of inference nor for the rules of substitution of the Hilbert-Bernays predicate logic (which are insufficient and therefore in need of a precise reformulation) has the heredity of the designated value with reference to these rules been demonstrated.

Scholz supplements this criticism with a reference to the critical passages in the two reviews by Arnold Schmidt just mentioned. The remark on the insufficient formulation of the rule of substitution relates to the insufficient version in the first edition of Hilbert-Ackermann (corrected in the second edition, not least on account of the criticism in Scholz's mimeo *Logistik* of 1932/1933). This correction was explicitly noted by Quine in his review of the second edition in the *Journal of Symbolic Logic* (Quine 1938) as well as in Bernays's preface to the first volume of *Grundlagen der Mathematik* (Hilbert and Bernays 1934, VI).

The only place in the subsequent logical literature where Dubislav's procedure, the problem of ambiguity and its partial clarification are discussed is Hans Reichenbach's *Elements of Symbolic Logic* (Reichenbach 1947). § 23 introduces "truth characters of one-place functions", i.e. quasi truth-values. In a footnote their interpretation as "necessity", "possibility" and "impossibility" in Russell's *Introduction to Mathematical Philosophy* (Russell 1919) is mentioned. Right after this we learn of the introduction of Dubislav's tables in the paper of 1929 with the application to "case analysis", completed by Reichenbach's use of the tables for the interpretation of modalities in 1932 (Reichenbach 1932).

After the "Definition of tautologies containing functions" in § 24, the subject of § 25 is "The use of case analysis for the construction of tautologies in propositional functions". This method, the introduction of which Reichenbach attributes to Dubislav, uses quasi truth-tables and because of this is restricted to monadic quantificational logic. To examine an expression correctly built up according to the rules of this logic means "going through all possible cases resulting for different truth characters of its constituents" (Reichenbach 1947, 131). Among the difficulties arising from ambiguities there are those that are harmless, and those that aren't. The harmless ones are those in which "the indeterminacy of the middle line drops out" (ibid.). This statement obviously refers to the combinations "disappearing by themselves" in (Dubislav 1932), although Reichenbach does not include a reference to that work. On the other hand, Reichenbach also gives several examples of cases which are not so harmless insofar as the ambiguity arising from the concurrence of two ambivalent pairs of values may be overcome only by forming a combination in which the first member of one of the pairs corresponds to the first member of the other, and likewise for the second members. This, however, can only be decided by "material thinking", i.e. contentual considerations, as indicated by Arnold Schmidt.

For his part, Reichenbach points out that in these sorts of cases we need to refer to an infinite amount of objects, though he does not regard such a maneuver as being illegitimate. Consequently, the situation is this: if the case analysis shows, without material considerations, that a formula is a tautology, then it certainly is a tautology. However, if the examination leaves the result underdetermined, the formula still may be a tautology. Thus, as Quine puts it within his review of Reichenbach's book in 1948, what is provided is "a partial test of validity in quantification theory, [...] adequate to a *portion* of *monadic* quantification theory" (Quine 1948, 162). While this limitation is clear to Reichenbach, he showed "no awareness that test methods have existed since 1915 for the *whole* of *monadic* quantification theory" (with reference to Löwenheim 1915; Quine 1945). Regardless of this weakness, the judgment is clear: Dubislav's procedure of case analysis by quasi-valuation is a sufficient, but not a necessary criterion for validity.

Let me finally illustrate this point by reference to three examples that, while leading back to the origin of my own occupation with Dubislav's procedure, also shed some light on the rather abstract explanation of the method discussed at the beginning of this paper. Let us first take Dubislav's case analysis in the Caius example:

(Λ_x	(Mx	$\rightarrow$	Sx)	$\wedge$	Mc)	$\rightarrow$	Sc
0	0	0	0	0	0	**0**	0
1	0	1	1	1	0	**0**	1
0	1	0	0	1	1	**0**	0
0	1	0	1	1	1	**0**	1
0	2	0	0	0/1	0/1	**0**	0
1	0	2	2	1	0	**0**	0/1
1	2	2	1	0/1	0/1	**0/1**	1
0	1	0	2	1	1	**0**	0/1
0/1	2	0/2	2	0/1	0/1	**0/1**	0/1

The compound formula to be tested is the result of a step-by-step construction beginning with the two elementary propositional functions Mx and Sx and continued by either instantiation, or quantification, or propositional connection. In the first step Mx and Sx are combined to form a conditional Mx $\rightarrow$ Sx which in the second step is universally quantified to yield Λ_x (Mx $\rightarrow$ Sx). In the third step we instantiate Mx and Sx for c ("Caius") to get Mc and Sc, respectively. In step 4, Λ_x (Mx $\rightarrow$ Sx) and Mc are combined within a conjunction Λ_x (Mx $\rightarrow$ Sx) $\wedge$ Mc, which in step 5 is subjunctively joined with Sc, yielding (Λ_x (Mx $\rightarrow$ Sx) $\wedge$ Mc) $\rightarrow$ Sc.

To calculate the final column, with the values of the compound formula, we start by assigning all possible values to the initial elementary functions Mx and Sx – i.e. first (0, 0), then (0, 1), etc. until we reach (2, 2). Writing the values under the components Mx and Sx of the compound formula on top, we get columns 2 and 4 of the schema. The further calculation precisely follows the steps of the construction of the compound formula, taking the values from Dubislav's quasi truth-tables (in which we have merely replaced + by 0, – by 1 and * by 2):

ax	at
0	0
1	1
2	**0/1**

ax	$\bigwedge_x$ ax
0	0
1	1
2	1

ax	$\bigvee_x$ ax
0	0
1	1
2	0

ax	¬ ax
0	1
1	0
2	2

ax ∨ bx

bx / ax	0	1	2
0	0	0	0
1	0	1	2
2	0	2	**0/2**

ax ∧ bx

ax / bx	0	1	2
0	0	1	2
1	1	1	1
2	2	1	**1/2**

ax → bx

ax / bx	0	1	2
0	0	1	2
1	0	0	0
2	0	2	**0/2**

The column reached in the final step (here printed in bold letters) contains, in the third line from bottom, the ambiguity 0/1. In view of the table for "sub" (→), the desired 0 could only be obtained if, in the ∧ – column of the same line, we had a 1 (otherwise the last step would result in $0 \rightarrow 1$, which yields 1). But to achieve this, also the right-hand component Mc of the conjunction would also have to be 0. Whereas this does not result unambiguously from the 2 of Mx since according to the table of cases we get 0/1. This shows that even Dubislav's own example does not work without recourse to material considerations.

As our second example we take the case analysis of the expression

			(Λ_x	ax	→	Λ_y	by)	→	Λ_z	(az	→	bz)
ax	bx											
0	0	:	0	0	0	0	0	**0**	0	0	0	0
0	1	:	0	0	1	1	1	**0**	1	0	1	1
1	0	:	1	1	0	0	0	**0**	0	1	0	0
1	1	:	1	1	0	1	1	**0**	0	1	0	1
2	0	:	0	2	0	0	0	**0**	0	2	0	0
0	2	:	0	0	1	1	2	**0**	1	0	2	2
2	1	:	0	2	1	1	1	**0**	1	2	2	2
1	2	:	1	1	0	1	1	**0**	0	1	0	2
2	2	:	0	2	1	1	2	**0/0**	0/1	2	0/2	2

The schema shows that we have only a single ambiguity, appearing in the last line in the column of the z-quantifier reached in the second last step. But here the ambiguity does indeed drop out as Dubislav had expected, since the antecedent of the conditional, reached in the last step, obtains the value 1, thereby assigning the value 0 to the conditional for each of the three possible values of the succedent.

With some embarrassment I confess that in the first draft of my logic script, my example was just this formula (which is indeed a tautology). As such, after demonstrating the smooth working of Dubislav's procedure, I claimed that his work did provide a *decision procedure* for classical monadic quantificational logic. However, the fact that it yields only a sufficient criterion of validity, and not a necessary one, soon became clear through counter-examples of a type that was also employed by Reichenbach (although I had not consulted his work at that time). Let me close with a very simple example, procured by one of the tutors of my logic course at Erlangen: the generalized tertium non datur.

Λ_x	(ax	V	¬	ax)
0	0	0	1	0
0	1	0	0	1
0/1	2	0/2	2	2.

In this case, the ambiguity remaining in the last line cannot be removed by way of mutual compensation of two ambiguities. For, in the pertinent last line, before reaching the final column, only a single ambiguity is extant. Thus, it seems that Dubislav's idea of a cancelling out of ambiguities cannot be saved, not even by Reichenbach's benevolent attempt at clarification, proffered in 1947.

References

Behmann, Heinrich. 1922. Beiträge zur Algebra der Logik, insbesondere zum Entscheidungsproblem. *Mathematische Annalen* 86: 163–229.

Dubislav, Walter. 1926. *Über die definition*. Berlin-Schöneberg: Hermann Weiß.

Dubislav, Walter. 1927. Über die definition. 2nd rev ed. Berlin-Schöneberg: Hermann Weiß.

Dubislav, Walter. 1928. Zur kalkülmäßigen Charakterisierung der Definitionen. *Annalen der Philosophie und philosophischen Kritik* 7: 136–145.

Dubislav, Walter. 1929. Elementarer Nachweis der Widerspruchslosigkeit des Logik-Kalküls. *Journal für die reine und angewandte Mathematik* 161: 107–112.

Dubislav, Walter. 1931–32. Les recherches sur la philosophie des mathématiques en Allemagne (Aperçu général). *Recherches Philosophiques* 1: 299–311.

Dubislav, Walter. 1931. Die definition. Third, completely revised and enlarged edition. Leipzig: Felix Meiner (*Beihefte der ‚Erkenntnis"*, 1).

Dubislav, Walter. 1932. Die Philosophie der Mathematik in der Gegenwart. Berlin: Junker und Dünnhaupt (*Philosophische Forschungsberichte*, Heft 13).

Dubislav, Walter. 1981. Die definition. 4th ed. With an introduction by Wilhelm K. Essler. Hamburg: Felix Meiner.

Fraenkel, Adolf A. 1929. [Notice of Dubislav 1929]. *Jahrbuch über die Fortschritte der Mathematik* 55 I (Heft 1, publ. 1931), 33.

Gödel, Kurt. 1930. Die Vollständigkeit der Axiome des logischen Funktionenkalküls. *Monatshefte für Mathematik und Physik* 37: 349–360.

Gödel, Kurt. 1931. Über formal unentscheidbare Sätze der *Principia Mathematica* und verwandter Systeme I. *Monatshefte für Mathematik und Physik* 38: 173–198.

Hilbert, David, and Wilhelm Ackermann. 1928. Grundzüge der theoretischen Logik. 2nd rev ed. Berlin: Julius Springer (*Die Grundlehren der mathematischen Wissenschaften in Einzeldarstellungen*, Band XXVII).

Hilbert, David, and Paul Bernays. 1934. Grundlagen der Mathematik. Erster Band. Berlin: Springer (*Die Grundlehren der mathematischen Wissenschaften in Einzeldarstellungen mit besonderer Berücksichtigung der Anwendungsgebiete*, Band XL).

Löwenheim, Leopold. 1915. Über Möglichkeiten im Relativkalkül. *Mathematische Annalen* 76: 447–470.

Post, Emil Leon. 1921. Introduction to a general theory of elementary propositions. *American Journal of Mathematics* 43: 163–185.

Quine, Willard Van Orman. 1938. Review of Hilbert/Ackermann 1938. *Journal of Symbolic Logic* 3: 83–84.

Quine, Willard Van Orman. 1945. On the logic of quantification. *Journal of Symbolic Logic* 10: 1–12.

Quine, Willard Van Orman. 1948. [Review of Reichenbach 1947]. *Journal of Philosophy* 45(6): 161–166.

Reichenbach, Hans. 1932. Wahrscheinlichkeitslogik. *Sitzungsberichte der Preußischen Akademie der Wissenschaften. Physikalisch-mathematische Klasse* 1932:476–488.

Reichenbach, Hans. 1947. *Elements of symbolic logic*. New York: Macmillan.

Russell, Bertrand. 1919. *Introduction to mathematical philosophy*. London/New York: Allen and Unwin/Macmillan.

Schmidt, Arnold. 1932. [Review of Dubislav 1931]. *Zentralblatt für Mathematik und ihre Grenzgebiete* 2(Heft 1): 1–2.

Schmidt, Arnold. 1933. [Review of Dubislav 1932]. *Zentralblatt für Mathematik und ihre Grenzgebiete* 5(Heft 4): 145.

Scholz, Heinrich. 1933. *Logistik. Vorlesung* [Münster i.W.] Winter-Semester 1932/33, Sommer-Semester 1933. Mimeographed lecture courses.

Scholz, Heinrich. 1934. [Review of Dubislav 1932]. *Jahresbericht der Deutschen Mathematiker-Vereinigung* 43:88–90 of the section "Literarisches" [italic pagination].
von Neumann, Johann J. 1927. Zur Hilbertschen Beweistheorie. *Mathematische Zeitschrift* 26: 1–46.
Wittgenstein, Ludwig. 1921. Logisch-Philosophische Abhandlung. *Annalen der Naturphilosophie* 14 (Heft 3/4): 185–262. Bilingual monographic edition: *Tractatus Logico-Philosophicus*. London: Kegan Paul, Trench, Trubner & Co. 1922.

Chapter 9
"Demonstrations", Not "Deductions": Walter Dubislav on Transcendental Arguments

Temilo van Zantwijk

Well known examples of transcendental arguments, like Aristotle's defense of the principle of contradiction, Descartes' cogito argument, and Kant's transcendental deduction of the categories, vary in many respects. Yet, while these lines of reasoning depend on quite different presuppositions, and have various argumentative scopes and force, they all share an important feature. Specifically, they are supposed to overcome skeptical doubts without exceeding the bounds of justifiable discursive commitments. It is on this basis that Strawson claimed, that if Kant's concept of transcendental deduction were freed from its association with idealism then it would be coherent to claim that the mere possibility of certain experiences should be held to depend upon a set of necessary conditions (Strawson 1959, 1966, 40). But, despite such maneuvers, the question of if and how transcendental arguments can be thought to be sound has not been settled yet. For, influential opposition to this brand of argumentation has emerged from scholars such as B. Stroud, who maintained that the idea of necessary conditions of possible experience stands in contradiction to the transcendentalists' declaration that they are not presupposing the truth of the propositions entailing these necessary conditions (Stroud 1968; Stern 1999; Schaper and Vossenkuhl 1989; Niquet 1999). The skeptic's point is that merely believing that a proposition which entails necessary conditions of possible experience, is true, provides a warrant for accepting the presuppositions underlying experience. As such, transcendental arguments are said not to stand up to skeptical doubt.

Special thanks go to Matthew Grellette and Katharina Eis for their help to improve an earlier draft of the text and for many thoughtful comments.

T. van Zantwijk (✉)
Institute of Philosophy, University of Jena, Jena, Germany
e-mail: Temilo.van.Zantwijk@uni-jena.de

N. Milkov and V. Peckhaus (eds.), *The Berlin Group and the Philosophy of Logical Empiricism*, Boston Studies in the Philosophy and History of Science 273, DOI 10.1007/978-94-007-5485-0_9,

At the same time these skeptical doubts are part of a moderate version of skepticism, which does not involve a complete rejection of sensible inter-subjective communication. For instance, it does not imply that “meaning” is a senseless expression and so forth. But, the skeptic can and will maintain that the conditions of possible experience, established via transcendental argumentation, may not necessarily be true. Thus, from the argument that a proposition S entails an essential condition of possible experience, it does not follow that S is true. For this result follows when one merely believes S to be true.

Stroud’s objection is interesting because it doesn’t question the form or validity of transcendental arguments. Nor does it reject the endeavor of seeking out the necessary conditions of possible experience. Rather, it only takes aim at the persuasive force of transcendental argumentation. For example, suppose you manage to state a valid transcendental argument, meant to justify the claim that a universal law of causation (“The same causes are related to the same effects under the same conditions”) is a condition of possible experience (not as a condition of “nature” as the very object of that experience in fact). As such, you are able to explain what you mean when speaking of the conditions of “possible experience”. What have you really gained? The use of transcendental arguments in philosophy presupposes the worth of a special kind of reasoning that draws upon the relation between conceptual content and the theoretical commitments implied by the practice of justifying experience. What is more, it holds that form of reasoning up as the means by which to identify the foundations of our actual experience. As such, transcendentally justified propositions are portrayed as being imbued with the very same persuasive force as immediately self-evident axioms. However, in Stroud’s opinion this approach involves a verificationist gap, since transcendental arguments only establish the reasonability of beliefs and not the truth of propositions.

It is important to note that Stroud’s criticism is open to counter objections. For example, one might ask if it makes any difference to share necessarily held beliefs with others, or to share the knowledge with others that a proposition must be true. However, such counter objections force their proponents to make many new assumptions. For instance, relying upon the idea of necessarily held beliefs requires one to construe the realm of human beliefs such that it is possible to make a clear cut distinction between the beliefs we necessarily have, and share with others, and our private beliefs. Now, this statement clearly amounts to a construction of the common ground of human experience as a set of necessarily held beliefs. As such, it seems that in order to raise this sort of objection, one would have to accept some strong version of transcendental idealism, i.e. a system of transcendental deductions in the sense of Fichte’s “Wissenschaftslehre” or Schelling’s “System des transcendentalen Idealismus”.

On top of the problem of fairly stating transcendental arguments, one would also have to ascribe consistency and completeness to your set of transcendental deductions. Here a new problem, regarding how to justify this much stronger additional claim, emerges. For their part, German Idealists couldn’t help themselves in reintroducing the concept of intellectual intuition (*intellektuelle Anschauung*) at this point. For, in doing so, they were able to ground their systems in self-evident appreciations of the construing mind. Whereas, Kant, having abandoned intellectual intuitionism from critical philosophy, took a different approach. For him, transcendental arguments must amount to logically valid pieces of argumentation, that are not based on intuition. At the same time, they must be strong enough to justify our knowledge by reference to our everyday and scientific experiences.

It's worthwhile to reflect on the criteria of justification that Kant commits himself to on this front. He holds the adequacy of our statements to be grounded in the terms of experience. Thus, the problem of adequacy arises, according to Kant, insofar as we necessarily make use of non-empirical expressions like "space", "time" and "causality" to make sense of our experiences. Accordingly he refers to the transcendental deduction of forms of perception and the categories as "an indispensable requirement" of theoretical reason, because these concepts are related to objects without any support from the senses. Since they also are synthetic concepts, with a meaning which cannot be figured out by analyzing content, we need a special form of argumentation in order to prove their adequacy or aptness at informing us about the world, as it appears to us. As such, assessments of adequacy are not made according to an appreciation of things in themselves, but to the categories in relation to appearances.

Dubislav's criticism of transcendental deduction is widely neglected in the contemporary debate on this subject. However, it is of special interest here because of its proof-theoretic approach to the subject. In his small treatise "On the Methodology of Criticism" (*Zur Methodenlehre des Kritizismus*) Dubislav gives an account of transcendental deduction by way of comparison to mathematical proofs. Treating Kant's transcendental deductions as a type of argument (*Begründung*) within a general theory of proof, Dubislav erects a theoretical framework by reference to which he can discuss the assertive force of transcendental arguments. In this manner, he took aim at the primary concern underlying Stroud's objection. Thus, Dubislav, drawing on the account of J.F. Fries and his followers up to L. Nelson, arrives at the conclusion that the case of transcendental deduction can be stated in an unobjectionable way. Specifically, he held that argumentative force of transcendental arguments should be held not to be a matter of logical proof. Rather, it should be understood to be a matter of a weaker type of demonstration (*Aufweisung*). As a result it seems inappropriate to supply a constitutional theory of experience on the basis of fundamental judgments (*Grundurteile*) founded on transcendental deductions.

Given this approach, Dubislav can recognize the value of transcendental arguments as being possessed of this limited scope, and at the same time avoid transcendental idealism as a constitutional theory of (possible) experience. After a short survey of Dubislav's theory, we'll address the question of what role such limited transcendental arguments play within the formalist philosophy of mathematics and natural science, especially against the background they have in Dubislav's account. With an eye to these considerations, I will argue that formalism is vitally interested in transcendental arguments as a means of attacking the notorious problem of the adequacy the formalist axiomatic construction of scientific disciplines.

9.1 Philosophical and Mathematical Method

Dubislav starts from the presumption that the difficulties of Kant's conception of transcendental deduction are the results of problems in Kant's general ideas about basic concepts, axioms and proofs in mathematics.

Mathematics, in Kant's view, depends upon a set of basic assumptions (*System von Grundvoraussetzungen*), which are understood to be self-evident truths, neither open to, nor in need of, justification. What is more, these truths entail nonempty basic concepts (*Grundbegriffe*), whose referentiality is supposed to be self-evident on the basis of pure perception (*reine Anschauung*) (Dubislav 1929, 6f.). On the other hand, in philosophy, intuitively accessible axioms are lacking. In their stead, a regressive method of gaining basic propositions by analyzing conceptual content is supplied. With regard to mathematics again, Kant's view is that advances are made by the progressive determination of concepts starting from basic concepts, as they are contained in fundamental axioms. As such, mathematical method is therefore "dogmatic". Though, not in the pejorative sense pertaining to rational psychology and those other branches of metaphysics whose concepts are obtained deductively without correspondence to possible experience, but in the sense of deductive, logical valid reasoning from true assumptions to true conclusions. A "doctrine" according to Kant, means a theory with the force needed to be accepted by every rational being as soon as it is understood. Thus, whereas mathematical theories are "doctrines", philosophical methods are, in Kant's view, analytic, in the sense that they extract content from conceptual expressions. For example, the notion of "spatial extension" from that of "body". Furthermore, they are regressive in the sense that they start from more specified content and arrive at more general conceptual expressions. Basic concepts, thus, are universal concepts. Fundamental truths (*Grundsätze*) are true propositions consisting of expressions referring to the different fields of human cognition: namely, to pure intuition as the field of mathematical knowledge, to empirical perception as the field of natural science, and to everyday experience with its basic universal concepts, like "change", "cause" and "effect". Kant supplies an example of a fundamental truth of all possible experience within the "Second Analogy of Experience", where he states: "All Change happens in accordance with the law of connection of causes with effects" (Kant, *Critique of Pure Reason*, B 233).

Dubislav gives a criticism of Kant's account of fundamental truths attacking the underlying idea of axioms borrowed from Kant's philosophy of mathematics.

In Kant's view, the core of scientific justification is, Dubislav argues, mathematical proof within an axiomatic theory. According to this account we trust in axioms when they fulfill two conditions: they must be necessarily true, and they must contain nonempty conceptual and referential expressions. Now, regressively established fundamental truths in philosophy, as measured by the same criteria of scientific justification, would deserve exactly the same unquestionable assent if and only if they were to follow from self-evident true nonempty propositions according to logically valid rules. Just as propositions obtained by derivation are true if and only if the propositions they are derived from are true and nonempty, which is the reason why—besides rules of valid reasoning—we need axioms as fundamental judgments in mathematics (in Kant's view), the fundamental propositions derived by conceptual analysis in philosophy are true if and only if the judgments from which they are extracted (by what means we will have to see) are true propositions containing nonempty referential and conceptual expressions obtained according to valid rules of derivation.

Not surprisingly, Dubislav launches a fundamental attack on the underlying assumptions concerning the "content" of mathematical concepts as Kant conceived of them, and on his idea of mathematical method as determining concepts by "construction". A more extended version of his account of Kant's philosophy of mathematics is given in his own work on that subject (Dubislav 1932). The first point of discussion therein concerns pure intuition (*reine Anschauung*) as the faculty of the mind involving mathematical construction (Dubislav 1932, 51). The objection that Kant didn't offer sufficient reasons to ascribe this faculty to humans doesn't carry too much weight on its own. However, in connection with the objection that the idea that geometry and arithmetic must depend on pure perception because they refer to exactly one object each—space as the very form of outer, time as the form of inner sense—it leads to consequences inconsistent with modern physics.

In fact, before Dubislav, Kant's theories of space and time had lead Kurt Grelling and Reichenbach to break with Kantianism more generally (Peckhaus 1994). The former rebelled against the Neo-Friesian School based on the work of Leonard Nelson (Nelson and Grelling 1974). While the latter, after having proposed a Kantian approach to the distribution of probabilities in his dissertation paper, had turned away from Kant after he dwelt more thoroughly on relativity theory (Reichenbach 1916, 1920).

The fact that Dubislav, as a formalist, did not accept the fundamental claims of Kant's philosophy of mathematics is not surprising. The interesting question is what caused him to refer to the work of Kant at all. In the course of his criticism Dubislav gives an important hint as to his motivations on this point.

Kant's ideas about scientific method are "irreconcilably" opposed to modern scientific method. This conflict is deepened and broadened, because Kant bases the pure intuition of space and time, as well as the categories, on the vested rights of reason set out in the fundamental propositions (*Grundsätze*). The problem with this move is that the corresponding basic judgments are closely related to axioms, in the sense Kant deploys with regard to mathematical theory: true propositions that can't be subjected to a test. Hence, they are neither subject to confirmation nor refutation. Rather, as Dubislav points out, within his former paper on critical method, they are demonstrated (*aufgezeigt*) via the use of the infamous (*berühmt berüchtigte*) transcendental method (Dubislav 1932, 52f.). We may conclude, therefore, that the question of transcendental deduction, in Dubislav's view, is not to be solved in isolation from methodological questions. As such, Kant's concept of science depends on a view of axioms that, according to Dubislav, has been overcome by formalism. Accordingly, the question of what transcendental deductions, or in a broader sense transcendental arguments (as we will see, Dubislav doesn't conceive of them as deductions), are supposed to be turns into the question how such an argument can be freed from the environment of scientific concepts and criteria of justification in which Kant embedded it.

9.2 How Are Ground Judgments to Be Justified?

Unlike later accounts of Kant's deductions, Dubislav completely neglects its references to legal practice. Kant explicitly introduces his concept of deduction by reference to the distinction between the question of stating the case (quid facti) and the question of justifying the judgment (quid iuris) in court. Presumably, Dubislav takes these statements to be nothing but metaphorical elucidations of a method that has to be formulated in logical terms only. At an interpretive level, it is surely questionable that he simply dismisses all the comments that Kant provides on the deductions as irrelevant to the question how a deduction is to be formulated. Indeed, he calls Kant's comments on this topic subtle or even captious (*spitzfindig*), and thinks of them as being formulated in a roundabout way. As a result, he doesn't grant that Kant provides anything like a reliable theory of transcendental deduction. Dubislav arrives at this opinion via the assumption that Kant, like himself, must have accepted the generally recognized division of forms of reasoning (*Begründungen*). On this understanding, there is deductive reasoning, which is held to be a matter of logically sound inferences. Second, there are empirical demonstrations, including incomplete inductions, which are to be tested by observation and experiment. Third, there are calculations of probabilities, including statistical inference. Finally, there can be lines of reasoning which combine any these types, such as one might find to be at play within any number of scientific explanations (Dubislav 1929, 18f).

Following Bolzano, Dubsilav categorizes these different forms of reasoning by the measure of assent they are thought to demand with regard to a given conclusion (which is to be understood as an assertion (*Behauptung*) bearing a truth value). According to this line of thought, an argument (*Begründung*) is an operation supplying a conclusion with at least some assertive force. In the case of a logically valid inference, the assertive force is absolute, hence providing for a state of certainty. Whereas, the other forms of argumentation only provide for the conditional acceptance of an assertion. For instance, when calculating probabilities in cases of rational choice, the assertive force of the argument, should be expressed by a quantitative degree of assent. (Bolzano 1989, 126ff., 1992, 81ff.) In the case of incomplete inductions, which in Dubislav's view are conclusions drawn by analogy, assent can't be measured quantitatively and has to be estimated by the power of judgment. Against this background the problem of transcendental deduction turns into a simple question: if we consider the four forms of reasoning to be complete, to which of these forms is transcendental reasoning is to be reduced?

When speaking of "transcendental deductions", Kant seems to be subject to a certain theoretical prejudice. Specifically, he appears to hold that transcendental arguments must supply absolute assertive force and thereby count as valid deductions.

This is due to the fact that the grounds of judgments are not axioms in the Kantian sense, for we do not have immediate access to, or certainty of, that content. The problem of transcendental deductions only arises because the grounds of judgments are supposed to have the epistemological status of axioms even though they are not.

Convinced that Kant does not provide an answer to this problem, Dubislav turned to Fries' analysis of transcendental deductions, which takes the word "deduction" in a proof-theoretic sense. Fries analysis of these arguments differs significantly from Kant's, both with respect to its underlying assumptions and with regard to its account of the logical structure of this sort of reasoning. Therefore, it is important to note that, in what follows, references to transcendental deduction in the Kantian sense will be referred to as "transcendental deduction (K)". Whereas, Fries' version shall be referred to as "transcendental deduction (F)". Analogously, a demonstration (Aufweisung) in the Kantian sense will be referred to as a "demonstration (K)", while Fries' concept of demonstration shall be indicated by the term "demonstration (F)".

Following Fries, Dubislav accepts the following analysis of the concept of transcendental deduction: Let U be a "basic judgment", which is the expression of a fundamental proposition in human experience. S(U) is an empirical psychological judgment, telling, us that U expresses a true proposition immediately (intuitively) evident to any rational being who makes use of concepts entailed in U. In Fries' view, as accepted by Dubislav, U does not rest on S(U), which solely is a second order statement about U. Accordingly Dubislav concedes that a transcendental deduction (F) is a fair reconstruction of a transcendental deduction (K). For, Fries' version allows one to state a transcendental deduction (K) as a form of proof:

The task now, is to consider just what to make of this analysis. Dubislav argues that Fries implicitly accepts a necessary validity condition which says that a transcendental deduction (F) is valid only if the empirical-inductive derivation of S(U) is valid and doesn't make use of U. But how can this condition to be fulfilled? Since his early writings, Fries admitted that, in many cases, available derivations of ground judgments are obviously circular. Hence, in such circumstances, valid transcendental deductions (F) are beyond our reach. Thus, the question arises of whether a Friesian transcendentalist can avoid the Humean trap, which says that experience simply presupposes experience and therefore cannot be justified by way of sound argumentation?

If we take a look at the transcendental deduction (F) of the ground judgment of causality, the Kantian version states that the same causes are followed by the same effects (which is not to be confused with the Leibnizian principle of causality, which tells us that nothing exists unless sufficient causes effectuate its existence). Now, Dubislav agrees with Fries on the assumption that, in their everyday practices, humans behave as if the ground judgment of causality were a self-evident truth. However, such a demonstration (F) of this ground judgment doesn't seem to satisfy the conditions of a justification, in the sense of a transcendental deduction (K). Therefore Fries recurs on his idea of inductive evidence supporting not immediately the ground judgment of causality but its immediate intuitive evidence, which is an unconscious ("dark") acceptance of the content expressed in the ground judgment.

From this account of transcendental deduction (F), Dubislav derives his main objection against the idea that transcendental arguments are a form of valid deductive reasoning. Specifically, in order to prove U by transcendental deduction (F), the grounds for S(U) must be drawn from everyday experience. Therefore, it remains unclear why a transcendental deduction (F) should have lead to a stronger measure of assent than a regressive demonstration (F) which has the same (small) argumentative force as an empirical deduction (K): "Denn weder durch die Deduktion noch durch die Aufweisung werden die betreffenden Einsichten ihrerseits irgendwie begründet" (Because neither by deduction, nor by demonstration the concerning insights are in any way validated) (Dubislav 1929, 34). As such, Dubislav held that Kant's attempt to deduce the categories of experience does not take him any further than the empirical deductions he criticized within the work of Locke and Hume (Dubislav 1929, 9).

9.3 Formalism and Transcendental Idealism: The Question of Time

Next, it will be shown that Dubislav, drawing on considerations put forward by Fries and Nelson, tried to overcome the framework of Kant's transcendental philosophy that informed the work of both of those scholars. As has been shown above, according to Dubislav and Reichenbach, Kant's theories of space and time, as given a priori forms of perception (*Anschauung*), and his understanding of causality, do not fit with the theory of relativity that governs modern physics. Now, the question arises of how this result is related to formalism, understood as a way to conceive of axiomatic systems. Formalism refers to the idea of founding natural science and psychology as structural sciences (*Strukturwissenschaften*). As such, it reduces scientific disciplines to a few basic relations that do without synthetic a priori judgments. Reduced to its logical structure, a theory allows of various interpretations. An interpretation is the application of a theory to a sphere of objects. Kantian transcendentalism, on the contrary, implies foundationalism. Any possible object belongs to one coherent sphere of appearances, constituted by the pure intuitions of space and time and a limited set of categories. In what follows, it will be suggested that the reason that formalism emerged from considerations within modern natural science (especially quantum physics) regards the fact that the question of adequacy of empirical theories that was addressed by the transcendental deductions can't be stated in a way that makes sense anymore. As a consequence there seems to be no place for transcendental deductions in scientific method anymore.

In order to make this point clear, it is helpful to take a look at the transcendental deduction of time. In accordance with Leibniz, Kant maintained a theory of time based on the concept of causality. As such, Kant's transcendental deduction of time, in the "Transcendental Aesthetics", aims at a justification of time perception that speaks to the directedness of that phenomenon. However, the idea that time has a direction, that is so fundamental to the psychology of our everyday experience, is incompatible with the way that time is conceived within modern physics. Indeed, even Kant was well aware of the fact, that from the succession of our perceptions, it doesn't follow that time must be directed from the past to the future. Rather, the psychological order of experience only shows, "that in the imagination one thing is earlier, another later, but not that one state of the experienced object procedes another" (*daß meine Imagination eines vorher, das andere nachher setze, nicht daß im Objekte der eine Zustand vor dem anderen vorhergehe*) (Kant, *Critique of Pure Reason*, B 233). Kant connects the psychological order of imagination with the concept of causation and holds a causal theory of time. The idea behind the causal theory of time is that causation constitutes a directed time order, such that an "arrow of time" is constitutive of human experience. Even within Newtonian physics, it is inappropriate however to think of time as being generally directed. For, the basic equations of classical mechanics by no means imply that motion is irreversible. As such, Reichenbach, observed that the "between"-relation is a reversible property of time that is invariant for a reversal of time direction. Accordingly, he maintained that the laws of mechanics do contain information about temporal relations, but not about the direction of processes determined by these relations: "Neither the laws of mechanics nor mechanical observables give us a direction of time, unless such a direction has been defined previously by reference to some irreversible process" (Reichenbach 1956, 35).

Nevertheless the causality-based theory of time had some merits. For, not every natural process is reversible. After all, since the nineteenth century, the study of thermodynamics has called the general time-symmetry of natural processes into question. Changes in temperature, pressure, or the bulk of bodies and gases seem to be irreversible, and are treated as such at the macro-level of physical observation (Kornwachs 2001, 32f.). With this in mind, it is clear that the concept of entropy allows one to assign a very high probability to certain a state of affairs as the irreversible result of a process. For instance, the regular distribution of ink molecules over a surface of water is highly probable, when compared to the likelihood of their being concentrated within a limited area of such a surface. Probable inference is not the same as deductive reasoning, insofar the conclusion of any probable inference is a probability statement, which must not be true. Therefore C.F. von Weizäcker objected that Boltzmann's solution is based on an invalid inference from a non-deterministic to a deterministic assertion about the future (Von Weizsäcker 1939, 274ff.).

What can be said in favor of the Kantian approach is that nothing like a physical explanation of the direction of time is available. Thus, prima facie, it makes sense to search for a philosophical solution. As a starting point, one can note that space and time have been the basis for theories of motion since antiquity. Indeed, Aristotle denied the reality of time and thought of it as a measure or quantity of movement with respect to different states of affairs of the things given to us in perception. While the idea of uniformly flowing, objective time occurs only as an unquestioned assumption in classical mechanics (Newton 1999, 408). After Einstein criticized absolute time as a variable independent from the objects and motions subject to physical investigation, the question of time became even more demanding (Einstein 1905). Following Einstein, A. Minkowski understood time as being related to the relative velocity of an inertial system and, thus, integrated it as a bound variable in the space-time continuum (Minkowski 1909). However, to conclude that physics was "temporalized" by these considerations seems disproportionate (Zimmerli and Sandbothe 2007, 8ff.). In fact, it even remains unclear if the normal meaning of the word "direction" is implied within the talk of "time dimensions" (Böhme 1966, 105). In this vein, K. Gödel remarked that time still was thought of as a reversible and symmetrical structure with strong analogies to the concept of space (Zimmerli and Sandbothe 2007, 9f.).

As Reichenbach and C.F. von Weizäcker have pointed out, it was quantum theory that is forced to give up the general concept of time as a reversible flow (Von Weizsäcker 1992). For, the motion of electrons can't be described with respect to spatio-temporal trajectories that fully determine the past and the future (under the assumption that a complete description of the presence is available). As such, the Schrödinger Equation measures the wave function that describes the motion of the quantum object. Thus, while the underlying conception of motion is deterministic and reversible, it only allows us to make statistical statements about the future distribution of quantum objects. On this understanding, the future is thought of as a sphere of possibilities not fully determined by the past. Accordingly, the observer or the measurement device registers which of these possibilities is actually realized. The idea of separating the subjective attitude of the observer, which is bound to the irreversibility of time, from the objective process, which is supposed to be reversible, amounts to something of a working solution. For, the fundamental question, of whether time has to be conceived of as being reversible or not, is left open (Zimmerli and Sandbothe 2007, 10ff.).

Dubislav aims at a formalist constitution of experience that does not depend upon synthetic a priori judgments about the structure of nature. In his "Philosophy of Mathematics" Dubislav clearly supports Hilbert's formalism, whom he considers to be his teacher, against Frege's logicism and Brouwer's intuitionism. As he states:

> In the face of the successes of formalism [...] at the one hand and considering the difficulties and ideological burdens that logicism and intuitionism have to bear each in it's own way only on the ground of formalism it seems to be possible to transfer pure logic and pure mathematics in the condition Gauss has referred to, as we mentioned in the beginning. (Dubislav 1932, 48).

In his "Natural Philosophy", this amounts to an analysis of time which maintains that, besides a set of definitions and axioms determining the way our statements about time are related to one another, i.e. relations like "before", "after", "at the same time", a series of conventions must be acknowledged that enable us to relate logical structures to our perception of time. Specifically the conventions dealing with time-measuring, especially with regard to its scale, unit and zero-point. For example, the congruence of two intervals measured at the same place must be determined by a convention about the counting of periodically returning events (Dubislav 1933, 147).

Alongside these initial moves towards a formalist account of time, which also draws on conventionalism, Dubislav launched a severe attack on the Kantian theory of time as a pure form of perception. In particular, he completely rejected the idea that theory of space and time is considered to be independent from the results of natural science. This rejection also extended to the, later, Neo-Kantian shape that Cassirer gave to this notion, implying that Kant's original idea must be transferred into a formal conceptual framework of valid a priori relations (Dubislav 1933, 141). Accordingly, Dubislav maintains that formal structures do determine content, but that content, at least in natural science, always has to be empirically testable in relation to measurable or observable facts.

Finally, we now have to ask how this result fits in with Dubislav's treatment of transcendental arguments, which consists of the suggestion that transcendental arguments should not be rejected outright, but cannot plausibly be portrayed as valid deductions. In his "Philosophy of Mathematics", Dubislav elucidated the method of formalizing a scientific discipline in three steps (Dubislav 1932, 12ff.). First, all the statements of the discipline are to be stated in a complete deductive chain of reasoning. In this process all non-logical assumptions are to be made explicit. Furthermore, the introduction of concepts is to be carried out by derivation from a set of fundamental concepts (*Grundbegriffe*). On this basis, all statements of a given discipline are to be divided between a class of axioms and a class of theorems. For, the complete reduction of the theorems to the axioms and definitions allows one "to take a formalist stand". That is, the given discipline can no longer be conceived of as a system of truths deduced from the basic concepts and axioms, but, rather, must be understood to express a network (*Netzwerk*) of conceptual relations. The third and last operation of formalization consists in the representation of the network with the means of the logical calculus (basically Dubislav thinks of classical propositional and predicate logic with identity, referring to it as "the calculus"; he also draws on type theory).

If we take a look at the role definitions play within the process of formalization, we have to distinguish between substitution instructions (*Substitutionsvorschriften*) and assignment instructions (*Zuordnungsvorschriften*). Dubislav thinks of the substitution instructions as arbitrary stipulations without any impact on the interpretation of the system. In fact, the question of how to interpret a model in the formalist view depends on the choice of certain assignment instructions According to Dubislav the assignment instructions have the same character as arbitrary stipulations (for a criticism of this view based on the idea that definitions are guided by interests compare Gabriel 1972, 53–55). But here the question of the correct determination of concepts by other concepts touches upon the question of an adequate interpretation of a system. For, how can we determine whether a relational network represents anything or not? Evidently, the question of adequacy can't be answered through deference to the internal logic of the system. As such, the method of stipulations enabling reductions comes to an end. For instance, to answer the question (to use the famous analogy of a formal system with a railroad map which can be found in Carnap) if the railroad map adequately represents the railroad network, it is not sufficient to know the relations between the junctions on the map. What we want to know is if the map is applicable to the real network. What the transcendentalists' point of view comes down to, in this case, is the idea that we cannot think of a real network unless we conceive of it in terms of a consistent set of relations between junction points. Therefore, in some practical sense we might say that the concept of a consistent set of relations is necessary for the possibility of the experience of a network. This does not mean that a specific set of relations that we use within our everyday thinking—for instance, the relations between Berlin, Leipzig and Munich as junction points—can never be falsified by experience. For, the set taking Munich to be in between Leipzig and Berlin is, in fact, is falsified by experience. However, it will not be possible to falsify the requirement of a consistent set of relations. For, using the railway example, in referring to a railroad network we are already conceiving of it in terms of such a consistent set. Thus, transcendental arguments are needed to show which propositions ought to be acknowledged as a priori true, although not invulnerable, in making use of a system of propositions and concepts. We may conclude therefore that within the formalist framework transcendental arguments, understood in the weaker sense of demonstrations (F) (*Aufweisungen*), are of vital importance to the question of adequacy and this result might partly explain Dubislav's strong concern with transcendental arguments after all.

References

Böhme, Gernot. 1966. *Über die Zeitmodi. Eine Untersuchung über das Verstehen von Zeit als Gegenwart, Vergangenheit und Zukunft mit besonderer Berücksichtigung der Beziehungen zum Zweiten Hauptsatz der Thermodynamik.* Göttingen: Vandenhoeck und Ruprecht.

Bolzano, Bernard. 1989. In *Wissenschaftslehre §§ 269–306*, ed. Jan Berg. Stuttgart-Bad Cannstatt: Frommann.

Bolzano, Bernard. 1992. In *Wissenschaftslehre §§ 349–91*, ed. Jan Berg. Stuttgart-Bad Cannstatt: Frommann.

Dubislav, Walter. 1929. *Zur Methodenlehre des Kritizismus.* Bad Langensalza: H. Beyer & Söhne.

Dubislav, Walter. 1932. *Die Philosophie der Mathematik in der Gegenwart.* Berlin: Junker & Dünnhaupt.

Dubislav, Walter. 1933. *Naturphilosophie.* Berlin: Junker & Dünnhaupt.

Einstein, Albert. 1905. Zur Elektrodynamik bewegter Körper. *Annalen der Physik* 17: 891–921.

Gabriel, Gottfried. 1972. *Definitionen und Interessen. Über die praktischen Grundlagen der Definitionslehre.* Stuttgart-Bad Cannstatt: Frommann-Holzboog.

Kornwachs, Klaus. 2001. *Logik der Zeit–Zeit der Logik.* Münster: LIT-Verlag.

Minkowski, Hermann. 1909. Raum und Zeit, 80. Versammlung deutscher Naturforscher und Ärzte. *Physikalische Zeitschrift* 10: 75–88.

Nelson, Leonard, and Kurt Grelling. 1974. Bemerkungen zu den Paradoxien von Russell und Burali-Forti. In *Die kritische Methode in ihrer Bedeutung für die Wissenschaft*, ed. Leonard Nelson, 95–127. Hamburg: Meiner.

Newton, Isaac .1999. *The principia. Mathematical principles of natural philosophy.* Trans. Bernard N. Cohen and Anne Whitman. Berkeley: University of California Press.

Niquet, Marcel. 1999. *Transzendentale Argumente. Kant, Strawson und die Aporetik der Detranszendentalisierung.* Frankfurt: Suhrkacbrsmp.

Peckhaus, Volker. 1994. Von Nelson zu Reichenbach: Kurt Grelling in Göttingen und Berlin. In *Hans Reichenbach und die Berliner Gruppe*, ed. Lutz Danneberg et al., 53–86. Braunschweig/Wiesbaden: Vieweg.

Reichenbach, Hans. 1916. Der Begriff der Wahrscheinlichkeit für die mathematische Darstellung der Wirklichkeit. *Zeitschrift für Philosophie und philosophische Kritik* 161: 209–239.

Reichenbach, Hans. 1920. *Relativitätstheorie und Erkenntnis a priori*. Berlin: Springer.

Reichenbach, Hans. 1956. *The direction of time*. In ed. Maria Reichenbach, foreword by Hilary Putnam. Berkeley: University of California Press.

Schaper, Eva, and Wilhelm Vossenkuhl. 1989. *Reading Kant. New perspectives on transcendental arguments and critical philosophy*. Oxford: Blackwell.

Stern, Robert. 1999. Introduction. In *Transcendental arguments. Problems and prospects*, ed. Robert Stern, 1–11. Oxford: Clarendon.

Strawson, Peter F. 1959. *Individuals: An essay in descriptive metaphysics*. London: Methuen.

Strawson, Peter F. 1966. *The bounds of sense: An essay on Kant's critique of pure reason*. London: Methuen.

Stroud, Barry. 1968. Transcendental arguments. *Journal of Philosophy* 65: 241–265.

Von Weizsäcker, Carl F. 1939. Der zweite Hauptsatz und der Unterschied von Vergangenheit und Zukunft. *Annalen der Physik* 36: 275–283.

Von Weizsäcker, Carl F. 1992. *Zeit und Wissen*. München: Hanser.

Zimmerli, Walther C., and Mike Sandbothe. 2007. Introduction. In *Klassiker der modernen Zeitphilosophie*, 1–30. Darmstadt: Wissenschaftliche Buchgesellschaft.

Walter Dubislav in 1931 in Berlin (by Willy Römer)

Chapter 10
Dubislav and Bolzano

Anita Kasabova

10.1 Brief Introduction

Walter Dubislav (1895–1937) was an active member of the Berlin Group of logical empiricism in the early 1930s. A philosopher, mathematician and logician, he shared the thematic focus of the Berlin Group on the natural sciences, mathematics and logic. He shared the methodological demand of the Berlin Group that philosophical method of inquiry should follow the rigor and precision of formal sciences in exposition and logical reasoning (Rescher 2006, 283). A rigorous methodology for philosophy was also required by Bernard Bolzano (1781–1848), the Prague mathematician, logician and philosopher. Was it Bolzano's efforts to separate logic from psychology in the *Theory of Science* (Bolzano 1837) or his reconstruction of mathematics in the *Contributions to a Better Founded Exposition of Mathematics* (1810) which attracted Walter Dubislav's attention?

Dubislav was not interested in Bolzano's early attempts to develop a mathematical method for expounding objective dependence relations which hold between judgments as grounds and consequences (Bolzano 1810, II, § 2). His research is focused on the later Bolzano (1837). In a series of papers published between 1929 and 1931, he deals with Bolzano's Kant-criticism and Bolzano's contribution to modern logic. More specifically, he examines what he calls Bolzano's propositional functions (*Aussage- oder Satzfunktion*), his notion of analyticity and analytic statements, as well as his notions of probability (*Wahrscheinlichkeit*) and derivability (*Ableitbarkeit*).

My thanks go to Nikolay Milkov for his helpful advice.

A. Kasabova (✉)
Department of Anthropology, New Bulgarian University, Sofia, Bulgaria
e-mail: anita.kasabova@gmail.com

N. Milkov and V. Peckhaus (eds.), *The Berlin Group and the Philosophy of Logical Empiricism*, Boston Studies in the Philosophy and History of Science 273, DOI 10.1007/978-94-007-5485-0_10,

10.2 From Kant to Bolzano: Dubislav on Bolzano's Kant-Criticism

Dubislav was drawn to Bolzano's Kant-criticism and his meticulous efforts to secure a pre-Kantian (or rather, a pre-transcendental idealist) position in philosophy, as is testified by his 1929 article "Ueber Bolzano als Kritiker Kants" and his planned edition of František Přihonský's *Neuer Anti-Kant* (1850) in collaboration with Heinrich Scholz. Přihonský was a member of Bolzano's school and in the *Neuer Anti-Kant oder Prüfung der Kritik der reinen Vernunft*, Přihonský systematizes Bolzano's criticisms of Kant which are scattered across many papers between 1818 and 1837. Dubislav and Scholz (1931c) prepared a critical commentary on Bolzano-Přihonský's dispute with Kant. The edition never appeared but their commentaries were published in Edgar Morscher's (2003) edition of the *New Anti-Kant*. Kantian philosophy was still extraordinarily influential in Dubislav's time and hence he approved the Prague philosophers' examination of the explanations and proofs Kant puts forward for the claims set down in his philosophical system. According to Dubislav, if one accepted the view that Kant's philosophical system could be appropriately evaluated by scientific means, a critical appreciation from a scientific perspective would ascertain which of Kant's doctrines are true and which are false (Dubislav 1931c, 203).

Dubislav (1929, 358) explains that the scientific value of Bolzano's Kant-criticism lies in the exposition of unsolved problems in systematic philosophy by rebutting Kantian metaphysics and logic. Thus he argues that Bolzano did not deal with Kant because he was a famous philosopher but because he believed that a critical examination of Kant's doctrines would provide a convenient access to a series of important philosophical problems. On this view, criticizing Kant is an instrument for clarifying philosophical problems that go beyond Kant—such as the question whether epistemology underwrites metaphysics and logic or the Kantian division between mathematics and philosophy supplemented by the latter's claim that exact definitions and strict demonstrations cannot occur in philosophical investigations (Kant 1789, B754–5, B759). Such claims were a thorn in the side of the Berlin Group as much as they were in Bolzano's. Hence Dubislav in turn reconstructs Bolzano's critical reconstruction of Kant because a proper display of Kant's doctrines would clarify not only Bolzano's views, but also Dubislav's own. For instance, Bolzano (1837, § 65) elaborates Kant's definition of analyticity and his distinction between analytic and synthetic judgments.[1] Dubislav, for his part, pays tribute to Bolzano's own notion of analyticity and analytic propositions. He thus conceives Kant-criticism as an exercise in defining and specifying the viewpoint

[1] Bolzano writes: "Kant penetrated this distinction the deepest and it is to him that the author of this book owes his correct view on this issue. [. . .] It suffices to grasp this distinction appropriately, on order to understand that there are attributes (*Beschaffenheiten*) which belong to an object and necessarily belong to it according to the concept we form of that object, without being presented as components of this concept." (1837, § 65.8, cf. also § 148, my translation—A. K.).

of those who criticize him (Dubislav 1929, 358). In this way, Bolzano, Přihonský and Dubislav avoid the paradox of criticizing a *Critique*. By reconstructing Kant's doctrines clearly and concisely, they present their own philosophical position. For example, Kant's account of how cognitions are produced is appropriately expressed by Bolzano's statement that "the possibility of cognizing *(Erkennbarkeit)* an object is the possibility of pronouncing a true judgment on it." (Bolzano 1837, § 26.4, my translation—A. K).

Dubislav (1929, 363–365) expounds Bolzano's critical examination of Kant's distinctions between analytic and synthetic judgments and Kant's notion of analyticity. I analyse Dubislav's reconstruction of Bolzano's Kant-criticism and Bolzano's own view, since Dubislav (1929, 1930a, b, 1931a, b, c, d) pays special attention to Bolzano's notion of analyticity. According to Kant, a judgment is analytic if its predicate-concept is (covertly) contained in or by its subject-concept. His well-known example of an analytic judgment is "all bodies are extended" where the predicate-concept does not add anything to the subject-concept that is not already contained in it (Kant 1789, B10–11, JL, § 36). On this view, analytic judgments are affirmative judgments and have the subject-copula-predicate form of categorical judgments: "All *A* are *B*". In addition, analytic judgments are epistemologically warranted by the principle of contradiction: the truth of an analytic judgment must be cognizable in accordance with the principle of contradiction (Kant 1789, B190).

Dubislav (1929, 364) remarks with Bolzano that this definition excludes hypothetical judgments which, according to Kant's table of judgment-forms, do not have the subject-copula-predicate form (Kant 1789, B95) and that Kant's analytic judgments are trivial and affirmative judgments only. When advancing the principle of contradiction as an epistemological warrant for analytic judgments, however, Kant also considers negative analytic judgments such as: "no unlearned person is learned" (Kant 1789, B192). If Bolzano (and Dubislav) omit Kant's negative analytic judgments, it is probably due to the fact that these latter do not modify the Kantian notion of analytic judgments. The predicate-concept of negative analytic judgments covertly or implicitly includes a partial negation of the subject-concept: "is learned" partially negates "no unlearned person". But this negation does not advance our knowledge of the subject-concept 'no unlearned person', since the negation is contained by the subject-concept.[2] Dubislav discusses Bolzano- Přihonský's objections that Kant's definition of analytic judgments is (i) too wide because it also applies to judgments such as: "the father of Alexander, King of Macedonia, was King of Macedonia" and "a triangle similar to an isosceles triangle is itself isosceles" which have the form "an *A* which has *B*

[2] *Pace* Y. Bar-Hillel (1950, 97), who notes Dubislav's (1926) "return to Bolzano's proposal" to accept analytically false as well as analytically true statements. Bar Hillel writes "[i]t is well known that a term corresponding to Bolzano's 'analytically false' lacked in Kant's terminology, that therefore Kant's classification of propositions into analytic and synthetic ones was by no means exhaustive." While Kant's classification may not be exhaustive, this is not because he did not accept analytically false statements but rather because he lacked Bolzano's innovative notion of statements with a variable component (Bar-Hillel 1950, 97).

is *A*". The predicate-concept merely repeats the subject-concept without clarifying the meaning of 'father of Alexander King of Macedonia' or 'triangle similar to isosceles triangle'. On the other hand, however, if the predicate-concept is contained as an essential mark in the subject-component, the definition of analytic judgments is (ii) too narrow because it excludes judgments such as "every object is either *B* or non-*B*" (1837 § 148; NAK, 34–35).[3]

10.2.1 Dubislav on Bolzano's Notion of Analyticity

Having examined Bolzano's criticism of Kant's analytic judgments, Dubislav (1926, 1929, 1930a, b, 1931a, b, c, d) reconstructs Bolzano's own notion of analyticity which is based on the method of variation of presentations (*Vorstellungen*) in a proposition (*Satz*) and the notion of validity (1837, §§ 147, 148.1). According to Bolzano, a declarative statement is analytic if and only if it contains at least one presentation which can be arbitrarily varied without disturbing its truth or falsity. In addition, a declarative statement is analytic if and only if all the statements which could be obtained by the arbitrary variation of this presentation, are either all true or all false, provided only their subject-presentation is objectual (*gegenständlich*) (1837, § 147). In other words, analytic statements produced by the process of variation must (i) be either all true or all false and (ii) have the same truth value as the original proposition. In addition (iii) the process of variation must produce an objectual (*gegenständliche*) statement, that is, a declarative statement which has an actual or possible referent. Hence the subject-presentation, as a component of the statement, has to have a referential relation to its object, regardless of whether that object actually exists. (1837, § 137). Dubislav (1931c, 224) accepts Bolzano's claim that the objectuality constraint holds for general assertions because, on his view, an assertion such as "all triangles have three angles" is applicable if and only if the respective presentation is objectual. He adds that although in mathematical logic this constraint is obsolete, Bolzano's claim is equivalent to the logicist interpretation: "it is the case for all *x*: '*x* is a triangle' implies '*x* has three angles'", and the statement in quotation marks is true if the sentential function '*x* is a triangle' is "always false".

Bolzano's explication of analytic statements (as well as his notion of probability) is also based on the notion of validity (*Gültigkeit*) (1837, § 148.1). Although

[3]Quine, in "*Two Dogmas of Empiricism*", has similar objections against the Kantian notion of analytic judgments (Quine 1951, 21). Cf. Edgar Morscher (2003).

N.B. Kant could rebut (ii) because, in addition, he accepts judgments as analytic if they rest on the principle of contradiction. He gives the following examples of analytic geometrical principles: "a = a", "the whole is equal to itself", "(a + b) > a, i.e., the whole is greater than its part" (Kant 1789, B 16). But discussing the relevance of Bolzano- Přihonský's Kant-criticism goes beyond the scope of this chapter. Let it suffice to say with Dubislav that via Kant, Bolzano worked his way into crucial problems of philosophy and discovered solutions which anticipate views in modern logic and philosophy.

Dubislav discusses validity in relation to probability and derivability, he does not mention it with regard to analytic statements.[4] Bolzano says that a declarative statement *P* is analytic if and only if it contains a replaceable presentation *R* which can be arbitrarily varied without disturbing the truth or falsity of *P*. The resulting variants of *P* are either all true or all false, so that *P* is universally valid or universally contra-valid in regard to *R*. In § 147, Bolzano introduces the notion of universal validity of declarative statements as such as follows: a proposition as such is universally valid in regard to the collection of true statements *T* if all *T*-variants of *P* are true. *P* has a degree of validity (*Grad der Gültigkeit*) which is defined as the ratio of the number of true variants to the total of variants. If all the variants are true, *P* is universally valid and its validity is 1. If all the variants are false, *P* is universally contra-valid and its validity is 0. If some variants are true and some are false, P's degree of validity is a fraction between 0 and 1. Bolzano's validity is a relative notion, since the degree of validity of a given statement is always relative to a given variable component, so that one and the same statement can have different validities.

In addition, Bolzano distinguishes between analytic and logically analytic statements by relating the notion of universal validity to his notion of analyticity when he defines logically analytic statements as logically and universally valid or logically and universally contra-valid. (1837, § 148.3). The difference between analytic and logically analytic statements is that the invariable presentations of the latter are logical concepts such as the copula 'is' or 'has', the concept of negation or the concept 'something'. Bolzano uses an epistemological criterion for distinguishing between analytical and logically analytical statements: he says that for assessing the analytical nature of this sub-species of analytical statements, only logical knowledge is necessary, whereas for assessing the truth or falsity of analytical statements in the wider sense, a completely different kind of knowledge is required, since extra-logical concepts are brought in. Moreover, Bolzano admits that the distinction between analytic and logically analytic statements is rather unstable, for the domain of concepts belonging to logic is not so sharply delimited that disputes could never arise.[5]

According to some commentators, Bolzano's analytic statements are propositional forms: if a sequence of presentations in a declarative statement is replaced by another sequence, by means of such a uniform variation of presentations in

---

[4]See on this Jan Berg (Berg 1999, 122–124).

[5]"[Logically analytic statements] differ from [analytic statements in the wider sense] in that for an assessment of the analytic nature of the former, only logical knowledge is necessary because the concepts which form the invariable part of those statements all belong to logic. The assessment of the truth and falsity of propositions of the former, however, require a wholly different kind of knowledge, since concepts alien to logic intrude. This distinction is admittedly unstable, for the domain of concepts belonging to logic is not that sharply delimited so that some controversy is inevitable." (1837, § 148.3, my translation—A. K.). Bolzano's distinction between analytic statements and logically analytic statements is famously discussed by Bar-Hillel (1950), Berg (1999), and Künne (2008).

a statement, a propositional form or invariant is produced which is analytic if and only if all its component presentations are either true or false and if and only if it remains true or false, despite the changes produced in some of those component presentations. (1837, § 148, NAK, p.36, Bar-Hillel 1952, 67; Morscher 2003, XLV).[6] Bolzano uses the notion "propositional form", or "sentential form", depending on our interpretation of "*Satzform*"—according to him, all propositions have the uniform subject-copula-predicate structure "*A* has *b*" (1837, § 127). Following Russellian usage, *Satz* is usually translated as *proposition* and, following Morscher, this translation is applied to Bolzano, based on Bolzano's claim that "*Sätze an sich*" are not linguistic expressions but the sense of those expressions (1837, § 19) which is independent of the mental or linguistic acts in which it is expressed. *Sätze* on the other hand, are declarative statements and that is the translation used in this chapter. Furthermore, in recent years (Textor 1997; Berg 1999; Sebestik 2007), Bolzano's "*Satzform*" has been considered as sentential, rather than propositional because it is an expression which becomes a sentence and it is obtained by considering some parts of a sentence variable. Propositions are either true or false and do not have variable parts because they are the sense expressed by a sentence or sentential form. The latter is indeterminate.

Bolzano does not have Frege's two-place casting mould of 'function-argument', in which expressions fit together as 'complete' and 'incomplete' and hence he also lacks the Fregean notion of a function is an incomplete expression which takes a number of names as arguments and produces one proposition as the value. Nonetheless, Bolzano did not explicate statements as connections of presentations related by the copula but assigned primacy to statements over presentations by suggesting that we do not use the concept of presentation for defining the concept of a statement, since presentations are merely those parts of a statement which are not themselves statements (1837, § 128). In this sense, Bolzano anticipated Frege.

There is no consensus amongst commentators about whether Bolzano's standard form of statements "*A* has *b*" is a propositional form or a sentential form (*Satzform*), let alone an abstraction of propositional forms or propositional function $A(b)$ as Dubislav claims. Some commentators (Berg 1999; Sebestik 2007; Textor 1997) hold (*contra* Dubislav 1929, 1930a, b, 1931b, c; Morscher 1999b, 2003; Siebel 1996, 1999) that the variable presentations are parts of declarative statements expressing a *Satzform* (sentential form). Thus "Caius is a man" and "Caius has wisdom" express the sentential form "*A* has *b*". On a semantic level, sentences express propositions (*Sätze an sich*), that is, linguistic senses. Bolzanian statements as such (*Sätze an sich*) are roughly equivalent to Fregean thoughts: they are linguistic senses or

[6]"But suppose a statement contains just a single presentation which could be arbitrarily varied without disturbing the truth or falsity of the statement; i.e. if all statements obtainable from it by arbitrarily substituting this presentation by others, are either all be true or all false, provided only they have objectuality (*Gegenständlichkeit*). This property of the statement is already sufficiently remarkable to differentiate it from all those statements for which this is not the case. Hence I allow myself to call statements of this kind *analytic*, borrowing an expression from Kant." (1837, § 148.1, my translation—A. K.).

possible contents of a sentence, expressible or thinkable, which are either true or false.[7] The standard form of sentences (*Satzform*) [*A* has *b*] is also either true or false (1837, §§ 19, 28, 126, 127) but, unlike Bolzanian statements as such, it is a linguistic expression obtained by considering parts of a sentence variable (Sebestik 2007; Textor 1997).

By contrast, Dubislav's colleague Heinrich Scholz (1931c, 208–210) suggests the notion of *perfect assertive-form* (*perfekte Aussageform*), or a form in which *all* components we explicate as variable are replaced by the appropriate signs. On Scholz's reading of Bolzano's § 148, a statement is analytic, if and only if it can be obtained through replacement *(Einsetzung)* from a perfect statement–form which always turns out either true or false. Scholz's distinction between a sentential form and a perfect sentential form corresponds to Bolzano's distinction between analytic and logically analytic statements.

10.3 Dubislav on Bolzano as a Precursor of Modern Formal Logic

Dubislav argues that Bolzano anticipated the thinkers of his time and that is why there was no fruitful interaction between his theories and theirs, because he was misunderstood or ignored by his contemporaries.[8] Dubislav attempts to bring Bolzano into the contemporary discussion by reformulating his discoveries in contemporary terms, so as to show the Prague philosopher's relevance to contemporary views, at the risk of misrepresenting the latter's claims. Thus Yehoshoua Bar-Hillel (1952, 337–338) acknowledges Dubislav's evaluation of Bolzano's contributions to logic whilst rejecting Dubislav's (1931b) claim that Bolzano anticipated modern *mathematical* logic.[9] Bar Hillel also rejects the claim

[7] Bolzano holds that if a proposition is true, it expresses the sense of a certain combination of words. Omnipotence can be predicated of God if and only if the subject "God" actually has this property, otherwise the proposition is false and has no sense (1837, § 28).

[8] If Bolzano was little known at his time, the main reason was political rather than scientific: a Roman Catholic priest and professor of theology, Bolzano was removed from his post at the German University of Prague in 1820, after nearly being excommunicated for criticizing the official theological manual. The mathematical discoveries of the young Bolzano, such as the 1917 theorem that, given any bounded sequence (a_n) of real numbers, there exists a convergent sub-sequence (a_{nj}) which was later called the "Bolzano–Weierstraß theorem", remained unnoticed until it was independently re-discovered by Weierstraß 50 years later. Dubislav (1931d, 344, 1931e) briefly mentions Bolzano's contributions to mathematics. Bolzano's logical and philosophical teachings were, however, propagated in the Danube Monarchy by his students R. Zimmermann and F. Přihonský and influenced philosophers such as Husserl and Meinong.

[9] "The expression 'mathematical logic' is not free of ambiguities, but if its component 'mathematical' is not to be devoid of any literal value, then we cannot assent to Dubislav when he calls Bolzano "a forerunner of *mathematical* logic". There seem to be among German logicians a certainly understandable tendency to praise Bolzano beyond his certainly great merits. Even if

of Scholz–Dubislav (Dubislav 1931c) that Bolzano anticipated modern semantic logic. Borrowing Carnap's terminology he argues that Bolzanian concepts such as statement as such or proposition (*Satz an sich*), presentation (*Vorstellung*) and variable presentation (*veränderliche Vorstellung*) "belong to the *non-semiotical* part of the meta-language of the object language dealt with, which was colloquial German for Bolzano and is ordinary English with us. They do not belong to the *semantical* part of the metalanguage, and in their definition no mention is made of any semantical concepts such as 'designate', 'express', etc." (1952, 324). *Pace* Bar-Hillel, Bolzano does not distinguish between meta-language and object-language as Carnap and Tarski did, though he did make semantic innovations, for example by introducing the distinctions between subjective and objective presentations and subjective and objective statements—the latter being the significations of mental or linguistic expressions. In addition, Bolzano was attentive to the role of signs and their signification which are key semantic and semiotic notions. He also introduced the notion of explication (*Verständigung*), a statement communicating the meaning or signification an interlocutor relates to a certain sign (1837, §§ 285, 668).[10] Nonetheless, Bar-Hillel has a point in saying that these innovations do not belong to semantic *logic*—or to formal semantics.

Although Bolzano did not formalize his theory, modern and contemporary logicians who are his commentators, did.[11] Bolzano's contribution to modern formal logic is his method of variation, his notion of analytic statements and what Dubislav (1929, 1930a, b, 1931b, d) and some other commentators (Bar Hillel 1952) call propositional functions. Whether or not this name is appropriate for Bolzano's analytic statements is discussed below. Another Bolzanian innovation is his notion of derivability (*Ableitbarkeit*) which is a precursor of the modern notion of logical consequence—or, as Dubislav (1930a, b, 1931b, d) holds, the notion of implication—as well as the notion of grounding (*Abfolge*) which contributes to the logic of explanation and to methods of deductive knowledge.

10.3.1 Dubislav on Bolzanian Propositional Functions

Bolzano's notion of analytic statements is arguably a predecessor of propositional functions in the sense that their analyticity depends on their containing at least one presentation which may be arbitrarily varied to produce either true or false variants of the original proposition. But, *pace* Dubislav, Bolzano's equivalent to propositional functions is not so much the notion of propositional forms but rather

---

he did not anticipate either semantics or mathematics, he did investigate topics far beyond his own time and created foundations for many disciplines of actual value." (1952, 337–338).

[10]See on this Kasabova (2006).

[11]Bar Hillel (1950, 1952), Corcoran (1993), Etchemendy (1999), Künne (2006, 2008), Dubislav (1931c), Siebel (1996, 1999, 2002), and Tatzel (2002).

the variation of presentations, even though the "*veränderliche Vorstellung*" is not, strictly speaking, a variable quantity which may assume any one of a set of values. "Variable" translates *veränderlich*, though neither Bolzano nor Přihonský used the word "variable" in its contemporary mathematical sense. In their usage, a *variable* is not a letter but refers to a constant which can be replaced and produce new statements.

Dubislav repeatedly praises Bolzano's "uncovery of those judgment-forms containing variables" ("*Aufdeckung von derartigen Variable enthaltenden Urteilsformen*") and the "so-called assertive or propositional functions" ("*sogenannte Aussage- oder Satzfunktionen*") (1929, 365, 1930a, 408, 1930b, 265, 1931b, 449–450, 1931c, 206, 1931d, 341, 1932e).

"By uncovering those judgment-forms containing variables, Bolzano made one of the deepest discoveries in the domain of elementary logic. These formations (*Gebilde*), which Bolzano designated as statements with variable presentations, are called propositional functions. These formations are such that, if the variables contained in them are replaced by their values according to a rule of substitution (*Substitutionsvorschrift*), one obtains statements in the usual sense of the word. So we can designate those assertive or propositional functions as casting moulds for sentences."[12] (1929, 365, my translation—A. K.)

By way of criticizing Kant's philosophical claims and reconstructing his views using a rigorous method of inquiry, Bolzano discovered important logical and philosophical principles. According to Dubislav, Bolzano's discovery of propositional forms is one of the reasons why (1929, 1931b) Bolzano is not only 'Kant's critic' but also 'a precursor of mathematical logic'. In words that Dubislav borrowed from the French logician and mathematician, Louis Couturat (1905), sentential or, as Russell says, propositional forms, are casting moulds of linguistic expressions.[13] Dubislav (1931b, 450–451, 1931c, 341) considers Bolzano's use of the method of variation of the component-parts of propositions a "classical discovery" of the "so-called assertive or propositional function": "He characterizes an assertive function as follows, which we render in the terminology used today: an assertive function is

[12]"Mit der Aufdeckung von derartigen Variable enthaltenden Urteilsformen hat nun Bolzano eine der tiefsten Entdeckungen auf dem Gebiete der elementaren Logik gemacht. Man nennt diese Gebilde, die Bolzano selbst als Sätze mit veränderlichen Vorstellungen bezeichnet hat, Satzfunktionen. Es sind also Gebilde so beschaffen, daß, wenn man die in ihnen enthaltenden Variablen nach einer Substitutionsvorschrift durch Werte derselben ersetzt, Sätze im üblichen Sinne des Wortes resultieren. Man kann also anschaulich derartige Satzfunktionen mit L. Couturat als Gießformen für Sätze bezeichnen."

[13]Unfortunately Dubislav gives no reference for Louis Couturat. In 1905 Couturat published *Les Principes des Mathematiques: avec un appendice sur la philosophie des mathématiques de Kant, L'Algèbre de la logique, Les définitions mathématiques* and *Définitions et démonstrations mathématiques*. The last two works are cited in the bibliographie of *Die Definition* (1931a), as well as the German translation of Les principes des mathématiques (1908). It is likely that Dubislav refers to Couturat (1905) when citing the expression "Gießformen".

a formation containing one or several place-holders such that a proposition (*Satz*) results, if a place-holder is filled in according to a rule of replacement." (Dubislav 1931b, 450–451, my translation—A. K.).[14]

In order to emphasize Bolzano's anticipation of modern and mathematical logic, Dubislav's reconstruction sometimes departs from the former's original account. Thus Bolzano does not use concepts such as "*Leerstelle*", ("place-holder") or "*Einsatzvorschrift*" ("*rule of substitution*") but "variable presentations" which are replaceable by other presentations (1837, 147). In footnotes to (1929, 449, 1931c, 340), Dubislav remarks that Bolzano designates what he himself calls propositional functions as (declarative) statements with variable presentations (*Sätze mit veränderlichen Vorstellungen*), a hint that commentators inevitably construct their own views by reconstructing the theory of an earlier author. Sure enough, Dubislav's (1931a, 116) notion of determination of concepts is very close to the 'propositional function' he imputes to Bolzano. This is interesting for two reasons:

1. Dubislav's notion of concept corresponds to his characterization of Bolzano's variable presentations: "Concepts in a logical sense are merely signs of a particular kind, namely signs in the shape of sentential- or (as they are also called) propositional functions of a variable. Such a propositional function [. . .] is taken as [. . .] a casting mould for statements [. . .]. A propositional function of a variable is produced if, in a statement, a sign is substituted by a variable [. . .]." (1931a, 116, my translation—A. K.).[15] The mathematical logician Kurt Grelling (1932, 197), an active member of the Berlin Group until 1937, objects against Dubislav that, on the latter's view, propositional functions are signs but Dubislav does not explain what they designate, i.e. what they mean or signify (*bedeuten*). Could Dubislav's propositional functions be the meanings or significations (*Bedeutungen*) of concepts? Apparently not, because Dubislav (ibid) also claims that a propositional function represents or stands for the concept "prime number". Hence propositional functions are signs and concepts are meanings or significations of signs.
2. Dubislav's (1931a, 116–117) view that concepts are meanings of signs is very close to Bolzano's account of *Vorstellungen an sich* (which may or may not be arbitrarily replaceable in a declarative statement).[16] Bolzano distinguishes between subjective presentations which are mental or linguistic acts and

[14]"Eine Aussagefunktion charakterisiert er folgendermaßen, wobei wir die heute übliche Terminologie benutzen : eine Aussagefunktion ist ein Gebilde, welches ein oder mehrere Leerstellen dergestalt enthält, daß, wenn man die Leerstelle nach Maßgabe einer Einsatzungsvorschrift ausfüllt, eine Aussage resultiert".

[15]"Begriffe im Sinne der Logik sind lediglich Zeichen besonderer Art. Und zwar Zeichen in Gestalt von Aussage- oder, wie man sie auch genannt hat, Satzfunktionen einer Variablen. Unter einer derartigen Aussagefunktion [. . .] versteht man [. . .] eine Gießform für Aussagen [....] Eine Aussagefunktion einer Variablen resultiert, wenn man sich innerhalb einer Aussage ein Zeichen durch eine Variable [. . .] ersetzt denkt".

[16]"Vorstellung [. . .] welche sich willkürlich abändern läßt" (1837, § 148.1)

objective presentations or presentations as such which are constituent parts of declarative statements as such or propositions (1837, § 50). He characterizes objective presentations (*Vorstellungen an sich*) as significations or meanings (*Bedeutungen*) of signs which are designated by signs (ibid., § 285). Dubislav (1931a, 117) comments: "If one were to strip Bolzano's explications of their mystical character, that which he wanted to be understood as a presentation as such can be determined as a propositional function of a variable which has precisely those attributes which Bolzano ascribed to his presentations as such." (my translation—A. K.).[17]

Hence Dubislav's view that concepts are significations or meanings of signs is Bolzanian but is it formalist? In *Die Definition* (1931a, 113) Dubislav classifies Bolzano's account of determination of concepts (*Begriffsbestimmung*) as idealist, distinguishing Bolzano's notion from his own which he calls formalist. Is Dubislav's view formalist? Grelling (1932, 198) says 'no'. On Dubislav's view, it seems that for a formalist, concepts are signs without meaning or signification and thus Grelling (1932) objects that Dubislav has not explained what a concept designates. If Dubislav holds that it designates a class, as he seems to (1931a, 116), what are the criteria for designating an object as belonging to one class rather than another? Grelling concludes his objection against Dubislav's 'formalist' definition of concepts with the deadly question which level of existence he would ascribe to a class?

"Physical reality can hardly be ascribed to [a class]—it is not something one can meet in the woods [. . .]. Then again one can hardly express it as a presentation so, if at all, one would have to ascribe ideal existence to it, which [. . .] amounts to having jumped out of the frying pan into the fire, or [. . .] having cast out the devil with Beelzebub." (1932, 198, my translation—A. K.).[18]

10.3.2 Dubislav on Bolzano's Notions of Derivability and Probability

Bolzano's notion of derivability (*Ableitbarkeit*) is characterized by a compatibility constraint (*Verträglichkeit*) and a substitutional criterion. Bolzano uses the method of variation—the idea that components of declarative statements can be varied or

[17]"Wenn man die Erläuterungen, die Bolzano für das, was er unter einer Vorstellung an sich verstanden wissen wollte, ihres mystischen Charakters entkleidet, dann ist festzustellen, daß eine Aussagefunktion einer Variablen im obigen Sinne gerade diejenigen Beschaffenheiten besitzt, die Bolzano seinen Vorstellungen an sich zuschrieb."

[18]"Physische Wirklichkeit kann man ihr nicht gut zuschreiben, man kann ihr nicht im Walde begegnen [. . .]. Da man sie auch nicht gut als Vorstellung aussprechen kann, so muß man ihr, wenn überhaupt, eine ideale Existenz zuschreiben, womit man also vom Regen in die Traufe gekommen ist, oder [. . .] den Teufel mit Beelzebub ausgetrieben hat."

substituted—in his account of derivability (*Ableitbarkeit*). Some statements P_1 to P_n are derivable from other statements Q_1 to Q_n with respect to common components i_1 to i_n if and only if all substitutions of presentations i_1 to i_n which produce only true statements in P_1 to P_n also produce only true statements in Q_1 to Q_n and the propositions are *compatible* relative to their variable components i_1 to i_n if at least one substitution for i_1 to i_n produces only true statements in P_1 to P_n (WLII, §§ 154, 155.2).

Dubislav (1930a, b, 1931d), points out that derivability is a special kind of compatibility (*Art der Verträglichkeit* 1931b, 451) and corresponds to the notion of formal implication in mathematical logic. Having asserted that Bolzanian derivability is a precursor of formal implication, Dubislav remarks that in the latter accounts the compatibility constraint is left out (Dubislav 1931b, 452). Berg (1999) and Siebel (1996, 1999) concur that Bolzano's account of derivability is not easily reconciled with modern logic. The reasons are twofold: first compatibility is a three-place relation between statements as such on one hand, and presentations, on the other, since two statements as such (or two classes of statements as such) are compatible with regard to their variable components (the presentations). (1837, § 154) Second, derivability is not a relation between linguistic signs but the semantic content or sense of linguistic signs.

Derivability, as Dubislav (1931b) explains, is a relation of implication which relates to the validity (*Gültigkeit*) of formal or logical implication as well as material or relative implication, for Bolzano examines the process of deduction leading from premises to conclusions and the validity of implicational statements or inferences. Thus Q is deducible from P in a step-by-step deduction showing that Q is true if P is true. Dubislav (1931b) also notes that Bolzano uses two kinds of derivability (formal and material). Bolzano's distinction between derivability in a broad and narrow sense is similar to his distinction between analytical statements and logically analytic statements. As Dubislav puts it, Bolzano "realized that there are two kinds of derivability relations: first, those which mere logical knowledge suffices to determine. Second, there are those relations which can only be determined by means of extra-logical knowledge".[19] (1931b, 452–453, my translation—A. K.)

Bolzano's derivability in a broad sense (material implication) holds for conditionals related by the '*if* ... *then*' conjunct where 'implies' relates parts of a sentence to make a more complex sentence. Bolzano considers such implications as conditionals "in the broad sense of derivability" which require knowledge outside the domain of logic (1837, § 223). He gives the following example: Caius is a man

[19]"Bolzano erkannte nämlich, daß es zwei Arten von Ableitsbeziehungen gibt. Erstens solche, zu deren Feststellung man lediglich logischer Kenntnisse bedarf, und zweitens solche, zu deren Feststellung außerlogische Kenntnisse herangezogen werden müssen." Apparently Bar-Hillel (1952) did not read this part of Dubislav's reconstruction of Bolzano because he claims that Bolzano "does not distinguish, strangely enough, between material and formal derivability, but he does so, for instance, with respect to a closely related concept, that of consequence (*Abfolge*)." (Bar-Hillel 1952, 86). *Pace* Bar-Hillel, Bolzano's logical derivability is close to the modern notion of *consequence*, whereas *Abfolge* is *grounding* (or ground-consequence).

implies Caius has an immortal soul, where 'implies' is relative to Caius. To accept or understand (*einsehen*) this, "we must know that all human souls are immortal". But in order to know that the implication is correct (*richtig*), it suffices to recognize it as an instance of the inference scheme 'for every *x*, if *x* is a man, then *x* has an immortal soul'. Cases where 'implies' denotes a material implication, that is, if the '*A* implies *B*' means that *A* is false or *B* is true are problematic because of counter-intuitive results—the so-called paradoxes of material implication: either the whole conditional is true whenever the antecedent is false or the whole conditional is true whenever the consequent is true.

Later commentators (Corcoran 1993; Etchemendy 1999) hold that Bolzano's notion of derivability is a (primitive) precursor of Tarski's (1936) notion of logical consequence. Bolzano's notion of logical derivability (derivability in a narrow sense) is close to Tarski's logical consequence, since Bolzano says that in cases of derivability such as 'A implies B or A implies not-B' all except the logical presentations have to be varied. Logical derivability is a relation using the *if ... then* construction denoted by the verb '*implies*' which can relate either sentence-schemas, as in 'B is a bachelor, implies B is unmarried' or parts of a sentence, as in conditional clauses.

But the analogy between Bolzano and Tarski is limited (Siebel 1996, 1999). Tarski's notion of logical consequence does not have a compatibility constraint, nor does it hold between the contents of statements. In addition, it concerns a meta-linguistic framework and an interpreted language and turns on truth-conditions or satisfaction conditions of propositional functions. A propositional function *Fd* is satisfied if and only if all properties of *F* are satisfied by a domain or set of individuals *d* which is defined by the properties of *F*. For Tarski, "[t]he sentence *X* *follows logically* from the sentences of the class *K* if and only if every model of the class *K* is also a model of the sentence *X*" (1936, 417). For Tarski, the replaceable elements are the objects falling under the variables of the non-logical constants, whereas for Bolzano, the replaceable elements are non-logical presentations which are parts of statements as such.

Dubislav does not mention Bolzano's notion of grounding (*Abfolge*): a statement that *p* is true because *q*. The grounding relation provides the semantic conditions for a deduction (*Ableitung, Herleitung*). These conditions are designated by the conjunct '*because*' which denotes the grounding relation, a very peculiar relation, by virtue of which some terms act as grounds to others (Bolzano 1837, § 162).[20] The explanatory force of the grounding relation lies in "drawing out the elements of an implicit deduction", by means of which we "obtain the key to new truths which were not clear to common sense".[21] Unlike derivability, the grounding relation holds

[20]"Ein sehr merkwürdiges Verhältnis, vermöge dessen sich einige derselben zu andern als Gründe zu ihren Folgen verhalten." Bolzano 1837, § 162; § 221.note: "der Begriff einer solchen Anordnung unter den Wahrheiten, vermöge deren sich aus der geringsten Anzahl einfacher Vordersätze die möglich größte Anzahl der übrigen Wahrheiten als bloßer Schlußsätze ableiten lasse".

[21]1810, *Beyträge* II, § 2; 1837, § 401.

only between true sentences of the form: '*p* because *q*' which are compatible as ground and consequence and its terms are either single sentences or collections of sentences.

Instead, Dubislav (1930a, 409, 1930b, 264–265, 1931d, 343) relates Bolzano's notion of derivability to the latter's account of probability (*Wahrscheinlichkeit*). He considers Bolzano's probability as containing (*enthalten*) a derivability relation.[22] Dubislav's reconstruction is based on the premise that Bolzano's notions of derivability and probability are characterized by the compatibility constraint. A derivability relation is a probability relation with the numerical value P =1 and the numerical values of this relation lie in the interval 0,1 (1930b, 264). More recently, Jan Sebestik (2007, 38–39) takes up Dubislav's point, adding that it is Bolzano's compatibility constraint which enables "the extension of deductive logic to inductive logic via probability." Sebestik praises what he calls Bolzano's extraordinary achievement in providing "the first logical definition of probability. For the first time deductive logic and inductive logic are united in a global theory and the former appears as a limit case of the latter." (ibid., 38–39).

Bolzano was hardly the first mathematician and logician to deal with probability—he Bolzano agrees with Laplace that probability is a relation in which the number of propitious (*günstige*) cases stands to the number of possible cases (1837, § 161, note 2). His innovations are (1) that he provided a systematized account of probability as a property (*Beschaffenheit*) of statements and (2) that he introduced the distinction between objective probability or the ratio of the number of true variants to the number of (collection of) statements and subjective probability or degrees of confidence and credibility. A statement *M* has objective probability $P = 1$ or certainty if *M* is derivable from a collection of statements *C* relative to variable presentations *i*. If *M* is not derivable from *C* relative to variable presentations *i*, it has the probability $P = 0$, that is, *M* and *C* are incompatible. Thus certainty and incompatibility are the limits of probability with the values 1 and 0. In addition, Bolzano introduces conditional probability: a statement *M* is conditionally probable if its probability is $0 < P < 1$ (Bolzano 1837, § 161.1). Dubislav leaves out Bolzano's account of subjective probability: the degree of confidence with which we judge that *p* (or take *p* to be true) is Bolzano's tool for determining the limits between a cognition (*Erkenntnis*) and an error (*Irrtum*) (Bolzano 1837, § 317).

---

[22]Twice Dubislav (1930a, 409, 1931d, 343), Dubislav cites the following passage in the *Wissenschaftslehre* (1837, § 161.1): "Let us consider certain presentations i, j ... in a single proposition A or in several propositions A,B,C,D, ... as variable, and in the latter case suppose that propositions A,B,C,D are in a relation of compatibility in regard to these presentations. Then it will often be particularly important to know the relation of the collection of cases in which propositions A,B,C,D ... all become true, stands to the collection of those cases in which an additional proposition M becomes true, and whether we should also take M to be true or not. For if the latter collection comes to half of the former, we can hold M to be true merely on account of the truth of propositions A,B,C,D ... and if this is not the case, then we cannot. So I permit myself to call this relation between said collections the *relative validity* of proposition M in regard to propositions A,B,C,D, or the *probability* proposition M attains from the *presuppositions* A,B,C,D." (my translation—A. K.).

10.4 Dubislav and Bolzano on Definition

Dubislav (1931a) distinguishes between the following accounts of definition (1) determination of essence or *definitio rei* (*Wesensbestimmung, Sacherklärung*); (2) determination of concepts or *definitio nominis* (*Begriffsbestimmung*); (3) setting conditions for the meaning or usage of a sign or *definitio lexicalis* and (4) stipulating the meaning of a new sign or the new usage of a familiar sign or *definitio stipulationis*. In addition, he expounds the concept of definition, determines the relation between *definiens* and *definiendum* and examines rules for use of concepts in definitions. He is particularly interested in (2) or the definability of concepts (*Begriffsbestimmung*).

As I mentioned in Sect. 10.2.1, Dubislav (1931a, 116) considers concepts as meanings of signs and signs as sentential forms of variables (*Aussagefunktionen von Variablen*).[23] However, Dubislav (1931a, 1932) does not rebut Grelling's (1932, 194) objection that he fails to distinguish between two kinds of dependence relations: (i) the relation between assumptions in a system of assumptions and (ii) the relation between the sense of statements and the definition of the signs occurring in them. For Dubislav (1931a, 117) claims that the *sense* of statements depends on the propositional functions which depend on a system of assumptions (*System der Voraussetzungen*)—that is, he aims at (i) whilst dealing with (ii). According to Grelling (1932, 193–194), the *sense* of a statement depends on the definition of the signs occurring in it. Thus the sense of the statement "the events *A* and *B* occurring at places *a* and *b* are simultaneous" can be quite different, depending on how simultaneity is defined, but this sense does not depend on a whole system of assumptions which mutually support each other.

Given Dubislav's investigations of Bolzano's contributions to modern logical theories, it is somewhat surprising that he did not heed the latter's views on definition in (Dubislav 1931a). Despite Dubislav's (1931a, 114–17) abridged reading of Bolzano as a Platonist idealist (which is wrong, because Bolzano did not postulate a 'third realm' of mind-independent entities), he does, as he puts it, extract the notion of propositional function from the latter's presentations as such (ibid., 117). Dubislav's (1931a) take on Bolzano appears slightly confusing: first, he classifies the latter's account of definition in (2) as an idealist determination of concepts (ibid., 117) and then he considers it as a case of (1)or determination of essences in the Aristotelian tradition (ibid., 133–134) which he subsequently rejects. In order to clarify Dubislav's confusion and to evaluate his objection, I reconstruct Bolzano's account of definition.

[23] In a reply to Grelling (1932) and Dubislav (1932, 203), concedes having tacitly accepted Pascal's characterization of definition that, he now admits, is too narrow because it does not allow for inductive definitions in which newly introduced signs are not eliminable.

10.4.1 Bolzano on Definition

Dubislav's account of Bolzano is actually correct: Bolzano subscribes to (1) but, in different parts of the *Wissenschaftslehre* he also elaborates (2) and (4).[24] On Bolzano's 'Aristotelian' or 'Porphyrian' view (1), a definition says what something is, based on (essential) predication (1837, §§ 127, 128, 136, 137). A thing is defined by predicates ascribing to it its essential properties. Thus a definition answers the question "what is it?" Bolzano agrees with Dubislav that definition as predication presupposes a (linguistic) system in which a presentation is segmented into a subject- and predicate-signification or, as Bolzano says, a subject-presentation (*Subjectvorstellung*) and predicate-presentation (*Prädicatvorstellung*) which he sometimes calls grammatical subject (*Unterlage*) and grammatical predicate (*Aussageteil*). Bolzano claims that declarative statements are reducible to or transformable into a canonical form [*A* has *b*] or [*A* has non-*b*]—a uniform structure which holds for all declarative statements of natural language.[25] In other words, "is *F*" is predicated of a grammatical subject *x* and the predicate either says or does not say what *x* is. In addition, Bolzano stipulates a truth condition for the canonical form: [*A* has *b*] is true if and only the statement that *A* has *b* is objectual (*gegenständlich*), that is, if it asserts of its object (*Gegenstand*) that which actually belongs (*wirklich zukommt*) to it (1837, § 28, 124). Bolzano's canonical form of declarative statements is thus a *definitio rei*, as Dubislav (1931a, 133–134) points out. But Dubislav criticizes Bolzano's version of (1) or the characterization of a thing's essential properties:

> We should ask how Bolzano can ground his claim that the given attribute-presentation *b* belongs to the objects in question by virtue of the mere concept under which we usually grasp them? He is obliged, for this purpose, to refer to his theory of truths and presentations as such [...] [and] to ascertain that such statements are valid not only relative to a system of basic presuppositions assumed as true, but *per se*. As a result, in our view his attempt at grounding becomes untenable. (Dubislav 1931a, 234, my translation—A. K.).[26]

---

[24]Bolzano gives an account of stipulative definition in part 4 of the *Theory of Science* and it is reconstructed in Kasabova (2006).

[25]Bolzano famously claims that all statements in natural language are expressible by a uniform structure: "that the following holds of all propositions in general. The concept of *having* [...] the concept signified by the word *has* occurs in all propositions. Besides this one component two others occur [...] in all propositions connected with each other by a *has* as indicated in the expression *A has b*. One of these components, namely the one indicated by *A*, stands as if it were to present the object dealt with in the proposition and the other, *b*, as if it were to present the *attribute (Beschaffenheit)* the proposition ascribes to that object. Therefore I permit myself to call [...] *A* the *supporting* or subject-presentation; [...] and *b* the *assertive part (Aussagetheil)* or *predicate presentation.*" (1837, § 127, my translation—A. K.). Cf. on this Textor (1997).

[26]"Wie kann aber Bolzano, dass ist zu fragen, seine These begründen, daß die genannte Beschaffenheitsvorstellung *b* den fraglichen Gegenständen vermöge des bloßen Begriffes zukommt, unter dem wir sie aufzufassen pflegen? Er ist genötigt, sich zu diesem Zwecke auf seine Lehre von den Wahrheiten an sich und Vorstellungen an sich zu beziehen [...], zu ermitteln, daß derartige

Pace Dubislav, Bolzano does not characterize a thing's essential properties by virtue of truths and representations as such, but by means of a definition with a grammatical subject-predicate structure—that is, by virtue of a linguistic system which determines the linguistic structure of definitions. This structure is syntactic: the order [subject-presentation—copula-presentation—predicate-presentation] and semantic: the sense of the statement [*A* has *b*] is what is expressible by [*A* has *b*] or what is meant. In other words, the components of a declarative statement are determined by the order of that statement. Dubislav may have been misled by Bolzano's truth condition: "a statement is true if it ascribes to its object something that belongs to it" (1837, § 124). But Bolzano does not merely provide a truth condition for the *definitio rei*; rather, his analysis concerns the structure of *Sätze an sich* and a semantic account of truth as a property of propositions (Textor 1997; Künne 2006).

At first blush it seems that for Bolzano, the essence of a thing is the collection (*Inbegriff*) of all properties derivable or inferrable from the concept (*Begriff*) of that thing (1837, §§ 111, 502) which corresponds to Dubislav's suggestion: Bolzano's *definitio rei* names a presentation such that we may infer from it all essential properties of the correlative object. Bolzano (ibid., § 502), however, revises the explanation given in § 111 and proposes to narrow it down by distinguishing between essential and derived properties of things. The first are necessary and the second are accidental. He would reply to Dubislav that an essential property of things must also be their necessary property and vice versa. For example an essential and necessary property of a triangle is that it is a system of three points. On the other hand, the property that the sum of all angles is equal to two right angles is a derived (*abgeleitete*) property of a triangle which objectively follows from (*abfolgt*) its essential property.[27] This latter property is not an essential property in the narrow sense. Thus the essence (*Grundwesen*) of a thing is the collection (*Inbegriff*) of only those properties yielded by its concept which are not inferrable (*herleiten*) from any other concept of that thing as consequences from a ground.[28] In addition, Bolzano might have asked Dubislav in regard to (1) whether a determination of essence (*Wesensbestimmung*) is equivalent to a *Sacherklärung* and whether a *definitio rei* determines a thing or a presentation of that thing. In the latter case, is the definition a *definitio rei* or a *definitio nominis*?

Aussagen nicht nur relativ zu einem als wahr unterstellten System von Grundvoraussetzungen gelten, sondern schlechthin. Damit wird aber sein Begründungsversuch für uns hinfällig."

[27]"It does not lie as a constituent in the concept of a triangle, but is only a consequence ensuing from this concept (*nur eine aus diesem Begriffe sich ergebende Folgerung*), that a triangle could be equilateral." (1837 § 55.10c, my translation—A. K.).

[28]"In this narrower meaning (*Bedeutung*) one takes the essence (*Wesen*) of a thing, also called the grounding essence (*Grundwesen*) to discern it better, as the collection of only those attributes ensuing from its mere concept, which cannot be objectively derived (*herleiten*) from any other concept of it (i.e. as consequences from their ground, § 198)." (1837, § 502, my translation—A. K.).

10.4.1.1 Bolzano on Nominal Definition

In addition to the traditional *definitio rei*, Bolzano works out an account of *definitio nominis*. He applies the distinction between constitutive (*constitutiven*) and derived (*abgeleiteten*) distinctive features (*Merkmale*) not only to things but also to presentations (1837, § 65.10, 120). He also introduces a crucial distinction between the components (*Bestandteile*) of a presentation (namely a property concept) and the attributes (*Beschaffenheiten, Merkmale*) of an object. Let us reconsider Dubislav's question: "We should ask how Bolzano can ground his claim that the given attribute-presentation *b* belongs to the objects in question by virtue of the mere concept under which we usually grasp them?" (Dubislav 1931a, 134). Dubislav is asking how Bolzano can justify his claim that a property concept belongs to the objects in question by virtue of the concept by which we grasp them.

Bolzano has a ready reply: a property concept is not composed of the features of its object but the properties of an object can be derived or inferred from the concept of that object (the property concept) without being thought as constitutive parts of that concept. "Wie es aber möglich sey, daß ein Gegenstand Theile habe, deren Vorhandenseyn aus unserer Vorstellung gefolgert werden kann, ohne daß ihrer darin gedacht wird; daß lässt sich freilich nicht eher wohl begreifen, bis man den Unterschied, der zwischen *Bestandtheilen* und *Merkmalen* obwaltet, deutlich eingesehen hat." (1837, § 65.8).

Bolzano expounds a general account of definition where non-essential properties and non-essential property-presentations are inferrable from the essential properties of objects. Only the latter are also constituents of presentations. Otherwise he would have to accept the erroneous claim (which he rebuts) that if an object has an infinite number of properties, the concept of that object would have to have an infinite number of constituents.[29] In order to protect this account against a conflation between properties of objects and constituents of presentations, Bolzano therefore has to reject structural isomorphism between objects and presentations.[30] Consequently, the first of Bolzano's conditions for a *definitio nominis* is that (i) there is no structural isomorphism between a presentation and an object: the components of a presentation are not to be confused with the attributes of its object (ibid., §§ 63, 64, 65).

The second condition, also related to the rejection of structural isomorphism, is (ii) the distinction between components or constituents (*Bestandtheile*) of

[29]"In my view it is by no means necessary that a concept ensuing that the object corresponding to it is composed of so and so many parts, should be composed of just as many parts (such as the presentations of those particular parts)" (1837, § 65.7, my translation—A. K.).

[30]A further reason for rejecting structural isomorphism are, as Bolzano points out, cases of complex objectless presentations such as [a regular 10-chiliagon (*Zehntausendeck*)], [round square], [blue yellow] or [golden mountain] which have no corresponding object, as well as objectual presentations comprising relative clauses, such as [a land without mountains] or [a book without copper] in which the attributive concept does not correspond to any property of objects falling under that concept but to properties the object is lacking (1837, §§ 63, 66, 70).

presentations and distinctive features (*Merkmale*) or attributes (*Beschaffenheiten*). An attribute of an object is not a component of the concept under which that object falls. Nor is the collection of properties belonging to and determining an object structurally isomorphic with the collection of the presentations (*Inbegriff*) of these attributes (ibid., § 64). In addition, attributes that necessarily belong to the presentation of an object are not presented as constituents of that concept. For instance, an equilateral triangle necessarily falls under the concept [equiangularity], yet [equiangularity] is not a necessary constituent of the concept [equilaterality]. Equiangularity is a necessary attribute of an equilateral triangle without necessarily being thought or presented in the attribute-concept [equilaterality] defining that object—it is inferrable from the concept [equilaterality] (ibid., § 64). The third condition (iii) for nominal definitions is thus the distinction between constituents or components of presentations on one hand and the presentations of attributes or features of objects, on the other (ibid., § 65.9). As Bolzano explains in the *Paradoxes of the Infinite*, in order to think a collection (*Inbegriff*) it is not necessary to think all the objects composing it (1831, § 14). For instance, I can think of an orchestra without thinking of all its players. In fact, I wouldn't need to think set-theoretically of all its members or mereologically of all its parts, because a collection is defined by what it does—thus the bass viol and the violin are essential and necessary constituents of the collection [orchestra] which I might think of, whilst the other constituents are inferrable.[31] Likewise, equiangularity and equilaterality are attributes of triangles which are inferrable from the concept [triangle] but they are not components of that concept (1837, § 65.10).

Bolzano's *definitio nominis* therefore allows for the inferrability of attributes from the concept of a given object without conflating concepts with their objects. The inferrable or derivable attributes are non-essential properties of the object and therefore they are not constituents of the concept of that object. Essential attributes such as "triangularity", however, are also constituents of the concept [triangle] since they belong to the nominal definition of a triangle. Thus we can interpret Bolzano's claim that "whatever one must necessarily think in order to have really thought a given presentation is also a constituent of the latter" (ibid., § 64.2, my translation—A. K.). Bolzano's account of nominal definition involves a clarification of the notion of intension or content of a presentation and of the relation between intension and extension. Having distinguished between the distinctive features or attributes of an object and the components of attribute-concepts, Bolzano rejects the structural isomorphism of objects and concepts which implies (i) that the content of a presentation is composed of the attributes of an object and (ii) that the content of a presentation is composed of subordinate presentations which stand under it. His clarification of the notion of intension involves criticizing Kant's notion of inclusion which is the latter's criterion for determining analytic judgements, discussed in Sect. 10.2 of this chapter. On Kant's view (shared by the young Bolzano, 1810,

---

[31] *Pace* Kneale and Kneale (1962, 364), "Bolzano seems to be in danger of confusing a whole of parts with a set of members."

§ 17), analytic judgments are those in which the predicate-concept (denoting the genus) is covertly contained in the subject-concept (denoting the species). On this view, analytic judgments function as nominal definitions because they relate *genus proximum* and *differentiam specificam*: the genus "extended" can be extracted from the species "bodies"—on the assumption that the content of a concept is composed of the sum total of partial concepts which are also attributes of the objects falling under that concept.

10.4.1.2 Bolzano and Dubislav on the Canon of Reciprocity

The later Bolzano (1837, § 120) explains that he was able to avoid the mistake of conflating the properties of an object with the components of its presentation and overloading (*überfüllen*) the intension of a concept by considering the content as a composition of its parts, due to Kant's distinction between analytic and synthetic judgments which compelled him to clarify the relation between the intension and extension of a concept. Bolzano (ibid., §§ 65, 120) and Dubislav (1931a, 12) note that Kant supported the so-called canon of reciprocity, namely that the intension and extension of a concept stand in an inverse relation: if the intension of a concept is conceived as a conjunction (*Knüpfoperation*) of attributes and its extension is conceived as the collection of objects comprised by this concept, then the more attributes or properties of objects are contained in its concept, the fewer are comprised (*umfassen*) by it or fall under it.

Dubislav (ibid., 12–13) pays tribute to Bolzano's critique of the erroneous canon of reciprocity,

> according to which extension and intension of a concept stand in a reciprocal relation. Furthermore, this theory of concepts is connected with the claim [...] that the so-called partial presentations of a concept are always also features of the objects falling under that concept [...], a claim Bolzano also proved as incorrect. Hence the confusion of the two states of affairs 'comprised by a concept' and 'falling under a concept', produced the bewildering terminology in which so-called partial presentations of a concept are called features of that concept since, according to the above-mentioned claim, partial components of a concept comprise those features under certain conditions.[32] (Dubislav, ibid., 12–13, my translation—A. K.)

Dubislav comments that Bolzano clarifies the confusion between intensive and extensive relations with concept-concept and concept-object relations: subordination or comprehension (*Umfassung*) are relations of inclusion between concepts,

[32]"wonach Umfang und Inhalt eines Begriffes sich zueinander reziprok verhalten sollen. Ferner wird mit dieser Begriffslehre die [...] ebenfalls von Bolzano als unrichtig erwiesende Behauptung verbunden, daß die sogenannten Teilvorstellungen eines Begriffes immer zugleich auch Merkmale der unter den Begriff fallenden Gegenstände [...] sein sollen. Daraus hat sich dann bei Verwechslung der beiden Sachverhalte "Von einem Begriffe umfaßt worden" und "Unter einen Begriff fallen" die verwirrende Terminologie entwickelt, die sogenannten Teilvorstellungen eines Begriffes Merkmale desselben zu nennen, weil unter der erwähnten Annahme die Teilvorstellungen eines Begriffes u. U. umfassen würden."

whereas subsumption, 'falling under', 'contained under' or 'contained by' is a relation of extension between concepts and objects. He does not mention that Bolzano's clarification is important for the notion of nominal definition.

Bolzano (1837, § 120) rejects both parts of the canon of reciprocity, as follows: (1) 'the intension of a presentation may be increased without increasing its extension.' (i) Consider redundant concepts such as [triangle which has the attribute equilaterality] in which the attribute of equilaterality is an added constituent of the concept [triangle] without increasing its extension. (ii) Consider auxiliary or adjunctive concepts which increase the content of the nominal concept without increasing its extension: the concept [round ball] has a larger content than the concept [ball] but their extension is the same. Bolzano's example is, however, problematic, as well as his rejection of the second part of the canon. (iii) By adding a new constituent to a concept, it is possible to increase its extension by increasing its intension. Bolzano also uses this condition for rejecting (2) 'the extension of a presentation may be increased without increasing its intension.' He gives the following example: the concept [a man who understands all European languages] is increased in extension by adding [living] to its intension. Unfortunately for Bolzano, his example shows the validity of the canon he rejects: the concept [a man who understands all living European languages] has an increased intension but a decreased extension, for [all European languages] are thus limited to the living ones, excluding the dead languages which are included in the former.

Bolzano offers a better argument for (2): (iv) a subordinate concept may be built (*bilden*), increasing the extension of the main concept without adding something to its content, since it is not necessary for a subordinate concept to be partly composed of the concept comprising it. The concept [actual] is not a component of the concept [possible] although [actual] is subordinate to and inferrable from [possible] (ibid., § 65.10).[33] As Dubislav (1931a, 12) says, "subordinated to" does not imply "a part of". Bolzano would add that, precisely for this reason, analyticity is not correctly defined as an inclusion of the predicate-concept in the content of the subject-concept, nor is a concept appropriately defined by decomposing it. Instead, analyticity is based on the method of variation of presentations and a concept is adequately defined (essentially as well as nominally), if we distinguish between its intension and extension.

Bolzano's contribution to the development of formal semantics is that his distinction between the content of a concept (*Bestandteile*) and its range of applicability over the particular objects it denotes (*Merkmale*) prefigures the distinction between intension and extension, the origin of which is officially attributed to Frege's famous distinction between *Sinn* and *Bedeutung* (Frege 1892).[34] In my view, however, (*pace*

[33]Cf. on this Künne (2008, 212–215).

[34]Roman Jakobson (1980) notes Bolzano's distinction between the meaning (*Bedeutung*) of a sign as such and the sense (*Sinn*) that this sign acquires in the context of the present circumstance. Unlike Frege, Bolzano uses *Bedeutung* to denote the presentation of a sign, which is why 'meaning' is the appropriate translation. Cf. Kasabova (2006).

Dubislav), Bolzano does not really refute the canon of reciprocity – the inverse relation of a concept's intension and extension is still valid for nominal definitions, for the explanation of what a word or concept means and how it is used does not rely on investigating or enumerating the attributes of the thing(s) denoted by this word or concept. As Bolzano's own example of adjunctive concepts such as [round ball] shows, on pain of circularity, the extension is not larger than the intension.

Unfortunately, Dubislav omits Bolzano's important contribution to the notion of stipulative definition, expounded in part 4 of the *Theory of Science*, in a chapter called: *Theory of Signs or Semiotics* (see also 1837, § 637).[35] Bolzano (ibid., § 668.9) prefigures Carnap by advancing the notion of explication (*Verständigung*) as definition. An explication improves the existing notion in a particular context by creating a new usage (ibid., § 284). Bolzano uses stipulative definition as a kind of explication for presenting the key notions of the *Theory of Science* (ibid., § 668.9). He introduces the notion of presentations and propositions as such by specifying the new usage of a familiar concept.

Bolzano's stipulative definition is based on the grounding relation (*Abfolge*): the property *isosceles* is an essential property of triangles because being triangular is inferrable from [isosceles], hence for Bolzano this kind of definition is inferential (ibid., §§ 111, 162, 198, 221.note).[36] In addition, inferential definition is important for determining infinite collections: a collection can comprise infinitely many items because it is determined by a generic concept and a classificatory principle: 'belongs to *A* or does not belong to *A*' (1831, § 14). Accordingly, Bolzano defines the concept [actual] as inferrable from [possible] (1837, § 65.10).

10.5 Conclusion

In this chapter I reconstruct Dubislav's perspective on Bolzano, relating it to more recent discussions amongst Bolzanians. At times the discussion is underpinned by Kantian notions the critique of which has long since become a philosophical commonplace. Dubislav's views on Bolzano—and Bolzano's views on notions such as analyticity, validity, variation, derivability, probability and definition—are of interest for historians of logic and philosophy.

---

[35] Jakobson (1980) points out Bolzano's contribution to semiotics, although logicians and philosophers usually neglect this fact. Bolzano considered the theory of signs as belonging to methodology or the theory of science proper. Logic taken in a wide sense is a theory of science and the theory of science proper is the *organon* which regulates our acquisition of knowledge and includes a didactic theory of signs because Bolzano subscribes to the view that the correct understanding and use of words are based on a correct understanding of signs. See on this Kasabova (2006).

[36] Jan Sebestik (1992, 139) notes that in Bolzano's notion of explication paraphrastic elucidations or contextual definitions appear for the first time in the history of logic. Cf. Kasabova (2006, 13).

References

Bar-Hillel, Yehoshua. 1950. Bolzano's definition of analytic propositions. *Theoria* 16: 91–117; and in *Methodos* 2: 32–55.
Bar-Hillel, Yehoshua. 1952. Bolzano's propositional logic. *Archiv für mathematische Logik und Grundlagenforschung* 1: 305–338.
Berg, Jan. 1999. Kant über analytische und synthetische Urteile mit Berücksichtigung der Lehren Bolzanos. In *Bernard Bolzanos geistiges Erbe für das 21. Jahrhundert*, ed. Edgar Morscher, 97–128. Sankt Augustin: Academia.
Bolzano, Berhard. 1810. *Beyträge zu einer begründeteren Darstellung der Mathematik (Contributions to a better grounded presentation of mathematics)*. Prague: Caspar Widtmann.
Bolzano, Berhard. 1831. In *Paradoxien des Unendlichen*, ed. F. Přihonský. Leipzig: Reclam. 1851.
Bolzano, Berhard. 1837. *Wissenschaftslehre*. Sulzbach: Seidel.
Corcoran, John. 1993. Meanings of implication. In *A philosophical companion to first-order logic*, ed. I.G.Richard Hughes, 85–100. Indianapolis: Hackett.
Couturat, Louis. 1905. *Les Principes des Mathematiques: avec un appendice sur la philosophie des mathématiques de Kant*. Paris: Felix Alcan.
Dubislav, Walter. 1926. *Ueber die sogenannten analytischen und synthetischen Urteile*. Berlin: Weiß.
Dubislav, Walter. 1929. Ueber Bolzano als Kritiker Kants. *Philosophisches Jahrbuch* 42: 357–368.
Dubislav, Walter. 1930a. Bolzano, Bernard: Wissenschaftslehre. *Erkenntnis* 1: 408–409 (book review).
Dubislav, Walter. 1930b. Diskussion über Wahrscheinlichkeit. *Erkenntnis* 1: 264–266.
Dubislav, Walter. 1931a. *Die definition*. Hamburg: Meiner.
Dubislav, Walter. 1931b. Bolzano als Vorläufer der mathematischen Logik. *Philosophisches Jahrbuch* 44: 448–456.
Dubislav, Walter. 1931c. In *Neuer Anti-Kant und Atomenlehre des seligen Bolzano*, ed. W. Dubislav, H. Scholz, and Franz Přihonský. Sankt Augustin: Academia. 2003.
Dubislav, Walter. 1931d. Bernard Bolzano in memoriam. *Unterrichtsblätter für Mathematik und Naturwissenschaften* 37: 340–344.
Dubislav, Walter. 1931e. Bernard Bolzano. Zum 150. Geburtstag des Philosophen. *Vossische Zeitung* 469, *Das Unterhaltungsblatt Nr. 233 vom 5.10.1931*.
Dubislav, Walter. 1932. Bemerkungen zur Definitionslehre. *Erkenntnis* 3: 201–203.
Etchemendy, John. 1999. *The concept of logical consequence*. Chicago: CSLI Publications.
Frege, Gottlob. 1892. Über Sinn und Bedeutung. *Zeitschrift für Philosophie und philosophische Kritik*, NF, 25–50.
Grelling, Kurt. 1932. Bemerkungen zu Dubislavs "Die Definition". *Erkenntnis* 3: 189–200.
Jakobson, Roman. 1980. A glance at the development of semiotics. In *Selected writings. Contributions to comparative mythology. Studies in linguistics and philology*, ed. Stephen Rudy and Linda R. Waugh, 199–218. Berlin: Walter de Gruyter.
Kant, Immanuel. 1789. In *Critique of pure reason*, ed. Paul Guyer. Cambridge: Cambridge University Press. 2007.
Kasabova, Anita. 2006. Bolzano's semiotic method of explication. *History of Philosophy Quarterly* 3(1): 21–39.
Künne, Wolfgang. 2006. Analyticity and logical truth. From Bolzano to Quine. In *The Austrian contribution to analytic philosophy*, ed. Mark Textor, 184–249. London/New York: Routledge.
Künne, Wolfgang. 2008. *Versuche über Bolzano/Essays on Bolzano*. Sankt Augustin: Academia.
Morscher, Edgar (ed.). 1999a. *Bernard Bolzanos geistiges Erbe für das 21. Jahrhundert*. Sankt Augustin: Academia.
Morscher, Edgar. 1999b. Logische Allgemeingültigkeit. In *Bernard Bolzanos geistiges Erbe für das 21 Jahrhundert*, ed. Edgar Morscher, 179–206. Sankt Augustin: Academia.
Morscher, Edgar. 2003. Im Spannungsfeld zwischen Kant und Leibniz—eine geistige Standortbestimmung. In Přihonský. xxi–lxxxiv.

Přihonský, Franz. 1850. In *Neuer Anti-Kant und die Atomlehre des seligen Bolzano*, ed. Edgar Morscher, Christian Thiel, Heinrich Scholz, and Walter Dubislav. Sankt Augustin: Academia. 2003.
Quine, Willard V.O. 1951. Two dogmas of empiricism. *Philosophical Review* 60: 20–43.
Rescher, Nicholas. 2006. The Berlin school of logical empiricism and its legacy. *Erkenntnis* 64: 281–304.
Sebestik, Jan. 1992. *Logique et mathématique chez Bernard Bolzano*. Paris: Vrin.
Sebestik, Jan. 2007. Bolzano's logic. *The stanford encyclopedia of philosophy* (Spring 2012 Edition, ed. Edward N. Zalta). http://plato.stanford.edu/entries/bolzano-logic/. Accessed 18 April 2012.
Siebel, Mark. 1996. *Der Begriff der Ableitbarkeit bei Bolzano*. Sankt Augustin: Academia.
Siebel, Mark. 1999. Bolzano über Ableitbarkeit. In *Bernard Bolzanos geistiges Erbe für das 21. Jahrhundert*, ed. Edgar Morscher, 147–178. Sankt Augustin: Academia.
Siebel, Mark. 2002. Bolzano's concept of consequence. *The Monist* 85: 580–599.
Tarski, Alfred. 1936. On the concept of logical consequence. In *Logic, semantics, metamathematics. Papers from 1923 to 1938*, 409–420. Indianapolis: Hackett. 1983.
Tatzel, Armin. 2002. Bolzano's Theory of ground and consequence. *Notre Dame Journal of Formal Logic* 43(1): 1–25.
Textor, Mark. 1997. Bolzano's Sententialism. *Grazer philosophische Studien* 53: 181–202.
William, Kneale, and Martha Kneale. 1962. *The development of logic*. Oxford: Oxford University Press.

Part V
Kurt Grelling

Chapter 11
The Third Man: Kurt Grelling and the Berlin Group

Volker Peckhaus

11.1 Introduction

The mathematician and philosopher Kurt Grelling (1886–1942) was one of Hans Reichenbach's closest collaborators in Berlin. He can be regarded as the third man in the Berlin Group besides to Reichenbach und Dubislav. He tried to keep the group running even after Reichenbach and Dubislav had left the city.

In 1991 there were several international meetings celebrating the 100th anniversary of Reichenbach's birthday. At two of these meetings I presented papers on Grelling which focused on his tragic fate (Peckhaus 1993, 1994, cf. also Peckhaus 1990, 142–149). On the basis of these papers, other material from my Grelling collection, and further research the New York psychologists Abraham S. Luchins and Edith H. Luchins wrote a biography entitled "Kurt Grelling: Steadfast Scholar in a Time of Madness." It was first published in the journal *Gestalt Theory* in 2000 and is also available online as an expanded version (Luchins and Luchins 2000). Contrary to these papers, the contribution presented here will predominantly discuss the scientific development of Grelling during his Berlin years, not so much going into details of his biography.

V. Peckhaus (✉)
Department of Philosophy, University of Paderborn, Paderborn, Germany
e-mail: peckhaus@hrz.upb.de

N. Milkov and V. Peckhaus (eds.), *The Berlin Group and the Philosophy of Logical Empiricism*, Boston Studies in the Philosophy and History of Science 273, DOI 10.1007/978-94-007-5485-0_11,

11.2 Grelling as a Neo-Friesian

Kurt Grelling was of Jewish origin. During the Nazi regime in Germany, however, he did not succeed to emigrate to the United States although he had been offered a position at the New School for Social Research in New York. In the end, he and his wife were murdered in Auschwitz. Their two children survived in Switzerland.

While he was still a student of mathematics, physics and philosophy at the University of Göttingen, Grelling became a transcendental philosopher in the Neo-Friesian tradition propagated by the young Göttingen philosopher Leonard Nelson. Grelling focused his work especially on logic and on the foundations of mathematics being in close contact to the Göttingen mathematicians around David Hilbert. Later in Berlin he found his way to logical aspects of Gestalt Theory, becoming a pioneer of what is called today "formal ontology".

In 1926 Hans Reichenbach was called to the Friedrich-Wilhelms University in Berlin. From the beginning Grelling took part in Reichenbach's seminars. He also joined Reichenbach's group in the "Gesellschaft für empirische Philosophie" led by Josef Petzoldt, a follow-up of the "Gesellschaft für positivistische Philosophie" founded by Petzoldt in 1912. Grelling had commented on this earlier creation in his column for the *Sozialistische Monatshefte* (cf. Grelling 1913). His comment shows that he was still not ready to accept the ideals of positivist philosophy at that time. Petzoldt had advertised the new society in a flyer enclosed to the first issue of the new journal *Zeitschrift für positivistische Philosophie*. In this flyer Petzoldt had expressed science's need for a philosophy emerging from science. This claim caused Grelling's polemics. In a historical view, he wrote, the direction of development was the other way around. Philosophy influenced science and culture from the outside. It was, e.g. philosophy that brought the emancipation from the ecclesiastical dogma.

> The egg thinks to be wiser than the chicken. How is it possible to prescribe a science it should emerge from another one? With this the autonomy of science is touched, a procedure which is not better than the demand of the Catholic Church, all science has to emerge from theology. (Grelling 1913, 1038)

Grelling admitted that some philosophical systems were deficient, but he denied that science had the competence to judge about this. A philosophy having emerged from science runs the risk "to make those mistakes in reasoning which had been overcome, with much ado, by autonomous philosophy long ago" (ibid.). Grelling also commented on Petzoldt's programmatic paper "Positivistische Philosophie" which opened the first issue of the *Zeitschrift für positivistische Philosophie* (Petzoldt 1913). There Petzoldt had rejected a foundation of experience on apriori functions. Grelling blamed him for making the same mistake Petzoldt himself accused the philosophers to commit, namely to "disregard the matters of fact, because the existence of apriori functions is a matter of fact, and not a theory made up out of thin air" (Grelling 1913, 1039). Grelling closed with the advice (ibid.):

> A bit more humility and respect for the brainwork of former centuries from the side of the spokesmen of the new society will be necessary if they want to get attention outside their inner circles.

Kurt Grelling's 1913 criticism shows that he was still quite far from adopting a naturalistic attitude towards scientific philosophy, as it was typical for the Vienna Circle and the Berlin group around Reichenbach. At that time Grelling was still a member of Leonard Nelson's Neo-Friesian School. The philosophers of the *Neue Fries'sche Schule* elaborated the approach of an anthropological critique of reason in the tradition of Jacob Friedrich Fries, a follower of Kant and opponent of his own contemporary Hegel (cf. Peckhaus 1990, ch. 5).

Grelling collaborated with Nelson especially in mathematical questions essential for Nelson who tried to keep contact to the Göttingen mathematicians, in particular to Hilbert. Grelling's name is best known for the so-called Grelling Paradox, or, due to the market power of Wikipedia, for the Grelling–Nelson Paradox[1] as it is called there, although there is no historical precursor for this name. The paradox is fruit of a discussion of Russell's paradox or, to be more exact, about the paradox of the concept "impredicable," which took place in the Neue Fries'sche Schule that lead to several attempts to solve these paradoxes since 1906 (cf. Peckhaus 1995, 2004). It was during Grelling's and Nelson's struggle with these "solutions" that Grelling and Nelson compiled material for a joint paper which was finally published under the title "Bemerkungen zu den Paradoxieen von Russell und Burali-Forti" in 1908 (Grelling and Nelson 1908). The authors looked for the basic logical conditions for the occurrences of the paradoxes and distinguished between

> the task of a proper "solution" of the paradox, i.e., the task of unveiling the underlying appearance, and the task of a "correction," i.e., the task of avoiding the paradox by introducing new, consistent concepts. Such a correction cannot be considered to be a solution, because the paradoxical objects, if they exist at all, are not eliminated by stopping work on them. (1908, 314)

Most importantly Grelling discovered new paradoxes, among them the semantic heterological paradox, Grelling's paradox, and it can be proved that it was only Grelling who constructed these new paradoxes. In its original version, it runs as follows (Grelling and Nelson 1908, 307):

> Let φ be the word that denotes the concept defining M. This word is either an element of M or not. In the first case we will call it "autological" in the other "heterological." ["Short," e.g., is autological, "long" heterological; "English" is autological, "German" heterological] Now the word "*heterological*" is itself either autological or heterological. Suppose it to be autological; then it is an element of the set defined by the concept that is denoted by itself, hence it is heterological, contrary to the supposition. Suppose, however, that it is heterological; then it is not element of the set defined by the concept that is denoted by itself, hence it is not heterological, again against the supposition.

This paradox was wrongly attributed to Hermann Weyl (1885–1955) by Frank Plumpton Ramsey (1903–1930, in Ramsey 1926). Weyl had mentioned it in *Das Kontinuum* as "a well-known paradox, essentially coming from Russell" and had discussed it as "scholasticism of the worst kind" (Weyl 1918, 2). This evaluation led to profound irritations between Grelling and Weyl.

---

[1] "Grelling-Nelson Paradox (2010)," Wikipedia, http://en.wikipedia.org/wiki/Grelling%E2%80%93Nelson_paradox (11 August 2010).

Two years after the joint publication with Nelson, Grelling made his Ph.D. in mathematics. Officially, he is listed among Hilbert's doctoral students, but the topic "Die Axiome der Arithmetik unter besonderer Berücksichtigung der Beziehungen zur Mengenlehre" (Grelling 1910a) was given to him by Ernst Zermelo who also wrote the report. This work was related to Zermelo's research on the transition from axiomatized set theory to axiom systems for arithmetic (esp. Zermelo 1909a, b). Hilbert simply followed Zermelo's report. The examiners in the oral examination were David Hilbert for Mathematics, Woldemar Voigt for Physics and Edmund Husserl for philosophy. Husserl attested Grelling "unusual philosophical knowledge" and an "understanding beyond average."[2]

11.3 Grelling and Reichenbach

Grelling's separation from Nelson's variation of critical philosophy was largely parallel, and not independent from Reichenbach's turn away from Kantianism. Reichenbach was 5 years younger than Grelling. They may have become acquainted to each other during Reichenbach's university studies at Munich in 1912/13. At that time Grelling was also in Munich studying National Economy and directing an independent philosophical colloquium. At the latest, Grelling met Reichenbach in Göttingen. Grelling had returned from Munich to Göttingen in the end of 1913, where Reichenbach studied from Easter 1914 to the beginning of 1915. Reichenbach affirms this in a testimonial for Grelling from 1940:

> I have known him personally since more than 25 years, and enjoyed his collaboration in particular during the years 1926–1933 when he attended my seminars in the University of Berlin and participated in discussions in the society for scientific philosophy.[3]

When Reichenbach was in Göttingen, he had not only a close relationship to Grelling, but also to Leonard Nelson. There is good reason to assume that it was on Nelson's advice that Reichenbach submitted his dissertation to the Philosophical Faculty of the University of Erlangen, where Nelson's uncle, Paul Hensel, had a chair for philosophy.

There were some overlaps in Reichenbach's and Grelling's political interests. When he came to Göttingen, Reichenbach was already a leading official of the free student's movement (*Freie Studentenschaft*). Grelling was a significant activist in the Göttingen group of this movement (cf. Linse 1974, 12).

However, there were also overlaps in scientific interests. In 1915 Reichenbach submitted his doctoral thesis *Der Begriff der Wahrscheinlichkeit für die mathematische Darstellung der Wirklichkeit* to the Philosophical Faculty of the University

---

[2]Ph.D. Files Grelling, University Archives Göttingen, Az. Phil. Fak., 1908–1914, G. Vol. II.

[3]Testimonial dated 3 October 1940, Hans Reichenbach Collection, University of Pittsburgh Libraries, Special Collection Department, [HR 037-28-10].

of Erlangen. In his thesis Reichenbach still followed a Kantian paradigm in trying to execute the critical programme for the theory of probabilities with the help of a transcendental deduction of the principle of distribution which was fundamental for Reichenbach's approach (cf. Reichenbach 1916a, b). Five years earlier, Grelling had published a paper on the philosophical foundations of the calculus of probabilities (Grelling 1910b). Both, Grelling and Reichenbach, relied on the same sources and Reichenbach quoted Grelling in a very benevolent way.

Both, Grelling and Reichenbach, changed their philosophical beliefs in the early 1920s under the impression of the revolutionary developments in modern physics. Both arrived at the conviction that Kant's theory of the aprioricity of space and time had been disproved by relativistic physics. This conviction is expressed in Reichenbach's *Relativitätstheorie und Erkenntnis a priori* (Reichenbach 1920) and in Grelling's lecture "Relativitätstheorie kritische Philosophie" presented at the meeting of the Fries Society in August 1921. Grelling closely followed Reichenbach's argumentation, but his conclusion was even more radical. In his thesis 14 he wrote: "Theory of Relativity and Critical Philosophy are incompatible."[4] This claim, as it was defended in a circle of critical philosophers, activated the final break between Nelson and Grelling.

In 1923 Grelling went to Berlin working as a school teacher at grammar schools. The change in his philosophical views became evident in his rapprochement to Monism and the philosophy of Bertrand Russell. Grelling translated four of Russell's books into German, Russell's *Analysis of Mind* (1921; German 1927a); *The ABC of Relativity* (1925; German 1928); *The Analysis of Matter* (1927b; German 1929), *An Outline of Philosophy* (1927c; German 1930).

11.4 Support for Scientific Philosophy

Grelling supported Reichenbach in making publicity for scientific philosophy. For the *Monist,* he took over, e.g., Reichenbach's job to write a report on what had been done in Germany in the field of exact philosophy in recent years. This report was published in 1928 (Grelling 1928a, b). On the first pages he discusses the relation between exact natural sciences and philosophy. In 1913 he had rejected the claim that the exact sciences needed a new philosophy emerging from within and not coming from outside science. According to his view, philosophy could deal with scientific problems as a meta-science. However, in 1928, he states that an intimate relation between exact natural sciences and philosophy was still lacking in Germany, partly due to the scientists: "Because their subjects are highly specialized, they easily lose a view of the whole and then face philosophical problems quite

[4]Kurt Grelling, "Relativitätstheorie und kritische Philosophie," typescript of the protocol of the meeting of the Fries Society on 15 and 16 August 1921, Bundesarchiv Berlin, Nachlass Nelson, 90 Ne 1, no. 388, fols. 243–246.

without comprehension" (Grelling 1928b, 393). Things were changing, however, in recent times. A lively philosophical interest among German mathematicians and physicist had appeared and some of them had done significant philosophical work on the borderline of their sciences. On the other side there was only a very light insight among leading philosophers into the philosophical problems emerging from the work of mathematicians and philosophers.

> Even today we have the spectacle of philosophers who lay claim to a guardianship over natural scientists. We still find philosophers rejecting scientific results which are verifiable only through experimentation or mathematical calculation because these results conflict with the allegedly apodictic truths of philosophy. Another favourite attitude of technical philosophers expresses itself in the assertion that the conclusions reached by natural science through painstaking experimental labour are but a confirmation of the results which the philosopher has long since attained through pure thought. (Grelling 1928b, 393–394)

Object of his report were those philosophers, who worked in close connection with the exact sciences, partly being representatives of these sciences, partly being philosophers who had included this subject in their studies and could therefore follow the development in this field.

Grelling stressed that these investigators may not really be called a school of philosophy, but he thought that all of them had certain fundamental principles in common:

1. They repudiate the premature construction of a philosophical system.
2. They respect the conclusions of natural science.

> This does not mean that they unconditionally recognize as valid everything which any scholar proclaims as the outcome of his investigations. But, in their view, the results of science must be confirmed or refuted, as the case may be, by the methods and devices of this science itself. (Grelling 1928b, 394)

Grelling, furthermore, wrote a long historical and systematic discussion of Reichenbach's *Philosophie der Raum-Zeit-Lehre* (Reichenbach 1928), published in the *Philosophischer Anzeiger* (Grelling 1930). He participated in spreading the philosophical contributions of other members of the Berlin group to a wider public. In 1931, e.g., Dubislav published the third, completely reworked and extended edition of his book *Die Definition* (Dubislav 1931). Grelling wrote a paper of 11 pages for *Erkenntnis* entitled "Bemerkungen zu Dubislavs 'Die Definition'" (Grelling 1932/33). He concluded his critical remarks by emphasizing how meritorious this book appears to him. Even where it provokes disagreement, it is inspiring (ibid., 189). Nonetheless, Grelling disagreed a lot. In particular he rejected Dubislav's interpretation of Frege which is fundamental because Dubislav posed his formalistic theory of definition against Frege's (ibid., 189–191). Dubislav saw the main feature of Frege's theory in the distinction between a designating sign and what is designated. According to him, there is a problem to name the combinations of signs which may occur in definitions. Grelling did not see any problem, at least not on Frege's side, because Frege was explicit in what kind of signs were allowed. Grelling, furthermore, claimed that, given Dubislav's characterization of a definition, the results achieved with his formalistic, game-theoretic approach can also be reached with the help of Frege's theory. Grelling criticized that Dubislav's concept of a definition of a formula was not founded because its justification implies an infinite regress. Dubislav's proof of the consistency of the logical calculus was accepted, but Grelling stressed that this proof was not purely syntactical, but presupposed a contentual interpretation of the calculus. He concluded that this question cannot be decided, if the consistency of logic as such (including the contentual side) is doubted. "Concerning the calculus of logic, however, Dubislav has proved its consistency under the presupposition that the contentual logic is consistent" (ibid., 192). Dubislav's conception of truth was also criticized with the fatal standard objection not to have distinguished between the definition and the criterion of truth. Besides this Dubislav's notion of a concept was said not to be sufficient, and Grelling accused Dubislav to have misconceived Heinrich Hertz's picture theory (ibid., 196–197) and to present an insufficient notion of a concept (ibid., 197–198).

Dubislav's response "Bemerkungen zur Definitionslehre" (Dubislav 1932) directly follows on the next pages. He emphasizes his agreement with Grelling in many questions of the philosophy of science. This sounds astonishing for the naive reader who is confused by the sharpness of the dispute. The reader might get the impression that this published exchange was a prepared demonstration of the dispute culture in the society. Although being in agreement that logic and the philosophy of science should form the core of philosophy, the members of the group disagreed in several central questions. Debates like this could initiate broader public disputes on these questions. They obviously wanted to set the topics for philosophical debates.

The Berlin Group and the Berlin Gestalt theorists were in close contact with each other, not only in the scientific programme, but also in institutional matters. Philosophers and psychologists were in the same faculty. Carl Gustav Hempel's doctoral thesis on the logical analysis of probability concepts (Hempel 1934) was mainly supervised by Hans Reichenbach, who quite suddenly departed to Turkey in 1933. Wolfgang Köhler stepped in substituting Reichenbach in the further procedure (cf. Hempel 1991, 8–9). The same took place in Olaf Helmer's case (Rescher 2006, 288).

The Nazi's seizure of power did not immediately stop the work of the Berlin Society for Scientific Philosophy. Two events, however, forced the society into a "big sleep" (*Dornröschenschlaf*) as Grelling wrote to Reichenbach in 1936.[5] The first incident was Reichenbach's move to the University of Istanbul, reformed and reopened by Kemal Atatürk in 1933, the second was Dubislav's imprisonment on remand 1935 when he was accused of criminal assault. After Dubislav had been released, he left Berlin for Prague where he killed in jealousy first his girl-friend then himself on 16 September 1937.[6] Grelling, in the meantime discharged from

---

[5]Grelling to Reichenbach, dated Berlin-Wilmersdorf, 16 January 1936, Hans Reichenbach Collection, Pittsburgh, [HR 013-14-06].

[6]Cf. the reports in the Prague Press, e.g. "Hochschulprofessor ersticht Malerin."

his position in school, became the main representative of the society in Berlin and tried to keep activities running. "It might interest you," he wrote to Reichenbach in March 1937,

> that I have established here two private logistic workshops several months ago, and that I have gathered a new Berlin Circle in which we discuss questions of logic and the philosophy of science.[7]

Grelling had created these workshops in the summer of 1936. Among the attendants were Franz Graf Hoensbroech of an old Dutch noble family who had published a paper on intension and extension of concepts in *Erkenntnis* (Hoensbroech 1931), Leopold Löwenheim, the mathematician and logician, famous for the Löwenheim-Skolem-Paradox, Jürgen von Kempski who acted as an editor of the *Archiv für Philosophie* from 1947 to 1964 which included the *Archiv für mathematische Logik und Grundlagenforschung* since 1950, a journal that made possible a new start of mathematical logic in Germany after the war, and the mathematician Luise Rothstein, a Polish Jew who had been a private student of Edmund Landau in the last years of his life. Carl Gustav Hempel reported about these activities on a postcard to the Finish linguist and logician Uuno Saarnio in January 1937:

> We met Dr. Grelling several times who now tries to create some sort of logistic center in Berlin; he has two seminars and a colloquium. The latter is attended by Löwenheim, the father of the Skolem–L[öwenheim] paradox, who was as it were "rediscovered" in Lichterfelde. L[öwenheim] was highly astonished when he heard from G[relling] that he and his paper had become famous in the meantime: although Löwenheim still works in logic, but he does not follow the literature.[8]

On 28 March 1933, Grelling was forced into retirement. He was able to compensate the financial losses with the revenues from an estate he managed since the death of his wealthy mother. These revenues even allowed him to keep a middle-class style of living as he wrote in a letter to Otto Neurath in July 1934. About his work "as a retired man of 48 years" he wrote in the same letter:

> I read a lot newer scientific literature, write reviews for the *Erkenntnis*, prepare an edition of the small writings of Gottlob Frege [never published], convene with Dubislav and some other acquaintances for philosophical discussions, the rest of my time I devote to my family, in particular to my two children of now four and seven years, who are my comfort when I am too much annoyed about the dumbness of man.[9]

---

[7]Grelling to Reichenbach, dated 14 March 1937, Hans Reichenbach Collection, Pittsburgh [HR 013-14-02].

[8]Hempel to Saarnio, dated Brussels, 21 January 1937, Saarnio Papers, Reino Saarnio, Helsinki.

[9]Grelling to Neurath, dated Menton, 21 July 1934, Neurath Papers, Vienna Circle Archives, The Hague.

11.5 Final Activities

This period was a time of intensive scientific activities. Grelling reactivated earlier fields of research like mathematical logic or the logical paradoxes. He also directed his interests to fields completely new for him like behaviouristic psychology and Gestalt theory. Besides this, Grelling had to organize a way of subsisting outside Germany. With the assistance of Otto Neurath he formulated some sort of scientific research programme which could be conveyed in France under the roof of Neurath's Mundaneum Institute. The projects included collecting historical and bibliographical material on the history of the logic of science, in particular the history of logistics (mathematical logic) in France focussing on the significance and impact of Louis Couturat. Grelling should, furthermore, write an elementary introduction to mathematical logic in French, which could serve as a substitute for Couturat's *L'algèbre de la logique* (Couturat 1905).[10] These plans could not be realized. Arrangements with Paul Oppenheim (1885–1977) were more successful. The philosopher and industrialist Paul Oppenheim had emigrated from Frankfurt to Brussels in 1933. He finally emigrated to the United States in 1939.[11]

Paul Oppenheim had been born in Frankfurt a.M. He studied in Frankfurt and Gießen, where he earned a doctorate in chemistry. In 1912 he married the Belgian Gabrielle Errera, a native of Brussels. Oppenheim first joined his father's jewellery firm, but left it 1924 to become a director of a chemical firm which became part of I.G. Farben. In his home town Frankfurt he had been very active in supporting the artistic and intellectual life. He later supported non-academic scientific research by financing intellectual collaborators. He hired, e.g., Carl Gustav Hempel as private scientific researcher. Hempel left Brussels for job hunting in USA. The arrangement was that Grelling could take over Hempel's position, if Hempel were successful in finding a job in the USA. Since 1937, Grelling first substituted Hempel, however, after the *Reichskristallnacht*, he emigrated to Brussels. After the war, Oppenheim continued this kind of support, working together with Olaf Helmer, John G. Kemeny, Nicholas Rescher, Nathan Brody, and others (Rescher 2006, 285).

Grelling's collaboration with Oppenheim led to a series of interesting papers on Gestalt theory which had been prepared by studies on psychological questions. Grelling delivered, e.g., a paper "Zur Theorie der Wahrnehmung" at the International Congress of Scientific Philosophy at the Sorbonne in Paris in 1935 (Grelling 1936). In this sketch of a theory of perception Grelling does not start from our strong beliefs in the reality of what we perceive, beliefs that might be explained by psychology, but that were justified or founded logically. He deals with perception as a matter of fact which he intended to explain in the usual scientific sense. He then presents a physiological theory of perception, finally discussing the

[10] Notes of a meeting of Grelling and Neurath at The Hague on 5 September 1937; cf. Neurath to Grelling, dated 6 September 1937, Neurath Papers, Vienna Circle Archives, The Hague.

[11] On Oppenheim cf. the Obituary of Paul Oppenheim (1977), Luchins and Luchins (2000), and Rescher (2006), 284–285.

question whether perception (*Wahrnehmung*) is knowledge (*Erkenntnis*). He arrives at basically negative results. Perception is no immediate knowledge, because the relation between the process of perception and the real object which might have been known is a mediated one. According to Grelling, intentional or immanent objects which appear to be given to us in every perception are not known (*erkannt*). The belief in the existence or reality of such intentional object is a blind belief that is not worthy to carry the name "knowledge". In a critical view the conviction in the existence of what has been perceived can at best be known as a plausible hypothesis.

Grelling clearly expresses an anti-realist position which keeps elements of Kant's critical philosophy. This is also obvious in his perspicacious evaluation of the scope of his theory. He finally remarks that the theory of perception just sketched is part of a wider theory of experience. Such a theory can never be a proof or a justification of experience, he writes. Everything it can do is to give a scientific explanation of the way experiences are made.

This paper on the theory of perception shows Grelling's interest in psychological questions which might be induced by the strong position of the Berlin Gestalt theorists, especially Wolfgang Köhler. Consequently, it is a short step towards a discussion of Gestalt theory which was regarded by Logical Empirists as an alternative to Behaviourism. One of the fruits of the collaboration between Grelling and Oppenheim was the joint paper "Der Gestaltbegriff im Lichte der neuen Logic," published in *Erkenntnis* (Grelling and Oppenheim 1937). Oppenheim and Grelling remark in the beginning that their paper is part of a more comprehensive study on order concepts as discussed in Hempel and Oppenheim's joint paper "Der Typusbegriff im Lichte der neuen Logik" of 1936 (Hempel and Oppenheim 1936). With this paper Grelling and Oppenheim intend to contribute to a discussion inaugurated by Wolfgang Köhler concerning the question whether it is justified to use the concept of Gestalt in exact sciences. They complain about the ambiguity of the concept, which is sometimes used without definition. The main objective of the paper is to give a suitable definition of the Gestalt concept. For this several auxiliary concepts have to be defined: A "classifier" is a characteristic function that relates a certain value to every element to which it can reasonably be applied (Grelling and Oppenheim 1937, 212). Such classifier gives an order of a domain by relating values to spots in the domain. The whole of such order relations, classifiers in a given domain, is called a complex. A correspondence is a relation between complexes. A transposition is a correspondence which transforms one domain into another. The concept of Gestalt is now the invariant of transpositions of a complex in relation to a correspondence (ibid., 216). This concept of Gestalt can be understood as a classifier whose arguments are complexes and whose values are Gestalt individuals.

The authors show that this concept of Gestalt comes close to the original notion introduced by Christian von Ehrenfels in 1890. It is basically equivalent to notions like "shape," "form," and "configuration." The authors suggest to restrict the use of the concept of Gestalt to this meaning, but they discuss a second main use suggested by Wolfgang Köhler. In this reading a Gestalt is a determinational system where "system with respect to a certain relation R" is understood as satisfying the following conditions (Grelling and Oppenheim 1937, 220):

> There is a division of the whole in such a way that each part of the division stands in the relation *R* to every other part, and that each object, standing at least with one other object in the relation *R* is itself part of the whole.

The relation in question is that of determination as defined by Carnap in § 37 of his *Abriss der Logistik* (Carnap 1929). In respect to their initial question the authors come to the result that both concepts apply to all real sciences, therefore also to exact natural sciences. Only the first, as they say, "*our* concept of Gestalt," does apply to formal sciences like logic and mathematics (Grelling and Oppenheim 1937, 203).

In a paper entitled "Logical Analysis of 'Gestalt' as 'Functional Whole'" submitted to the Unity of Science Congress in Cambridge, Mass., only published after Grelling's death (Grelling and Oppenheim 1939), Grelling and Oppenheim deepened their analysis of the second meaning, providing new definitions of "dependence," "interdependence," and "independence." The notion of dependence is further discussed and investigated in a formal way in a second paper that Grelling alone submitted to the Cambridge Congress, entitled "A Logical Theory of Dependence" (Grelling 1939). These papers make Grelling and Oppenheim important proponents of Mereology and Formal Ontology as been discussed today.

11.6 Conclusions

The records represent Kurt Grelling as a scholar full of inspiration and energy. The time and the circumstances prevented him to reach his goal of a scientific career. He had to cut down his own ambitions being number three in the Berlin Society for Scientific Philosophy. As a mathematician the starting point of his research was mathematical logic and set theory, disciplines which remained his permanent points of reference. From there he extended his interest into related fields like Semantics, but also other parts of scientific philosophy like Gestalt Theory and Mereology. In all these domains he had something to say which is still heard today. But he never had the chance to play the role of a leader, neither in institutional nor in scientific matters. He always remained a valuable collaborator.

References

Carnap, Rudolf. 1929. *Abriss der Logistik*. Vienna: Springer.
Couturat, Louis. 1905. *L'algèbre de la logique*. Paris: Albert Blanchard.
Dubislav, Walter. 1931. *Die definition*, 3rd ed. Leipzig: Meiner.
Dubislav, Walter. 1932. Bemerkungen zur Definitionslehre. *Erkenntnis* 3(1932/33): 201–203.
Grelling, Kurt. 1910a. Die Axiome der Arithmetik mit besonderer Berücksichtigung der Beziehungen zur Mengenlehre. Ph.D. thesis, Dieterichsche Universitäts-Buchdruckerei, Göttingen.
Grelling, Kurt. 1910b. Die philosophischen Grundlagen der Wahrscheinlichkeitsrechnung. *Abhandlungen der Fries'schen Schule* n.s. 3(3): 440–478.

Grelling, Kurt. 1913. Positivismus. *Sozialistische Monatshefte* 19(II): 1038–1039.
Grelling, Kurt. 1928a. Philosophy of the exact sciences: Its present status in Germany. *The Monist* 38: 97–119.
Grelling, Kurt. 1928b. Philosophy of the exact sciences. In *Philosophy today. Essays on recent developments in the field of philosophy*, ed. Edward Leory Schaub, 393–415. La Salle: Open Court, [Repr. Freeport: Books for Libraries Press 1968].
Grelling, Kurt. 1930. Die Philosophie der Raum-Zeit-Lehre. *Philosophischer Anzeiger* 4: 101–128.
Grelling, Kurt. 1932. Bemerkungen zu Dubislavs 'Die Definition'. *Erkenntnis* 3(1932/33): 189–200.
Grelling, Kurt. 1936. Zur Theorie der Wahrnehmung. In *Actes du congrès international de philosophie scientifique. Sorbonne Paris 1935, vol. 5: Logique & expérience*, Actualités scientifiques et industrielles; 392, 69–79. Paris: Hermann & Cie.
Grelling, Kurt. 1939. A logical theory of dependence [Paper sent in for the fifth international congress for the unity of science, Cambridge, MA, 1939]. In *Foundations of Gestalt theory,* Philosophia resources library, ed. Barry Smith, 1988, 217–228, Munich/Vienna: Philosophia Verlag.
Grelling, Kurt, and Leonard Nelson. 1908. Bemerkungen zu den Paradoxieen von Russell und Burali-Forti. *Abhandlungen der Fries'schen Schule* n.s. 2 (1907–1908) no. 3: (1908), 301–334.
Grelling, Kurt, and Paul Oppenheim. 1937. Der Gestaltbegriff im Lichte der neuen Logik. *Erkenntnis* 7(1937/38): 211–225.
Grelling, Kurt, and Paul Oppenheim. 1939. Logical analysis of 'Gestalt' as 'Functional Whole' [Paper sent in for the fifth international congress for the unity of science, Cambridge, MA, 1939]. In *Foundations of Gestalt theory*, Philosophia resources libraryed, ed. Barry Smith, 1988, 210–216. Munich/Vienna: Philosophia Verlag.
Grelling–Nelson Paradox. 2010. http://en.wikipedia.org/wiki/Grelling%E2%80%93Nelson_paradox. 11 Aug 2010.
Hempel, Carl Gustav. 1991. Hans Reichenbach remembered. *Erkenntnis* 35: 5–10.
Hempel, Carl Gustav. 1934. Beiträge zur logischen Analyse des Wahrscheinlichkeitsbegriff. Ph.D. thesis, Universitätsbuchdruckerei G. Neuenhahn, Jena/Berlin.
Hempel, Carl Gustav, and Paul Oppenheim. 1936. *Der Typusbegriff im Lichte der neuen Logik Wissenschaftstheoretische Untersuchungen zur Konstitutionsforschung und Psychologie.* Leiden: A.W. Sijthoff's.
Hoensbroech, Franz Graf. 1931. Beziehungen zwischen Inhalt und Umfang von Begriffen. *Erkenntnis* 2: 291–300.
Linse, Ulrich. 1974. Hochschulrevolution. Zur Ideologie und Praxis sozialistischer Studentengruppen während der deutschen Revolutionszeit 1918/19. *Archiv für Sozialgeschichte* 14: 1–114.
Luchins, Abraham S., and Edith H. Luchins. 2000. Kurt Grelling: steadfast scholar in a time of madness. *Gestalt Theory* 22: 228–281; expanded version http://gestalttheory.net/archive/kgrelbio.html. 10 Aug 2010.
Obituary of Paul Oppenheim. 1977. *New York Times*, 24 June 1977, Section D, 13.
Peckhaus, Volker. 1990. *Hilbertprogramm und Kritische Philosophie. Das Göttinger Modell interdisziplinärer Zusammenarbeit zwischen Mathematik und Philosophie*, Studien zur Wissenschafts-, Sozial- und Bildungsgeschichte der Mathematik; 7. Göttingen: Vandenhoeck & Ruprecht.
Peckhaus, Volker. 1993. Kurt Grelling und der Logische Empirismus. In *Wien–Berlin–Prag. Der Aufstieg der wissenschaftlichen Philosophie. Zentenarien Rudolf Carnap–Hans Reichenbach–Edgar Zilsel*, Veröffentlichungen des Instituts Wiener Kreis; 2, ed. Haller Rudolf and Stadler Friedrich, 362–385. Vienna: Hölder-Pichler-Tempsky.
Peckhaus, Volker. 1994. Von Nelson zu Reichenbach. Kurt Grelling in Göttingen und Berlin. In *Hans Reichenbach und die Berliner Gruppe*, ed. Danneberg Lutz, Andreas Kamlah, and Lothar Schäfer, 53–73. Braunschweig/Wiesbaden: Friedr. Vieweg & Sohn.
Peckhaus, Volker. 1995. The genesis of Grelling's paradox. In *Logik und Mathematik. Frege-Kolloquium Jena 1993*, Perspectives in analytical philosophy; 5, ed. Ingolf Max and Werner Stelzner, 269–280. Berlin/New York: Walter de Gruyter.

Peckhaus, Volker. 2004. Paradoxes in Göttingen. In *One hundred years of Russell's paradox. Mathematics, logic, philosophy*, de Gruyter series in logic and its applications; 6, ed. Godehard Link, 501–515. Berlin/New York: de Gruyter.

Petzoldt, Josef. 1913. Positivistische Philosophie. *Zeitschrift für positivistische Philosophie* 1: 1–4.

Ramsey, Frank Plumpton. 1926. The foundations of mathematics. *Proceedings of the London Mathematical Society* 25(2): 338–384.

Reichenbach, Hans. 1920. *Relativitätstheorie und Erkenntnis apriori*. Berlin: Springer.

Reichenbach, Hans. 1916a. Der Begriff der Wahrscheinlichkeit für die mathematische Darstellung der Wirklichkeit. i.e. Ph.D. thesis Erlangen 1915 [offprint from *Zeitschrift für Philosophie und philosophische Kritik*], Leipzig.

Reichenbach, Hans. 1916b. Der Begriff der Wahrscheinlichkeit für die mathematische Darstellung der Wirklichkeit. *Zeitschrift für Philosophie und philosophische Kritik* 161 (1916): 209–239, 162 (1917): 98–112, 222–239, 163 (1917): 86–99.

Reichenbach, Hans. 1928. *Philosophie der Raum-Zeit-Lehre*. Berlin/Leipzig: Walter de Gruyter.

Rescher, Nicholas. 2006. The Berlin school of logical empiricism and its legacy. *Erkenntnis* 64: 281–304.

Russell, Bertrand. 1921. *Analysis of mind*. London/New York: George Allen and Unwin/The Macmillan Company.

Russell, Bertrand. 1925. *ABC of relativity*. London: George Allen & Unwin.

Russell, Bertrand. 1927a. Die *Analyse des Geistes*. übersetzt von Kurt Grelling. Leipzig: Teubner.

Russell, Bertrand. 1927b. *Analysis of matter*. London: Kegan Paul, Trench, Trubner.

Russell, Bertrand. 1927c. *An outline of philosophy*. London: George Allen and Unwin

Russell, Bertrand. 1928. *Das ABC der Relativitätstheorie*. übersetzt von Kurt Grelling. Munich: Drei Masken Verlag.

Russell, Bertrand. 1929. *Philosophie der Materie*. deutsch von Kurt Grelling. Leipzig/Berlin: Teubner (*Wissenschaft und Hypothese*; 32).

Russell, Bertrand. 1930. *Mensch und Welt. Grundriß der Philosophie*. Munich: Drei Masken Verlag.

Weyl, Hermann. 1918. *Das Kontinuum. Kritische Untersuchungen über die Grundlagen der Analysis*. Leipzig: Veit & Company.

Zermelo, Ernst. 1909a. Sur les ensembles finis et le principe de l'induction complète. *Acta Mathematica* 32: 185–193.

Zermelo, Ernst. 1909b. Über die Grundlagen der Arithmetik. In *Atti del IV Congresso Internazionale dei Matematici, Roma, 6–11 Aprile 1908*, vol. 2, 8–11. Rome: Accademia Nazionale dei Lincei.

Kurt Grelling in the mid-1930s in Berlin

Chapter 12
Gestalt, Equivalency, and Functional Dependency: Kurt Grelling's Formal Ontology

Arkadiusz Chrudzimski

Kurt Grelling is best known as a mathematician and logician. He received his doctorate in mathematics, and what first comes to mind when one hears Grelling's name is the semantic paradox named after him.[1] Another well known point of association pertains to his defence of Gödel's incompleteness theorem.[2] However Grelling's writings also contain an interesting work in formal ontology. In a series of papers (the most important having been written together with Paul Oppenheim) he attempted to give a precise logical basis to the notoriously unclear concept of "Gestalt". It is this formal-ontological analysis that will be the topic of this paper.

12.1 The Emergence of the Concept of Gestalt

In its early days, scientific psychology seemed to be a relatively straightforward enterprise. The general picture it offered was this: Our mental life begins with sensory stimulation; each stimulus generates a corresponding impression; and

[1]The Grelling (or Grelling-Nelson) Paradox, formulated in 1908 by Grelling and Leonard Nelson is a semantic self-referential paradox resembling closely Russell's paradox. Let's call an adjective "autological" if it is applicable to itself and "heterological" if it is not the case (hence "short" is autological and "long" heterological) and ask if the very adjective "heterological" is a heterological word. Following the pattern of reasoning very similar to Russell's paradox or the barber paradox, it's not difficult to see that both possible answers, "yes" and "no", lead to a contradiction. See Grelling and Nelson (1908).

[2]Grelling argued against the reading of Gödel's Theorem as a paradox. See Grelling (1937).

A. Chrudzimski (✉)
Department of Philosophy, University of Szczecin, Szczecin, Poland
e-mail: arkadiusz.chrudzimski@univ.szczecin.pl

N. Milkov and V. Peckhaus (eds.), *The Berlin Group and the Philosophy of Logical Empiricism*, Boston Studies in the Philosophy and History of Science 273, DOI 10.1007/978-94-007-5485-0_12,

the stream of impressions is ordered by a couple of rudimentary associative mechanisms, impressively outlined by Hume. Nowadays we are very far from this ascetic elegance. For not only have we rejected the (in)famous constancy hypothesis, but even the fundamental concept of an elementary impression has come to appear rather hazy and arbitrary.

During the last decades of the nineteenth century many thinkers worked hard on deconstructing the traditional Humean picture. However, if we were to pick one person particularly responsible for its decline, our choice would probably fall on Christian von Ehrenfels. In 1890 Ehrenfels published the famous paper, "On Gestalt Qualities", where he drew our attention to a certain interesting phenomenon involved in the perception of organised wholes. Consider Ehrenfels' favourite example: the hearing of a melody. In a certain sense a melody clearly *consists of* tones (and rests) ordered in time. Within the framework of the traditional empiricist picture it is tempting to think of tones (or maybe of their aspects like pitch, timbre, and intensity) as something resembling Humean simple impressions and of the melody as a product of some relatively simple associative mechanism that puts the tones (or their aspects) together. Indeed at first blush it seems that there could be nothing in a melody over and above this "auditory material". Thus when we have the complete collection of tones we have, *eo ipso*, the melody.

But this picture is deeply mistaken. Consider what happens when a melody has been transposed into another key. In this case all the composing tones change in their pitch but in spite of this we still hear *the same* melody. It is even possible that a musically uneducated hearer notices no difference. For, the object at which he or she is primarily directed is not the isolated tones but rather the very melody in question. Thus, it seems that in our perception of organised wholes we have, before our minds, objects that can appear to be invariant, despite of massive changes in the "simple impressions" of which, the traditional empiricist picture says, they are composed. Moreover, it seems that these objects have been "directly given" to us. A melody is something *we hear in the first place* and not something that we need first to "abstract" from a collection of performances in different keys, on different instruments etc. Hence, if such things as melodies, spatial figures or exotic tastes are in any sense constituted on the basis of some "simple sensory material", their constitution must involve significantly more than a simple Humean association.

Ehrenfels' thesis was that a melody that remains the same while played in various keys is to be construed as an *extra quality* that cannot be identified either with any quality of the underlying tones, or with the totality of them.[3] He termed such extra moments "Gestalt qualities"—which proved to be a very happy choice. Similar phenomena attracted at this time also the attention of other philosophers belonging to the so-called "Austrian" tradition. Carl Stumpf, in his *Tonpsychologie*, spoke of

[3]The later conception of the Berlin school denied this claim and interpreted the Gestalt not as an extra quality, but rather identified them with the complex of underlying data. As Barry Smith puts this, according the Berlin school, "a collection of data [...] does not *have* Gestalt: it *is* a Gestalt" (Smith 1988a, 13).

the "phenomena of fusion" (*Verschmelzungsphänomene*) that are responsible for the constitution of unitary chords on the basis of tones (Stumpf 1890, 126 and 128 ff.) and in Edmund Husserl's *Philosophy of Arithmetic* we read of the "second order qualities" (Husserl 1891, 201) that constitute the unity of such entities as an avenue of trees or a line of soldiers. Husserl calls them "figural moments" (Husserl 1891, 203). Now, "fusion" and "figural moments" are also nice names but not nearly as appealing as Ehrenfels' "Gestalt". That's why we nowadays still talk about Gestalten and not about second order qualities, figural moments or fusions (Cf. Simons 1988, 160).

But what exactly is meant when we talk this way? The first examples came from auditory experience. As just mentioned, Ehrenfels' favourite example is a melody and Stumpf, in his explanations of the phenomena of fusion, concentrates on musical chords. Both are structures composed (diachronically and synchronically) of tones, but according to these authors they don't reduce to mere collections of tones. However, the auditory experience is not the only area where we encounter Gestalten. For, uncountable Gestalt qualities like spatial figures, threatening movements, and friendly faces dominate our visual experience. Other examples are smells and tastes. Some more complex Gestalten, like "wetness", span different sensory modalities. In the realm of inner experience moods and emotions can, arguably, be interpreted as Gestalten. And, as soon as we allow for higher order Gestalt qualities, even the most sophisticated cultural products can be construed this way (Cf. Smith 1988a, 16). The concept of Gestalt as introduced by Ehrenfels and employed in Gestalt-psychology, stand as a very general device that is able to structure practically all of our fields of experience.

Ironically enough, it is exactly this generality of the concept of Gestalt that brings its explanatory power and scientific legitimacy into question. Indeed, in the hands of some partisans of Gestalt-theory almost everything became a Gestalt; and the vague programmatic claim that a Gestalt is generally "something more" than the underlying material, can, when taken dogmatically, actually prevent any attempt at a serious reductive analysis. Here is the standard objection: First of all, if each and every complex structure we can encounter in our experience can be interpreted as a Gestalt, then the very concept of Gestalt is informatively empty and scientifically useless. We need at least some tentative criteria allowing us to distinguish between Gestalten and non-Gestalten. Second, as long as we haven't heard anything more substantial about Gestalt qualities being "non-reducible" to the underlying material, we are justified in dismissing all the Gestalt-talk as a piece of—suggestive, to be sure, but at the end of the day fruitless—psycho-ontological voodoo.

12.2 Ehrenfels on Gestalten

For his part, Ehrenfels tried to address at least the first part of the standard objection. The characteristic features of Gestalten are, according to him, the following: (A) ontological dependence on the foundation, (B) automatic generation,

(C) supersummativity, and (D) transposability. (Cf. Simons 1988, 164–167). Let me explain these points.

(Ad. A) The ontological dependence of Gestalten on their foundations means that there can be no Gestalt without some underlying plurality of elements. If we hear a melody, there must be a certain plurality of tones that "constitutes" it; if we see a spatial figure, there must be a plurality of lines arranged in a certain way, etc.

(Ad. B) The claim of automatic generation of Gestalten was one of Ehrenfels' main dogmas. According to his account, a corresponding Gestalt is given automatically when an appropriate foundation is given. In particular, the emergence of a Gestalt doesn't require any mental activity on the part of the involved conscious subject. We do not "produce" it, like some alternative theories have claimed.[4] Hence, if a certain plurality of tones that is able to constitute a certain melody is given to us, then we can't help but hear the melody.[5]

(Ad. C) A firm conviction that Gestalten are "something more" than a "mere sum" of the elements constituting their foundation, is a kind of trade mark of the whole Gestalt-theory. Precisely *because of that* Gestalten are psychologically and ontologically interesting. This claim is therefore very important. This said, it is also extremely vague. For, we do not know what "a mere sum of elements" is supposed to be. Is it a set, a mereological whole, or something else? Nor is it explained what "being something more" (than a mere sum) means exactly.

(Ad. D) The same Gestalt can appear on the basis of very different materials. The idea here is that the same melody can be played in various keys and on different instruments, and the same figure can appear at various places, in various sizes, and colours. This transposability is probably the most striking feature of Gestalten and the one that turned the philosophers' attention to the Gestalt-phenomena in the first place. It is also crucial for Grelling's analysis.

If we look at these four points, it appears to be a very natural move to construe a Gestalt as *a relation obtaining between the elements of its foundation*. A particular melody *M* would be, according to this proposal, nothing other than a rather complicated relation involving all the intervals and temporal arrangements of tones constituting a certain particular performance of *M*. If this relation remains unaltered, the melody remains the same, independently, in particular, of the absolute pitch of the beginning tone. That's why we can play the same melody in different keys.

In our logical notation we even have a nice device for extracting such relations in the form of the, so called, lambda-abstraction. To see how this works, just take an arbitrary complex description of a particular melody involving, say, four tones t_1, t_2, t_3, t_4. Let this be symbolised as "... t_1 ... t_2 ... t_3 ... t_4 ...." The expression "... t_1 ... t_2 ... t_3 ... t_4 ..." is to be understood as a (rather long), true sentence describing exactly how the four tones stand with regard to each other concerning

[4]On the "Production Theory" defended by Meinong and his followers see Smith (1988a, 26 ff).

[5]Ehrenfels writes: "Thus we can conclude that Gestalt qualities are given in consciousness simultaneously with their foundations, without any activity of mind directed towards them" (Ehrenfels 1890, 112 [152]).

their temporal order and interval sequence. Now, if we replace the names of the tones by variables and have them bound by the lambda-operator ("λ") we come to the expression "$\lambda\, x_1, x_2, x_3, x_4\, (\ldots\, x_1\, \ldots\, x_2\, \ldots\, x_3\, \ldots\, x_4\, \ldots)$", which behaves, syntactically, like a predicate. Concatenated with the appropriate number of names, it yields a sentence. The sentences: "$\lambda\, x_1, x_2, x_3, x_4\, (\ldots\, x_1\, \ldots\, x_2\, \ldots\, x_3\, \ldots\, x_4\, \ldots)$ Earth, Saturn, Mars, Venus", or "$\lambda\, x_1, x_2, x_3, x_4\, (\ldots\, x_1\, \ldots\, x_2\, \ldots\, x_3\, \ldots\, x_4\, \ldots)$ Paris, Chicago, Moscow, Barcelona" will be of course false. However, if we instead use four names of particular tones, we can get a true sentence. In particular the sentence:

$$\lambda\, x_1,\, x_2,\, x_3,\, x_4,\, (\ldots\, x_1\, \ldots\, x_2\, \ldots\, x_3\, \ldots\, x_4\, \ldots)\mathrm{t}_1,\, \mathrm{t}_2,\, \mathrm{t}_3,\, \mathrm{t}_4$$

is true, and means exactly the same as:

$$\ldots\, \mathrm{t}_1\, \ldots\, \mathrm{t}_2\, \ldots\, \mathrm{t}_3\, \ldots\, \mathrm{t}_4\, \ldots$$

Now consider another collection of four tones (t_5, t_6, t_7, t_8). Assume that they are exactly one octave higher than the original t_1, t_2, t_3, t_4, and that they are played in the same temporal arrangement. It should be obvious that the sentence:

$$\lambda\, x_1,\, x_2,\, x_3,\, x_4,\, (\ldots\, x_1\, \ldots\, x_2\, \ldots\, x_3\, \ldots\, x_4\, \ldots)\mathrm{t}_5,\, \mathrm{t}_6,\, \mathrm{t}_7,\, \mathrm{t}_8$$

will be also true.

For the sake of illustration, I used the much celebrated example of a melody, but the lambda device is of course absolutely general. Independently of the nature of the described objects and the complexity of the involved discourse, we are always able to extract the precise relation connecting the individuals we are talking about. As noted above, the lambda abstraction shouldn't be construed as producing names of relations. From a syntactical point of view, lambda terms are predicates. So it would be inappropriate to say that they "refer to relations", unless we are willing to treat predicates as referring expressions. Nonetheless, it is very natural to assume that they "have" relations as their "semantic values".

So, should we assume that repeatable Gestalten are in fact nothing other than repeatable relations extractible by a lambda abstraction? In fact this proposal seems very attractive, insofar as it subsumes the concept of Gestalt under a familiar ontological category.[6] However, Ehrenfels didn't accept it. True enough, he construed the very category of relation as a special case of Gestalt (Ehrenfels 1890, 101 f. [143]), but he also protested against the general identification of Gestalt qualities with the complex relations uniting the elements of their foundations. On this point, he wrote:

> It will not do, however, to identify the relation with any of the Gestalt qualities so far considered and to assert, for example, that a melody is nothing other than the sum of the similarities and differences of its individual tones, the square nothing other than the sum

[6]This was Meinong's and Marty's view. Cf. Meinong (1894), 323 f.; Marty (1908), 110. Grelling and Oppenheim report that also Ajdukiewicz proposed this interpretation of the concept of Gestalt. Cf. Grelling and Oppenheim (1937/38a), 196.

> of the spatial similarities and differences of its components. The melody can be heard, the square seen, not however any similarity and difference of two tones or two spatial determinations. And there is a further respect in which the relation is distinguished from other Gestalt qualities: it cannot come into being without some contribution of our part, without the specific activity of comparison. (Ehrenfels 1890, 102 [143])

It seems that Ehrenfels' arguments for not construing Gestalt qualities as relations—their accessibility to sensory perception and the fact that they appear without any conscious activity—are rather weak. The answer to the question whether we can "perceive" things like melodies and spatial figures or not, depends heavily on our concept of perception, and nowadays we are very far from the atomistic picture that Ehrenfels relied upon as the standard view.[7] Nor are we inclined any longer to regard relations as something particularly "subjective" and essentially connected with our "activity of comparison".

But it is not my goal to analyse Ehrenfels' arguments here. For the purpose of this paper all that must be noted is that, according to him, a Gestalt quality is definitely "something more" than the relation obtaining between the elements of its foundation. A Gestalt is a certain supplementary "positive quality" of the complex in question. In his own words:

> By a *Gestalt quality* we understand a positive content of presentation bound up in consciousness with the presence of complexes of mutually separable (i.e. independently presentable) elements. That complex of presentations which is necessary for the existence of a given Gestalt quality we call the *foundation* [*Grundlage*] of that quality. (Ehrenfels 1890, 93 [136])

It is not easy to understand what this supplementary quality (over and above the mentioned relation) is supposed to be, and this is very important. For, it seems that Gestalt ideology, as such, stands or falls with this claim. To see how this is so, suppose for a moment that we agree to identify Gestalten with relations of some sort. What would happen to the very concept of Gestalt? Would we still have any use for it? We might, but it would certainly loose its most attractive and mysterious aspect. That is, it would no longer be a special ontological category *sui generis* and the Gestalt theory would become a mere province of the theory of relations.

12.3 The General Ontological Framework

From the preceding sections we can see that the idea of Gestalt is far from being clear. Grelling was, in fact, deeply dissatisfied with the vagueness with which the Gestaltists "explained" the central concept of their theory. On the other hand, he was also convinced of the scientific merits of the Gestalt-theory. Thus, he set out to give the concept of Gestalt a firm logical basis.

---

[7]In fact we are so far from this picture mainly *because* of the work of Ehrenfels and his followers…

The first, and possibly crucial, aspect of the vagueness of the concept of Gestalt stems from the fact that it involves both psychological and ontological dimensions. Indeed, in the original writings of the Gestalt psychologists these two dimensions are systematically confused and extremely difficult to disentangle. The observation initiating the Gestalt movement was that our *perception* of organised wholes seems to involve something more than the perception of their foundation, but after this psychological observation we are almost immediately persuaded that the *composite objects* that we perceive must therefore be something more than mere collections of their components.

These, of course, are two different points! Ontologically speaking there neither are, nor were any witches. But, from the psychological point of view, the concept of witch proved to be quite important. Indeed, certain beliefs involving this concept have had a tremendous causal impact on our history. Thus, to this extent witches are psychologically real while ontologically fictitious. Couldn't Gestalten be like witches?

Another point illustrating the oscillation of the concept of Gestalt between its ontological and psychological dimensions is the aforementioned controversy between the automatic generation view and the production theory. As long as we consider Gestalten to be a primarily ontological category, the automatic generation view seems to be the only sensible option. However, as soon as we switch to the psychological considerations, the theory of production begins to look quite attractive. For, even if Gestalten are "directly given" to our mind, there can be unconscious psychological mechanisms producing them. From a psychological point of view the question of the existence and nature of such mechanisms is absolutely fascinating, whereas, for an ontologist, it's rather boring, because such a scholar is only concerned with a structural analysis of what is given.

Of these two dimensions, only the second one is relevant to Grelling's analysis. For, he concentrates solely on the ontology of Gestalt, while totally neglecting its psychology. As such, the questions that he wants to clarify do not concern the psychological mechanisms constituting Gestalt qualities before our minds. Rather, he asks what kind of entity a Gestalt quality is, what distinguishes a Gestalt quality from other kinds of qualities, and how it depends on its foundation. The obvious assumption of this analysis is that Gestalt is in fact an ontologically important category. However, it is important to note that even the possible discovery, that from an ontological point of view, Gestalten are like witches, wouldn't automatically make Grelling's work pointless. The ontology of a world *as it appears to us*—a kind of "phenomenological ontology" or "Erscheinungslehre"—can be philosophically interesting even if it should turn out that "in reality" there is no such a world.[8]

---

[8]"Erscheinungslehre" is the name used by Hedwig Conrad-Martius. Cf. Conrad-Martius (1916). Also Franz Brentano, who actually didn't believe in the existence of qualities given to us by our "outer experience", has in his philosophy an important place for such a kind of "as if" ontology. Cf. Brentano (1982), 14 ff.

Be this as it may, only the ontology of Gestalten will be the topic of the remainder of this paper. Before moving to the details of Grelling's explanation, let me clarify one further general point. Normally, in answering questions of this kind we assume a certain general ontological framework. When we ask whether there are negative facts or not, we presuppose a certain general ontology of facts, when we ask if a 4 week old embryo is the same entity as a child born 8 months later, we presuppose a certain general theory about the identity of objects enduring over time. So if we are going to clarify the special status of Gestalt qualities it is reasonable to suppose that we already have a certain ontological theory telling us what qualities, in general, are. And, in fact, both Ehrenfels and Grelling accepted certain "standard" theories of qualities. But, interestingly, their "standards" were quite different.

Practically the whole Brentanian tradition construed qualities (or properties in general) as individual moments that are ontologically dependent on their bearers. According to this view, qualities are what we nowadays call "tropes".[9] Each red apple has its individual redness which comes to being, and passes away, together with this apple, and what makes all individual moments of redness into instances of redness is the relation of similarity obtaining between them.[10] While Ehrenfels doesn't explicitly make this point, it is this concept of quality that surely stands as the background to his Gestalt paper.

A natural alternative to the trope account is a theory construing properties as universals. This view, in Aristotelian or Platonic form, became prevalent in twentieth century analytic philosophy, even thought, just like the tropist background of Brentanists, it was very seldom made explicit. According to this construal, properties are entities repeatable in many individuals, and what is ontologically dependent on the individual concrete bearer is not the property itself, but rather its particular exemplification. On this view, when I decide to eat an apple, "its" red colour doesn't disappear from the universe.[11] Exactly the same redness is still exemplified by many other things (and on the Platonic version of the theory, this redness even exists independently of its exemplifications). What ceases to exist is only a certain particular exemplification of redness.

Now, Grelling's ontological tools are neither tropes nor universals, but sets. This has to do with one of the most celebrated fetish of twentieth century logical philosophy—that of "extensionalism." Unsurprisingly, an extensionalist only wants to use explanations expressed in extensional discourse; and a piece of discourse is extensional if and only if the extension of it is a function of the extensions of its parts. Entities, traditionally stipulated as extensions, are: an individual for a proper name, a set for a monadic predicate, a set of ordered *n*-tuples for an *n*-ary predicate, and the truth values for sentences (see Carnap 1960, 48). This is why the majority of logically minded philosophers tend to assume that the properties and relations just

[9]This name has been introduced in Williams (1953).

[10]This is the standard version of trope theory. I cannot go here into details of the metaphysics of individual properties. Cf. in particular Campbell (1990), Loux (1998).

[11]See e.g. Armstrong (1978), Loux (1978), Chisholm (1989).

couldn't be anything else than some set theoretical constructions out of the entities that the common sense treats as being the bearers of properties (or terms of relations) in question. On this construal, the property *redness* is thus nothing other than the set of all red objects and the relation of *being bigger than* is the set of all ordered pairs where the first element of each pair is bigger than the second.

Now, extensionalism understood as a *methodological postulate* is a quite reasonable position. For, extensional discourse has some nice logical features, and there is surely nothing wrong in using it everywhere it suffices as a tool of analysis. However, the crucial question is, of course, whether in all the various cases it really *is* a sufficient tool. There are many classical arguments showing that the extensional construal of properties and relations leads to overtly absurd consequences. The sad truth is that, if properties were really nothing over and above sets of their bearers, then being a rational animal would be exactly the same thing as being a featherless biped. If you want to be extensionalist, you have to be prepared to swallow this result.

I believe that such counter-intuitive consequences suffice to warrant the dismissal of extensionalism, at least in this crude form, as a fundamentally wrong metaphysics. But, as it would end our journey with Grelling before it even began, let me try to play an *advocatus diaboli*, and say how extensionalism can be improved. An obvious move is to allow for the set theoretical constructions out of *possibilia*. That is, if our rational-animal-set and featherless-biped-set contained not only *actual* rational animals and featherless bipeds but also *possible* ones, then these sets would no longer coincide. For, there are, of course, some possible rational animals that have feathers, and some possible featherless bipeds that aren't rational. The price of this improvement—the introduction of possible objects in our ontology—isn't low. But, I believe that it is the only way to have a set theoretical metaphysics which is not obviously inadequate.

Now Grelling doesn't introduce possibilia. Thus, his set theoretical framework operates exclusively with actual entities and, as such, it is clearly inadequate. Nonetheless, all his results could be easily transposed into a possibilist framework. Just take one of the numerous ontologies of possible worlds which are these days widely (and typically unreflectively) accepted as the semantics for modal logic.

12.4 The Analysis of the Concept of Gestalt

In Sect. 12.2, above, I listed some features of Gestalten pointed out by Ehrenfels. From these features it was the *transposability* that Grelling and Oppenheim took to be crucial. According to their analysis, Gestalt qualities are transposable characteristics of some organized wholes that they call "complexes". In short, they think that Gestalt can be described as "invariant of transpositions." Accordingly, the definition that they accept as "fitting better with the logistic way of speaking" reads:

> Gestalten are equivalence classes of correspondences. (Grelling and Oppenheim 1937/38a, 196)

Needless to say, this definition "fits better with the logistic way of speaking" because it fits better with the ideology of *extensionalism*. That is, Gestalten have to be interpreted as equivalence classes of some kind, because *all* qualities are equivalence classes.

But let us return to the details of the Grelling-Oppenheim analysis. To formulate this definition, they introduce many auxiliary concepts. The first ones are that of *classifier* and *state-classifier* (S-classifier). Classifiers are general categories of families of mutually exclusive properties like "pitch of a tone" or "state of matter". They ascribe to their *arguments* (e.g. a tone or a physical body) certain *values* (like 440 Hz or solid). S-classifiers are classifiers of a certain special kind. They are classifiers whose arguments are "positions" in a "domain of positions" (Grelling and Oppenheim 1937/38a, 193). An example of a domain, understood in this sense, is a set of the points of space-time. It is assumed that a domain of positions is ordered by a certain "positional relation" (Grelling and Oppenheim 1937/38a, 195). Given these commitments, the concept of *complex* can be defined in the following way:

> In general [...] a complex is a relation between a class of S-classifiers and a domain of positions, such that every S-classifier assigns a value to each position in the domain. (Grelling and Oppenheim 1937/38a, 193)

Such a complex could be, for example, a sequence of musical tones. It involves a certain domain of positions (a time continuum), and ascribes to each argument, from the domain, a set of classifiers like pitch, timbre, intensity (a rest can be understood as zero intensity value). According to this description, a complex is a rather complicated set theoretical construction. For, remember, a relation is itself a set of ordered *n*-tuples.

The next important notion is that of *course of values*. Grelling and Oppenheim explain this concept through the example of a temperature chart (Grelling and Oppenheim 1937/38a, 193). A temperature chart is a graphic representation of the course of values of temperature in time. Temperature is, here, a single S-classifier, while the time continuum functions as a domain of positions. The crucial point is that beside isolated pairs, consisting of moments of time and values of temperature (at this time), we can track the "development" of temperature over time. Indeed, there are various "characters" of such a development (like e.g. "malarial"). For another example, we can return to a melody. Here the domain of position is also a time continuum, while the relevant S-classifier is the pitch of a tone. On this example, various "characters" of courses of values precisely amount to what we call "melodies".

Given these examples, we can see that it is sometimes important to speak of particular "characters" of courses of values. Thus, it is unsurprising that Grelling and Oppenheim introduce the next kind of classifier: the *course of values classifier*, or C-classifier for short. Such concepts as "interval sequence" (which characterizes a melody) or "character of a temperature chard" are C-classifiers in this sense.

With this development, we are now not far from the concept of Gestalt. The next point of observation is that some complexes are similar, in the sense that they embody similar courses of values. This means that a certain relation, which Grelling and Oppenheim call "correspondence", obtains between them. Indeed, in their eyes, it is this relation that is the key to understanding of the concept of Gestalt. In the simplest case this relation obtains between two complexes if:

> (1) Between the domains of positions there is an *isomorphism* with respect to their positional relation, i.e. there is a mapping of the one domain onto the other such that relative positions are preserved. (2) The S-classifiers are pairwise identical. (3) The course of values of corresponding, or, one might say "homologous" S-classifiers are equal. (Grelling and Oppenheim 1937/38a, 195)

Thus, this kind of correspondence obtains between two exactly similar performances of a piece of music (such as when we play the same CD at two different times) or between two exactly similar pictures placed at two different places (e.g. two ideally accurate prints from one and the same negative in the same scale).

But the relation of correspondence, as defined above, is far too strong to be used directly in analysis of the concept of Gestalt, at least as it is traditionally understood. For, this relation is not preserved when we print one of the pictures in black and white and the second one in colour, and it is not preserved when we print them both monochromatically, but change the reproduction scale. Now, there is doubtless an important sense in which a bigger picture corresponds to the smaller one, and the monochromatic print corresponds to the colour reproduction of the same motif.

Therefore, Grelling and Oppenheim propose to weaken the concept of correspondence. The main idea is that, for a given correspondence, *not all* S-classifiers and C-classifiers need to be taken into consideration (Grelling and Oppenheim 1937/38a, 195). Here are some examples. If we drop all the S-classifiers concerning the colour differences as irrelevant, then a monochromatic copy can be regarded as corresponding to the colour original. If we focus solely on the S-classifiers concerning the pitch of tones, then two performances of the same melody played on different instruments (but in the same key) can be regarded as corresponding. And if we consider only the classifiers concerning the sequence of intervals (and not the absolute pitch), then the correspondence will also obtain between two performances of the same melody played in different keys.

With this preliminary explanation, the generalized concept of transposition, as employed in the Gestalt theory can be introduced:

> We shall call "*transposition*" with respect to a given correspondence the operation which takes one complex into another which stands in a given correspondence to it. (Grelling and Oppenheim 1937/38a, 195)

It is this concept, according to Grelling and Oppenheim, that stands at the very heart of the notion of Gestalt. They claim that Gestalt is nothing other than an "invariant of transpositions."

Equivalence, so defined, is *inter alia* also an equivalence relation in the technical sense of the word (which means that it is a relation which is reflexive, symmetric, and transitive). As such it can be used to build the so-called equivalence classes.[12]

---

[12]On equivalence classes and their relations to other mathematical entities cf. Grelling (1969, 1970).

Accordingly, the concept of Gestalt can be finally defined as an "equivalence class of correspondences" (Grelling and Oppenheim 1937/38a, 196).

What has been defined so far, is the concept of a "Gestalt-individual", as these authors call it. An example of such a Gestalt-individual would be *the Blue Danube-waltz-melody*. But the Blue Danube-waltz-melody is, of course, a kind of melody (it falls under the concept of melody); and melodies are (together with spatial figures, dances, tastes etc.) Gestalt-qualities. Thus, in the idiosyncratic terminology of the Grelling-Oppenheim papers, we need to see a new kind of classifier. And indeed, they go on to distinguish between "(1) the classifier 'melody', (2) its arguments, individual tone sequences, (3) its values, the melodies of these tone sequences" (Grelling and Oppenheim 1937/38a, 196). Finally, we read that "in general 'Gestalt' can be represented as a classifier whose arguments are complexes and whose values are Gestalt-individuals" (Grelling and Oppenheim 1937/38a, 196 f.).

Admittedly this concentration of rather unusual terminology can significantly disturb our understanding of these otherwise clear ideas. A table illustrating the introduced concepts, with examples from various fields of our life, that we find in the short paper "Supplementary Remarks on the Concept of Gestalt" (Grelling and Oppenheim 1937/38b, 208), would therefore be very useful. Here I reproduce a part of it.

Fundamental concepts	Everyday life	Music	Psychology
Complex	A house	A sequence of tones	Black dots on white ground
State-classifier	"Material"	"Pitch"	"Colour"
Argument of the state-classifier	Place in space	Place in the sequence of tones	Place in the visual field
Value of the state-classifier	Stone	C	Black
Correspondence	Relation between model and house	Equality with respect to melody	Equality in phenomenal grouping
Transposition	Change of scale of measurement	Transposition	A certain change in the distances of the points
Gestalt ("Gestalt-quality")	Plan of the house	Blue Danube-waltz-melody	The phenomenal grouping
Quality of Gestalt	Symmetrical	Triple time	Stable

12.5 Dependence Systems

As mentioned many times, the concept of Gestalt tended to become all embracing. In particular, at the time of the Grelling and Oppenheim papers, it was also widely used to designate organized systems of causally interconnected aspects. However, Grelling and Oppenheim rightly noted that this notion of Gestalt is quite different from Ehrenfels' original concept, and therefore proposed to call such causally organised wholes *functional wholes* or *determinational systems* (*Wirkungssysteme*). The last part of the paper "The Concept of Gestalt in the Light of Modern Logic" and the whole paper "Logical Analysis of 'Gestalt' as 'Functional Whole'" is devoted to the analysis of this notion.

An example of such a functional whole, that they take from Köhler (1920, 54 ff.), is the distribution of charge on the surface of an isolated conductor. The feature that Köhler focused upon is that "the density of charge at any point determines the density at all others" (Grelling and Oppenheim 1939, 210). With regard to this example Grelling and Oppenheim claim that "whenever modern Gestaltists use expressions such as 'functional whole', 'organized whole', 'dynamic unity', they ascribe this property of 'interdependence' to their respective designata" (Grelling and Oppenheim 1939, 210). Such functional wholes are contrasted with merely "aggragative" wholes (*summative Ganzen, Und-Verbindungen*) that lack such interdependencies. An example, taken again from Köhler, are "three stones lying in three different continents. [...] The characteristic of such an aggregate may be called 'independence'." (Grelling and Oppenheim 1939, 210 f.)

Grelling and Oppenheim begin their analysis with the notion of *dependence*. At this point, they talk about the dependence of *functions*, and more precisely, of dependence of a function on a class of functions. Roughly, a function is said to be dependent on a class of functions if its value varies systematically with the values of functions belonging to this family. An example of such dependence can be the familiar law of economics saying that price is a function of supply and demand. The main idea is that if the values of demand and supply, at the time t_1, are exactly the same as at the time t_2, then the price also must be equal at t_1 and t_2. Here is the definition of dependence from Grelling's and Oppenheim's joint paper:

> [A] function *f* will be said to *depend* on a class φ of functions, when and only when *f* has the same value for any two arguments for which each element of φ has equal values (Grelling and Oppenheim 1939, 211).

A formal version from Grelling's own paper, "A Logical Theory of Dependence", reads (in a slightly altered notation) as follows:

Equidep $(f,\varphi) =_{df.} \forall x \forall y \{\forall g\, [g \in \varphi \supset g(x) = g(y)] \supset f(x) = f(y)\}$ (Grelling 1939, 218).[13]

Grelling's and Oppenheim's idea is that the notion of a functional whole, as employed by Gestaltists, can be defined by means of the concept of functional dependence as defined above. A functional whole can be understood as an organized complex in which the values of elements are mutually interdependent. This concept is contrasted with the non-connected aggregates (like the mentioned three stones).

---

[13]Grelling distinguishes in this paper between *logical* and *causal* dependence by means of Carnap's notions of *L*-truth and *F*-truth. He writes: "I want to suggest the following formulation: we speak of logical dependence if the definiens [...] is an *L*-true sentence, and of causal dependence if it is an *F*-true sentence" (Grelling 1939, 225).

To capture these notions Grelling and Oppenheim moved to introduce the following concepts of *interdependence* and *independence*:

> [A] class of functions, φ, will be called '*interdependent*' when and only when every element f of φ *depends* on the 'complementary class' consisting of all elements of φ except f (Grelling and Oppenheim 1939, 211).
>
> [A] class φ of functions will be called '*independent*' when and only when no element of φ *depends* on the complementary class. (Grelling and Oppenheim 1939, 211)

It is important to note that this opposition is not contradictory. Some class of functions are neither dependent nor independent.[14]

At this point, it is time to introduce the concept of a system, understood here as a class of functions, considered with respect to a certain relation. Grelling and Oppenheim write, that a class of functions φ is a system, with respect to a relation R, if and only if "this relation holds between each element of φ and the complementary class" (Grelling and Oppenheim 1939, 212).[15]

Now, if we substitute the relation of dependence, as defined above, for the relation R, with respect to which the concept of system has been defined, we get the notion of a *dependence system*. Grelling and Oppenheim believe that what the Gestaltists are referring to, when they talk about functional wholes as opposed to mere aggregates, are in fact just *systems of functions with respect to dependence*.[16]

12.6 Conclusion

The ontological clarification of the concept of Gestalt as presented above is doubtless an excellent piece of philosophical analysis. Already a clear distinction, between Gestalten as structural aspects available to transposition and Gestalten as causally self-regulating wholes, can save us from many dangerous confusions. I am not sure if the concept of a functional system as defined by Grelling and Oppenheim is able to capture *all* intuitions involved in Gestaltist's talk about organized wholes or dynamic unities, but it clearly formalizes at least *a part of them*, and suggests that other intuitions should be treated analogously.

---

[14]"In terms of the preceding analysis the opposition between aggregative and functional whole turns out not to be contradictory. For a class of functions can happen to be neither dependent nor independent: indeed *some* of its elements may depend on their complementary classes and others may not" (Grelling and Oppenheim 1939, 215). Grelling and Oppenheim suggest that with the help of the notion of probability it will be also possible to speak of "more or less dependent" functions (Grelling and Oppenheim 1939, 212).

[15]They introduce also the notion of a closed system: "A system which is not a part of a larger one with respect to the same relation may be called '*closed*'" (Grelling and Oppenheim 1939, 212 f.).

[16]"Now it looks plausible to translate the complete expressions 'functional whole' and the like in terms of 'system of functions with respect to dependence', or, shortly, 'dependence systems'" (Grelling and Oppenheim 1939, 213).

However, there remains a serious worry as to whether the concept of Gestalt, defined as an equivalence class of a certain correspondence, really corresponds to the ideas of Ehrenfels and his followers. For, at first sight, it seems that in the light of Grelling's and Oppenheim's analysis, Gestalten came to be identified with some *relations* obtaining between the elements of their foundations. However, we need to remember that Ehrenfels decisively rejected such identification.

True enough, Grelling and Oppenheim also explicitly reject the general identification of the concept of Gestalt with that of relation. However, they do not make this claim because they want to see Gestalten as a *sui generis* ontological category. Rather, as they write: "We take this definition to be too broad, since then every relation would be a Gestalt, and the latter expression would be completely dispensable" (Grelling and Oppenheim 1937/38a, 196). This suggests that they would accept this definition in some sort of a restricted form. And, indeed, Gestalten are, according to their proposal, *relations of a certain special kind.* (Cf. Simons 1988, 184) That is, they are relations that can function as invariant in operations of transposition with respect to certain equivalences.

But is this kind of relations really so special? When we take a look at Grelling's and Oppenheim's generalized conditions of equivalence, it becomes clear that absolutely *every relation* can function this way. In fact, Simons (1988, 184) shows that every relation can be interpreted as a course of values, and every course of values can, of course, be repeated, which is all that must be secured to have a Gestalt according to Grelling's and Oppenheim's explanation.[17]

But, does this mean that Grelling's and Oppenheim's analysis of the concept of Gestalt is ultimately unsuccessful? Not necessarily. First of all, it can just be the case, that there is indeed *nothing* in Gestalt qualities over and above certain relations obtaining between the elements of their foundations. In this case, the analysis of Grelling and Oppenheim would surely stand as a good job, freeing us from a perplexing philosophical myth. But their analysis can be interesting even if Ehrenfels & Co. are right, and Gestalten really contain something more than the formal traits identified by Grelling and Oppenheim. In this case, Grelling's and Oppenheim's analysis can be interpreted as identifying at least *some important features* of Gestalten, and as showing us that if Gestalt qualities should really be something more than relations then they have to contain something more *than this.*

It also needs to be noted that it is no accident that the concept of Gestalt becomes, in the hands of ontologists, far too general. For, if the class of Gestalten were restricted, in conformity with the original intention of its inventor, then it would be necessary to take into consideration certain *psychological facts.* Gestalten were introduced from the very beginning as something that plays a unifying role in our mental life. Gestalten are what we perceive in the first place, they are central to our aesthetic and moral concepts (think of such things as "a harmonic composition",

[17]Provided only we are able to correlate their terms with a "domain of positions" ordered by a certain "ordering relation". Simons (Simons 1988, 184) shows that it is a relatively simple job if we take natural numbers.

"a charming lady", or "a good life"), they are easy to remember and to re-identify. However, these psychological aspects are precisely what Grelling and Oppenheim want to put aside. What they want to realize is a formal, ontological analysis, not a psychological investigation. But then it is no surprise that the concept of Gestalt, so defined, has hardly any psychological relevance. Thus, it is here that the aforementioned psycho-ontological ambiguity of the concept of Gestalt returns to take its revenge.

Acknowledgments The work on this paper was supported by the *Austrian Foundation for the Promotion of Scientific Research* (FWF) and the *Foundation for Polish Science* (FNP, "Master" programme scheme, directed by Tadeusz Szubka).

References

Armstrong, David M. 1978. *Universals and scientific realism*, vol. 1–2. Cambridge: Cambridge University Press.

Brentano, Franz. 1982. *Deskriptive Psychologie*, ed. Roderick M. Chisholm and Wilhelm Baumgartner. Hamburg: Meiner.

Campbell, Keith. 1990. *Abstract particulars*. Oxford: Blackwell.

Carnap, Rudolf. 1960. *Meaning and necessity*. Chicago: The University of Chicago Press.

Chisholm, Roderick M. 1989. *On metaphysics*. Minneapolis: University of Minnesota Press.

Conrad-Martius, Hedwig. 1916. Zur Ontologie und Erscheinungslehre der realen Außenwelt. *Jahrbuch für Philosophie und phänomenologische Forschung* 33: 45–542.

Ehrenfels, Christian von. 1890. Über Gestaltqualitäten. *Vierteljahrsschrift für wissenschaftliche Philosophie*. 14:249–292. Repr. in Ehrenfels 1988. 128–155. English Trans. in Smith 1988b. 82–117. Cit. acc. to the English Trans. The page numbers of the German original (acc. to the ed. 1988) are given in square brackets.

Ehrenfels, Christian von. 1988. In *Philosophische Schriften*, Bd. II: *Psychologie, Ethik, Erkenntnistheorie,* ed. Reinhard Fabian. München/Wien: Philosophia Verlag.

Grelling, Kurt. 1939. A logical theory of dependence. In Smith 1988b, 217–226.

Grelling, Kurt. 1969. On definitions by equivalence classes and by group invariants. *Methodology and Science* 2: 116–122.

Grelling, Kurt. 1970. On the logical relations between groups and equivalence relations. *Methodology and Science* 3: 5–17.

Grelling, Kurt, and Leonard Nelson. 1908. Bemerkungen zu den Paradoxien von Russell und Burali-Fori. *Abhandlungen der Fries'schen Schule. Neue Folge* 2:301–334. Repr. in Nelson 1974. 95–127.

Grelling, Kurt, and Leonard Nelson. 1937. Gibt es eine Gödelsche Antinomie? *Theoria* 3: 297–306.

Grelling, Kurt, and Paul Oppenheim. 1937/38a. The concept of Gestalt in the light of modern logic. In Smith 1988b. 191–205 [English Trans. of: Der Gestaltbegriff im Lichte der neuen Logik. *Erkenntnis* 7 (1937/38):211–225].

Grelling, Kurt, and Paul Oppenheim. 1937/38b. Supplementary remarks on the concept of Gestalt. In Smith 1988b, 206–209 [First publ. in *Erkenntnis* 7 (1937/38):357–359].

Grelling, Kurt, and Paul Oppenheim. 1939. Logical analysis of 'Gestalt' as 'Functional Whole'. In Smith 1988b. 210–216.

Husserl, Edmund. 1891. *Philosophie der Arithmetik*. Halle 1891 (Husserliana XII. ed. L. Eley. Den Haag 1970). Cit. acc. to the Husserliana ed.

Köhler, Wolfgang. 1920. *Die physischen Gestalten in Ruhe und im stationären Zustand. Eine naturphilosophische Untersuchung*. Braunschweig: Vieweg.

Loux, Michael J. 1978. *Substance and attribute*. Dordrecht: Reidel.
Loux, Michael J. 1998. *Metaphysics. A contemporary introduction*. London/New York: Routledge.
Marty, Anton. 1908. *Untersuchungen zur Grundlegung der allgemeinen Grammatik und Sprachphilosophie*. Halle: Niemeyer.
Meinong, Alexius. 1894. Beiträge zur Theorie der psychischen Analyse. In Meinong 1969–78. Vol. I:305–395.
Meinong, Alexius. 1969–78. *Gesamtausgabe*, ed. Rudolf. Haller et al. Graz: Akademische Druck- und Verlagsanstalt.
Nelson, Leonard. 1974. *Gesammelte Schriften III. Die kritische Methode in ihrer Bedeutung für die Wissenschaften*. Hamburg: Meiner.
Simons, Peter M. 1988. Gestalt and functional dependence. In Smith 1988a, 158–190.
Smith, Barry. 1988a. Gestalt theory: An essay in philosophy. In Smith 1988a, 11–81.
Smith, Barry (ed.). 1988b. *Foundations of Gestalt theory*. München/Wien: Philosophia Verlag.
Stumpf, Carl. 1890. *Tonpsychologie*, vol. II. Leipzig: Hirzel.
Williams, Donald Cary. 1953. The elements of being I. *The Review of Metaphysics* 7: 3–18.

Part VI
Paul Oppenheim and Carl Hempel

Chapter 13
Paul Oppenheim on Order—The Career of a Logico-Philosophical Concept

Paul Ziche and Thomas Müller

13.1 Paul Oppenheim—The Co-operative Philosopher

Paul Oppenheim (1885–1977), initially trained as a chemist but "addicted to philosophy" from early on, and annoying scientists by asking "tedious" questions concerning his philosophical attempt to systematize the sciences,[1] is a curious member of the so-called "Berlin Group." Though being associated with this group in a number of ways, he does not regularly appear in listings of its prominent members. For instance, when listing authors that were close to the Vienna Circle, the manifesto of the Vienna Circle does name some members of the Berlin Group—Reichenbach, Dubislav and Grelling—, but omits any reference to Oppenheim (Hahn et al. 1929, 328). Oppenheim's status with regard to those groups of scholars is as difficult to assess as it is with regard to the development of twentieth-century philosophy of science as a whole. His name is connected with a number of landmark papers in the development of what came to be orthodox philosophy of science from the 1940s onward—but in these papers he invariably figures as a co-author, and he will probably remain unrivalled as the greatest philosophical co-author of the twentieth century.

[1] Max Born wrote concerning Oppenheim: "The Oppenheims had a son, who was a business partner of his father, but addicted to philosophy; he wrote a book concerning which he involved me in many a tedious discussion" (quoted after Rescher 1997, 337).

P. Ziche (✉) • T. Müller
Department of Philosophy, University of Utrecht, Utrecht, The Netherlands
e-mail: paul.ziche@phil.uu.nl

N. Milkov and V. Peckhaus (eds.), *The Berlin Group and the Philosophy of Logical Empiricism*, Boston Studies in the Philosophy and History of Science 273, DOI 10.1007/978-94-007-5485-0_13,

Besides his long-lasting and fruitful collaboration with Hempel that resulted in the canonical text on scientific explanation (Hempel and Oppenheim 1948a),[2] he also worked together with Hilary Putnam, John Kemeny, Olaf Helmer, Nicholas Rescher, Kurt Grelling, Nathan Brody, Hugo Bedau and Siegwart Lindenberg.[3] Throughout his career, Oppenheim described his own role in producing these joint papers in very modest terms.[4] Together with the scarcity of biographical information, and the wide range of issues that apparently interested Oppenheim—from methodological issues in the philosophy of science to the philosophy of economics and to problems in the philosophy of quantum mechanics—this modesty has contributed to obfuscating Oppenheim's own agenda.

Compared to the quantity and the great prominence of his collaborative work, Oppenheim's work as a single author is much less known, and, at least at first sight, is restricted to just one topic. As far as we could ascertain, Oppenheim's autonomous work can be chronologically grouped into two sets of texts, besides his dissertation as a chemist (Oppenheim 1908) and a short personal piece in honour of Hempel (Oppenheim 1969). In the 1920s and 1930s, he published one book (Oppenheim 1926), followed by two short summaries of its basic ideas (Oppenheim 1928, 1930) and a short article (Oppenheim 1937) that is already marked as coming from a collaboration with Hempel. After quite some time, he offered another substantial article (Oppenheim 1957), together with a short discussion note (Oppenheim 1959), both echoing the titles and the content of his works from the 1920s. None of these works has attracted much attention. They are sparsely cited,[5] and they have not yet

[2]This paper has been reprinted in several important anthologies (e.g. Feigl and Brodbeck 1953, 319–352). On the role of these anthologies in shaping a standard view of logical empiricism, see Giere (1996, 338). On the role of this paper within the history of theories of explanation, see Salmon (1990).

[3]Up to now, no complete list of Oppenheim's (co-)publications is available; but compare the data given in the biographical sources in note 12. Without claiming absolute completeness, the present paper tries to fill some gaps in the bibliography of Oppenheim.

[4]For typical examples, repeated in almost stereotypical fashion in virtually all of Oppenheim's joint papers, see e.g. Helmer and Oppenheim (1945), 25 note 1; Hempel and Oppenheim (1948a), 135 note 1; Kemeny and Oppenheim (1952), 307 note 1. On the style of his collaborative projects, see Rescher (1997), 341: According to Rescher, it was Hempel who recruited "a long series of collaborators" for Oppenheim.

[5]Even if referred to, Oppenheim's early texts seem to be little used: Beth (1959), one of the few works to have Oppenheim (1926) in the bibliography, does not refer to the book in his text at all.—The reception of Oppenheim's texts would merit closer study; J.H. Woodger (himself, being a biologist with no formal training in mathematics, looking for new logical tools to analyze theory forming in biology) views the Grelling–Oppenheim and Hempel–Oppenheim texts from the 1930s as models for "applications of the new logical ideas" (Woodger 1939, 81; cf. also Woodger 1952, 326). On Woodger—who is frequently referred to in Oppenheim's texts—see Gregg and Harris (1964).—Nelson Goodman in 1946 (Goodman 1946) devotes a review to the papers by Hempel and Oppenheim and by Helmer and Oppenheim on the concept of confirmation; the review appeared just weeks after he first presented his own seminal paper on counterfactual conditionals. See also the reference to the Kemeny–Oppenheim paper on factual support in Goodman 1953, 69 note 6.—See also Kemeny (1951) on the relationship between the Helmer–Hempel–Oppenheim

been brought into systematic contact with Oppenheim's well-known collaborative pieces. In this article we embark on a first attempt to fill this lacuna. We shall try to give a voice to Oppenheim speaking for his own, and to connect the themes and methods of his autonomous work with related topics in his collaborative works.

The picture that emerges is interesting in many ways. Oppenheim's main aim can be stated clearly: He wants to understand the *order of the sciences*. This program builds forth upon one of the focal issues of late-nineteenth century reflections on science, insofar as the attempt to order the sciences has been one of the predominant occupations of late nineteenth-century thinkers (cf. Ziche 2008), and is oriented as much to traditional standards and problems in philosophy as to recent developments in the sciences. Oppenheim's approach is characteristic through its application of innovative formal methods to questions and topics that belong, at least at first sight, to rather different contexts.

"Order" here does not imply that the sciences should be placed in any sort of hierarchy. Rather, what Oppenheim is looking for is a description of the various forms of sciences that there are, and a comprehensive analysis of the relations that hold between them. He wants to work out an open, continuous, multi-dimensional ordering with the help of formal methods. This ordering is open in the sense that it can accommodate all sorts of sciences, or rather "*Wissenschaften*," and is not limited to the natural sciences (in what follows, the term 'science' will always be used in this broad sense). While in his early autonomous work formal methods are limited to algebra and analytic geometry, mentioning logic only as one of the sciences to be ordered in his greater scheme, his later work is profoundly influenced by the formal methods of the new logic embraced by his logical-empiricist collaborators. Indeed, in his last major autonomous publication, "Dimensions of Knowledge" from 1957, we find him championing a syntactic approach to the analysis and ordering of scientific publications, while at the same time he remains faithful to his open conception of science.

The broad and tolerant view as to what counts as "*Wissenschaft*" remains characteristic for Oppenheim's approach throughout his career. His first publications include an analysis of disciplines ranging from metaphysics via mathematics to philology, history and geography, and his first joint publications with Hempel[6] and Grelling[7] are devoted to issues in the psychology of types of personalities. In these texts, he references authors such as Erich Jaensch, Ernst Kretschmer, Kurt Lewin and William Stern. It is this combination of an open and tolerant concept of "*Wissenschaft*" on the one hand, and the increasing awareness of and

ideas and Carnap's account of probability, and Kempski (1952) for an application of the Grelling–Oppenheim analysis of *Gestalt* concepts in the social sciences; on similar issues, see also Kluge (1999). In general, Oppenheim's ideas proved remarkably fertile for studies in the field of the social sciences; see, e.g. Znaniecki (1952, 182), with an affirmative reference to Hempel's and Oppenheim's ideas on the concepts of type and order.

[6]On Hempel's philosophical development, see e.g. Friedman (2000) and Wolters (2000).

[7]On Grelling, see Peckhaus (1993) and Luchins et al. (2001).

involvement with formal methods and with the orthodox ideas of logical empiricism and the newly emerging philosophy of science, on the other, that poses challenging questions, both historically and systematically. How does this broad conception of "*Wissenschaft*" fare when brought into contact with the emerging orthodoxy of logical empiricism? How clear is it that logical-empiricist orthodoxy necessarily implies a tidying up of the range of sciences that we may legitimately talk about? To put it succinctly: How does Oppenheimian tolerance survive?

13.2 Elements of an Intellectual Biography of Oppenheim

Historically, it is sufficiently clear that Oppenheim's program is related to ideas current in the Berlin Group. Oppenheim co-operated with Hempel and Grelling, there is a—though much less clearly visible—interaction with Reichenbach, and he also shares interests with Kurt Lewin whom Oppenheim quotes frequently, and who reviewed Oppenheim's book on the *Natürliche Ordnung der Wissenschaften* (Lewin 1929).[8]

The earliest hints at Oppenheim taking a lively interest in philosophy—apart from Born's indignant statement—come from biographical accounts dealing with Hans Reichenbach. The contact between Oppenheim and Reichenbach can be traced back to as early as 1921 (Rescher 1997, 338).[9] The concrete role of

[8]It is of great interest to compare Oppenheim's study with Lewin's *Comparative Theory of Science* (Lewin 1925). Lewin wants to study the "living roots" and the future perspectives of the traditional philosophical term "Wissenschaftslehre," and finds both in the "practice of the special sciences" (Lewin 1925, 49). His own method is one of "*comparative description*" (Lewin 1925, 70). Although he explicitly refers to terms from early logical empiricism (e.g. "*Einheitswissenschaft*," Lewin 1925, 57), and although his text is presented at the Erlangen conference on scientific philosophy that Carnap organized in 1923, he does not adopt a logic-based method. On Lewin's project see Köchy (2010).

[9]The Reichenbach papers in Pittsburgh would have to be consulted for further information. According to Hempel, he himself was introduced to Oppenheim by Reichenbach (Hempel 1991, 8). —In his classic *The Rise of Scientific Philosophy* (Reichenbach 1951), Reichenbach does not mention Oppenheim, although Reichenbach places quite strong emphasis on the formative role of the nineteenth century: He views the relevant developments of this period, however, exclusively in terms of the logical empiricists' account of science and philosophy.—Of great interest is Carnap's account of the conference on scientific philosophy at Erlangen in 1923 where he met Reichenbach for the first time, and where Lewin presented his idea of a comparative theory of science; on this conference see Thiel (1993). Some of the issues that we find back in Oppenheim's work have been treated at this conference: "pure logic," including "relational structures," but also "applied logic, e.g., the relation between physical objects and sense-data, a theory of knowledge without metaphysics, a comparative theory of the sciences, the topology of time, and the use of the axiomatic method in physics" (quoted from Reichenbach 1978, 40; in this volume, one also finds a considerable amount of biographical information on Reichenbach). The comprehensive volume on Reichenbach (Salmon 1979) refers to Oppenheim only via the Hempel–Oppenheim paper; the biographical account in Kamlah (1993) does not mention Oppenheim.

Reichenbach, however, is hard to pin down. According to Rescher, Reichenbach "helped Oppenheim formulate the ideas that formed the focus of his first publications," and he had a hand in getting those texts published (Rescher 1997, 338). Reichenbach probably comes closest to acknowledging a direct interaction with Oppenheim's ideas (the direction of influence may be left open for the moment) in his 1929 handbook-article *The Aims and Methods of Physical Knowledge*. Here, Reichenbach discusses questions of demarcation regarding "physics and the other natural sciences," and assembles a rather surprising list of reference authors. Any comparative characterization of the various methods in science presupposes, in Reichenbach's view, "a very precise analysis of conceptual formation within both sciences, and this belongs in the field of comparative studies of science, once again the object of intense research." With regard to this issue, he refers to Kurt Lewin, to the neo-Kantian Heinrich Rickert and the "cricital realist" Erich Becher, and to Oppenheim, of whose *Natürliche Ordnung* from 1926 he gives a short, but fitting one-sentence summary: "Oppenheim wants to order the sciences into a continuous two-dimensional schema, comparing the logical aspect of his system with the periodic system of the elements" (Reichenbach 1929, 127, 218).[10]

The breadth in scope and the tolerant attitude with regard to apparently rather different conceptions of science forms an important aspect of the practice of the Berlin Circle. Anecdotal evidence can support this point: After Reichenbach had to emigrate, it did not seem to present a problem for Hempel to switch to the *Gestalt* psychologist Wolfgang Köhler as his *Doktorvater* (see, e.g., Wolters 2000). It seems that by that moment (experimental) psychology's early problems with being accepted by the philosophical community were overcome, at least within the context of the Berlin Group.[11]

As regards Oppenheim's biography, both personally and intellectually, disappointingly little can be said.[12] In particular, his move from being a chemist and a manager in the chemical industry to embarking on a (non-professional; he never held a position at a university, although there are reports of his giving lectures at the university of Frankfurt in 1927)[13] career as philosopher of science cannot as yet be studied on the basis of documentary evidence. What can be said, is that after having studied chemistry, Oppenheim first worked in the family firm, that he was employed during World War I at the war ministry, and that since 1926 he held a post within the I.G. Farben corporation of chemical industries. In 1933, Oppenheim emigrated

[10]This text was originally published as "Ziele und Wege der Physikalischen Erkenntnis" in *Handbuch der Physik*, vol. 4: *Allgemeine Grundlagen der Physik*.

[11]Cf. Ash (1995). See also Ash (1994) and Cat (2007) for highly interesting discussions of the relationship between *Gestalt* psychology and logical positivism, including the links to the Berlin Group.

[12]The most extensive information on Oppenheim's biography can be found in Rescher (1997) and Rescher (2006). See also Schröder–Heister (1984) and Luchins et al. (2001); a nice anecdote on Einstein, Gödel and Oppenheim in Tucker (1985).

[13]Rescher (1997, 338). The archive of Frankfurt University, however, could not verify these reports on the basis of archival documents.

to Brussels, where his wife came from, and worked as a private scholar, with the aid of Hempel and Grelling. He continued this existence as a private scholar at Princeton where he arrived in 1939; he supported his colleagues, most notably again Hempel, in their plans to emigrate to the US. At Princeton, Oppenheim, together with his wife, ran what may be thought of as the modern equivalent of an eighteenth-century *salon*: a meeting-place for intellectuals, scientists and philosophers from all over the world, including, amongst many others, Einstein, Gödel and Quine. Reichenbach used to stay with the Oppenheims when visiting Princeton (Reichenbach 1993).

13.3 Oppenheim's Program: "Order"

Oppenheim's first publications carry strangely and strongly Goethean overtones. The ideas of "natürliche Ordnung," "Gestalt," and "Typus" that he employs are all well-known from traditional approaches to biological issues. Indeed, Oppenheim later names Goethe explicitly, though without committing himself in the least to Goethe's ideas (Oppenheim 1930). Still, these notions clearly place his project in an ambiguous historical framework. Oppenheim seems eager to take up older concepts and ideas, while at the same time emphasizing that his project is meant to take part in a decisively modern development that relies on modern logic as a necessary precondition. The notion of 'order' indeed has a comparably broad function within a large array of discourses in the period around 1900, featuring in modern logic as well as in a newly emerging philosophy of nature, and in more metaphysically minded projects such as in Hans Driesch's biology-inspired "Ordnungslehre."[14] Oppenheim's connection to the past is further reinforced by the fact that the very project of ordering, classifying and systematizing the various sciences is deeply rooted in the nineteenth century, and—just as the broad range of uses of the notion of "order" and Oppenheim's own tolerant attitude indicate—was usually pursued with a very open mind as to the acceptability of significantly different forms of "*Wissenschaften*" (Ziche 2008, ch. VII.2). Similar traditions can be traced for the Berlin Group which has been described as being deeply rooted in nineteenth-century traditions. Indeed, in a statement by Reichenbach that was taken up by Neurath, "Kantianism," "Friesianism" and "the influence of Cassirer and Nelson" are named as shaping the Berlin Group (Milkov 2008).

Oppenheim himself clearly identifies the problem of order as the central concern of his work. Probably the best summary of this issue is to be found in his first joint publications with Hempel, and his co-publication with Grelling on the concept of "Gestalt" is based upon the same ideas. Therein, Oppenheim simultaneously

[14]Cf. Ziche (2011); on the difficulty of characterizing Driesch's role precisely—as an opponent of logical empiricism, or rather as introducing a sort of "family conflict" within this group—see Danneberg (1993).—On the "Varieties of Order" within *Gestalt* psychology see Smith (1988a), 61–65.

comments on his earlier writings and sketches an agenda for future work. In their 1936 booklet on the *Typusbegriff*, Hempel and Oppenheim summarize Oppenheim's earlier publication from 1926, *Die natürliche Ordnung der Wissenschaften*, as introducing a difference between "statical" and "dynamic" subsumptions, which are, in turn, connected with two forms of concepts: scalable and classificatory concepts. In focussing on the role and the logical form of scalable concepts, Oppenheim intends to overcome the shortcomings of a traditional view of concept formation that uses concepts purely as means for classification. At the same time, he intends to make use of the innovations of modern logic, which had finally been able to provide a formal account of relations. A formal calculus adapted to the logic of relations is required for arranging concepts in serial form, and thus for any account of scalable concepts (Hempel and Oppenheim 1936a, 120). Oppenheim even goes so far as to view his theory of scalable concepts as a form of "*applied logistics*" (Hempel and Oppenheim 1936a, 121). The paper on *Gestalt* concepts[15] pretty much discusses parallel ideas, adding the idea of invariants to that of relationally defined scalable order concepts. Oppenheim links this paper via cross-references closely to the paper on typology. This already indicates Oppenheim's intention to maintain a high degree of coherence among his papers, independent of the different co-authors, and he again spells out his project explicitly: His investigations on *Gestalt* concepts are "part of more comprehensive considerations that have the 'concepts of order' as their object" (Grelling and Oppenheim 1937a, 211 note 2).

The problem of ordering the sciences in a natural way is worked out in great detail in Oppenheim's publications from the 1920s. His 1926 book—during the composition of which he so annoyed Max Born—on *Die natürliche Ordnung der Wissenschaften. Grundgesetze der vergleichenden Wissenschaftslehre* (*The natural order of the sciences. Basic laws of the comparative study of science*), also seems to be his first publication after his dissertation in chemistry in 1908. In this work, Oppenheim starts with a confession of faith that also uses terms from the *Gestalt* traditions: "*Science is a living whole*," and he strives for an integration of that holistic attitude with rigorous logical standards: "notwithstanding the strictest *logical rigour*, the final order must present itself as a *living whole*" (Oppenheim 1926, 1–2). The result of this integration must be intuitively transparent and simple, and this simplicity can best be achieved by employing a "*mathematical symbolism*," in agreement with this logical attitude. These programmatic claims presuppose, as he explicitly states, an attitude of "*tolerance*" towards all sciences that there are (Oppenheim 1926, 3–4).

Although his tolerant attitude requires him to start from the sciences as they are given in the institutions and practices of his time,[16] he selects a broad list of

[15]This paper was presented at several international conferences: In 1938, at the 4th conference on the unity of science in Cambridge, and again in 1939 in English at Harvard; see Stadler (1997), 427, 431.

[16]Oppenheim (1926), 8; in the German original, Oppenheim speaks of the "gegebenen Wissenschaften."

representative disciplines, ranging from mathematics and the natural sciences to economy, philology, history, geography and metaphysics.[17] The criteria underlying a "natural order" of these disciplines have to be developed inductively, and he remains somewhat vague as to how he wants to justify his choice of basic criteria. However, the form of classification he seeks eventually becomes very explicit. He aims at a "*continuous*" order (Oppenheim 1926, 16), which requires that quantitative concepts have to be applicable. In order to achieve this, he envisages a two-dimensional order, the *Denkfläche* (plane/surface of thought) on which individual sciences can be located, and within which the relations between disciplines become visible in a sort of topological ordering that defies the more traditional, classificatory idea of thinking in terms of yes-or-no-decisions that could result in a genealogical tree.[18]

Two pairs of polar terms span up this plane, and serve to define the first set of coordinate axes for localizing the sciences: the dimensions of types versus individuals, and of concrete vs. abstract. The typical mathematical techniques applicable to coordinate systems can be transferred to this ordering system.[19] One can define lines along which the individual coordinates (or combinations of coordinates) remain constant, and one can introduce various transformations between coordinate systems. New coordinates can be defined, also including polar hyperbolical coordinates that are made possible by generating relative measures such as the density of concepts or of properties ("Begriffsdichte"/"Merkmaldichte"). Moving along the various lines that can be described in these coordinate systems allows one to analyze historical processes in the genesis of disciplines, and to study controversies within or between disciplines.

Within this work, Oppenheim discusses the existing disciplines in great detail. Whereas the concrete location he ascribes to each discipline does not provide for any great surprises,[20] he also introduces some concepts that are of great interest given his later work with Hempel on explanation and on lawlike statements. In particular, these are the notions of a "degree of explanatory power" ("Erklärungsgrad"),

---

[17]Oppenheim stresses explicitly that "with this form of definition [in terms of a tendency towards the 'typical' resp. towards the 'individual'], there no longer is a break between the natural sciences and the humanities" (Oppenheim 1926, 25).

[18]In 1936, Kurt Lewin presents his conception of psychology in topological form, devoting quite a lot of attention to the mathematical basis of this mode of presentation (Lewin 1936). Oppenheim is not mentioned in this work. Regarding "problems of coordination," Lewin refers to Reichenbach, A.E. Blumberg and Feigl (Blumberg and Feigl were co-authors of an early paper on logical positivism in 1931; Lewin 1936, 59).

[19]Some of the illustrations in Oppenheim's *Natürliche Ordnung* are reproduced in Ziche (2008).

[20]Mathematics lies close to the pole of generality, in virtue of the low density of concepts and properties encountered in mathematical statements (only metaphysics comes closer to being absolutely general); history marks the extreme pole of concreteness, with a maximum of property density, but low conceptual density; geography also has high property density, and lies thus at the concrete side of the coordinate space, but it also displays a rather high conceptual density and thus differs fundamentally from history (Oppenheim 1926, 257; the coordinates he uses are explained on pp. 235, 237).

which is directly related to the degree of systematization brought about by a particular law of nature—an idea that will prove important in Oppenheim's later texts on explanation and on the unity of science,—and of a "degree of lawlikeness" ("Gesetzlichkeitsgrad;" Oppenheim 1926, 215). His two-dimensional formalism helps him to show that these dimensions do not coincide. He also starts to look for precise quantitative measures for the number of concepts or properties in a given context (Oppenheim 1926, 219).[21]

Remarkable are the further perspectives that he sketches at the end of his paper. He claims that his strictly logical framework makes his approach invulnerable to all charges of psychologism, and thus allows for further projects discussing the order of the sciences from the perspective of a "*psychology of thought*" (Oppenheim 1926, 279). These arguments can be directly put to use in order to clarify his work on types of personalities in the 1930s. What is more, they again bear a resemblance to ideas presented in the context of the Berlin Group: Wilhelm Ostwald, another champion of classifying the sciences with methods inspired by modern logic, lectured in 1930 in Berlin about his own brand of a psychology of the scientist and about a classification of psychological types of scientists.[22]

This picture of an ordering of *Wissenschaften* that is based on two variously describable dimensions is repeated in Oppenheim's next publication, *Die Denkfläche* (Oppenheim 1928). In this booklet, Oppenheim purports to give a short summary of his earlier book, and indeed, the text is best read as a précis of the *Natürliche Ordnung*. In a further short piece in *Kant-Studien* (Oppenheim 1930), Oppenheim broadens the basis of his ordering scheme and changes his terminology, moving from a (two-dimensional) plane of thought to a *space* of thought, the so-called *Denkraum*, in order to distinguish between synchronic ("Beschreibung") and diachronic ("Erzählung") descriptions; a move that is motivated by reflections on space and time triggered by relativity theory. The idea of a three-dimensional logical space as the natural habitat for the sciences makes its return, transformed in some ways, but even more surprisingly rather unchanged in its basic formal characteristics, in Oppenheim's publications of the 1950s, thereby exemplifying both the development and continuity of his work over his lifespan.

13.4 Hempel's Review and the Development of Oppenheim's Ideas: Towards 'Orthodox' Philosophy of Science

The papers on the notions "Gestalt" and "Typus" clearly display the multi-sided nature of Oppenheim's ordering project. Remarkably, however, none of his co-authors seems to have taken issue with this endeavour. Still, one can detect some clear developments in Oppenheim's publications and co-publications from

[21]On "Begriffszahl" and "Merkmalszahl" see Oppenheim (1926), 236sqq.

[22]Cf. Danneberg and Schernus (1994), 461. On Ostwald, see Ziche (2008), ch. IV.7; Ziche (2009).

the 1930s. These papers become more narrowly focused on central terms in the philosophy of science, such as laws and theories, and even seem to foreshadow the Hempel–Oppenheim account of scientific explanations. They discuss

> the so-called *explanation*, more precisely with regard to the formulation of *laws* that connect certain empirical data with each other. In the formulation of the laws there occur the concepts that describe the connected data, and if these concepts are formally inadequate, the same has to hold for the laws that are formulated by means of these concepts (Hempel and Oppenheim 1936a, 1).[23]

In a similar vein, operations within Oppenheim's coordinate systems can be described by the concept of reduction and, in turn, give a more precise meaning to this concept: Under which conditions, so one can ask on the basis of Oppenheimian coordinate systems, can a multidimensional order be reduced to one dimension? (Hempel and Oppenheim 1936a, 74sqq.)

Despite the emphasis on the role of laws within explanations, and on the possibility of describing processes in Oppenheim's coordinate systems via the notion of reduction, the "Gestalt" and "Typus" articles preserve the tolerant attitude of the 1920s. For, not only do they take issues in the psychology of personality as the intended applications of their theoretical account, but they also offer a surprisingly liberal view of logic and scientific method that can refer to, without the slightest hesitation, rather traditional forms of logic (Chr. Sigwart) as well as to the work of methodologists and philosophers of psychology such as G.E. Müller and G.F. Lipps (Hempel and Oppenheim 1936a, 53 note.). This surprising degree of tolerance surfaces again when the authors emphatically accept the methodological tool of "ideal types" (Hempel and Oppenheim 1936a, 83).

Again, Sigwartian logic and (Weberian?) ideal types seem perfectly compatible with a rigorously logical ideal, as stated in the plans for future publications that Hempel and Oppenheim sketch at the very end of the book on typological concepts: "It is planned to develop a general theory for the formation of ordering concepts in later publications, in a formally more rigorous manner, and to employ it for the logical analysis of yet further areas of science" (Hempel and Oppenheim 1936a, 121 note).[24] The whole project thus amounts to a "thorough and detailed proof for the *logical unity of science*" (Hempel and Oppenheim 1936a, 125).[25]

That the combination of breadth in scope and of logical clarity in the methods proves so persistent is quite surprising, as is, in a way, the harmonious cooperation between Oppenheim and Hempel. In 1931, Hempel—with whom Oppenheim up to

---

[23]The original runs: "[. . .] die sog. *Erklärung*, genauer gesagt für die Aufstellung von *Gesetzen*, die empirische Daten bestimmter Art miteinander verknüpfen. In der Formulierung der Gesetze nämlich treten ja die Begriffe auf, die die verknüpften Daten beschreiben, und sind diese Begriffe formal inadäquat, so muß dasselbe auch für die mittels ihrer formulierten Gesetze gelten."

[24]In German: "Es ist geplant, in späteren Veröffentlichungen eine solche allgemeine Theorie der ordnenden Begriffsbildung ausführlich und in formal strengerer Weise zu entwickeln und sie für die logische Analyse weiterer Wissenschaftsgebiete nutzbar zu machen."

[25]In German, this program is stated as looking for "einen vertieften, ins Einzelne gehenden Nachweis für *die logische Einheit der Wissenschaft*."

that moment had not been in contact; they only met in 1933 (Oppenheim 1969, 1)—published a thoughtful review of Oppenheim's (1926) book on the *Natural order of the sciences*. This piece included some rather trenchant critical remarks concerning Oppenheim's work, such as a request to give clearer criteria for scientificity. It seems obvious that this review led to a crucial revision in the development of Oppenheim's ordering schemes. In his brief review, Hempel applauds Oppenheim's rich collection of data and his conclusion that many apparent divergences in questions of fact can better be explained psychologically as differences in the logical point of view one assumes. Hempel, however, also voices two important points of critique. First, he urges Oppenheim to narrow down his notion of *Wissenschaft* in order to exclude metaphysics and normative disciplines as "systems of pseudo-sentences" (Hempel 1931, 473), in line with the logical empiricists' doctrine. Second, he criticizes Oppenheim for sticking to a traditional conception of logic and suggests the use of better formal tools, implicitly invoking Carnap's call for a logical analysis of scientific concept formation via the new mathematical logic.

In particular, Hempel has difficulties accepting the tolerant attitude of Oppenheim's work, which at the same time shows that he views precisely this type of tolerance as an essential characteristic of Oppenheim's earlier work: "Of course, this tolerance leads to the disadvantage that among the sciences ["Wissenschaften"] under scrutiny we also find metaphysics—although already restricting it, in essence, to epistemology—and certain normative sub-disciplines, for instance in economics and history." (Hempel 1931, 473)[26] Thus, he continued, Oppenheim's awareness of "the fundamental logical differences" between the disciplines

> does not lead, given the more descriptive tendency of the book, to the consequence that, according to me, is inevitable: Namely that 'normative' disciplines, and fields logically related to these disciplines are eliminated from the range of genuine sciences that are to be studied, because of their being systems of pseudo-sentences. After all, we are not talking about issues of terminology here. (Hempel 1932, 473)[27]

Hempel's critique clearly did not impede the cooperation between Oppenheim and Hempel. But to what extent does Oppenheim revise his fundamental program within these joint publications? What happens with Oppenheimian tolerance, when this attitude has to face the more orthodox ideas of logical-empiricist philosophy of science that were forcefully presented in Hempel's review? It has already been shown that their joint work on order concepts can be viewed as a continuation of Oppenheim's earlier ideas, and that, from the 1930s onwards, they start to develop and clarify some of these orthodox notions. In fact, Oppenheim's reaction

[26]In German: "Freilich ist diese Toleranz mit dem Nachteil verknüpft, daß in der Reihe der zur Untersuchung gelangenden Wissenschaften auch die Metaphysik—die allerdings im Wesentlichen schon auf die Erkenntnistheorie beschränkt wird—und gewisse normative Teildisziplinen, etwa der Nationalökonomie und der Geschichtswissenschaft auftreten."

[27]In German: "führt bei der mehr deskriptiven Tendenz des Buches nicht zu der m.E. unvermeidlichen Konsequenz, daß 'normative' und logisch verwandte Disziplinen als Systeme von Scheinsätzen aus der Reihe der zu untersuchenden echten Wissenschaften—es handelt sich hier ja nicht um eine Frage der Terminologie—gestrichen werden."

to Hempel's review plays out at two levels. Oppenheim reacts directly by accepting the charge to develop and adopt new formal techniques, to give more precision to his concepts and to introduce a reliable basis for the quantitative measures he adopts. Yet, at the same time he manages to circumvent the issue of pseudo-sentences and anti-metaphysics. Thus, the spectrum of sciences that are deemed acceptable is not narrowed down at all.

In a brief autonomous paper from 1937, Oppenheim rehearses his earlier claim that order concepts are more useful in describing the landscape of the sciences than traditional classificatory concepts, but he also makes some efforts to connect his notion of ordering with the theory of relations in mathematical logic. Yet, in doing so, he refrains from making any use of formal symbolism (Oppenheim 1937, 70 note 2). He also makes a move towards accepting Hempel's second point of critique. But in discussing the relative merits of classificatory vs. order concepts, he circumvents the need to explicitly declare metaphysics to be misguided. Rather, choosing between the two frameworks, i.e. classificatory or ordering concepts, becomes a pragmatic choice:

> Which of the two forms of linguistic presentation is to be preferred? The answer that we should prefer that type of language that in its form correctly represents the structure of reality has to be refuted because of its being metaphysical. In principle, none of the two forms of language can be 'right' or 'wrong,' but certainly the one can be more useful than the other, depending on the context (Oppenheim 1937, 75).[28]

This sounds very much like an acceptance of a Carnapian principle of tolerance which emphasizes the ultimately pragmatic standards of a language choice in science (see, e.g., Richardson 1994), and is compatible with a critique of metaphysics that remains subject to the pragmatically tolerant choice of linguistic frameworks. Did Oppenheim, then, renounce his open conception of the sciences and their order in favour of the orthodoxy of logical empiricism? In fact, his collaborative work from the 1940s on has been taken as one of the pillars supporting the development of that orthodoxy—we have already remarked that some of his co-authored papers have achieved the status of canonical texts of logical empiricism. A closer look at his autonomous work, however, reveals interesting and, as we think, highly relevant tensions within this picture.

These become directly visible when one focuses upon another important aspect of Hempel's review, namely the role of formal modes of presentation. Unusual as Oppenheim's formalizations are in 1926, they do not induce Hempel to issue a directly critical comment. Rather, he simply states that Oppenheim's ideas are formulated "in a symbolism that differs from the usual formalizations of logic." Although Hempel appreciates the "highly interesting points of view and problems" that Oppenheim can raise via his symbolism, he still charges Oppenheim with being

[28]In the German original: "Welche der beiden sprachlichen Darstellungsformen ist nun vorzuziehen? Die Antwort, vorzuziehen sei diejenige, welche ihrer Form nach die Struktur der Wirklichkeit richtig wiedergibt, ist als metaphysisch abzulehnen. Keine der beiden Sprachformen kann überhaupt 'richtig' oder 'unrichtig' sein, wohl aber kann je nach dem Zusammenhang eine zweckmässiger sein als die andere."

too close to the "concept formations of traditional logic" ("Begriffsbildungen der traditionellen Logik"), where "Begriffsbildungen" does not stand for the dynamic process of forming concepts, but rather for the type of concepts one employs. As such, "they partly require a more precise justification in the form of a thoroughgoing logical analysis of scientific concept formation." Later, Oppenheim himself places emphasis on modern logic resp. "logistics" (with authors such as Dubislav, Couturat and Carnap as points of reference). This does not stand in the way of broad content, and he himself also keeps using aspects of his topological and geometrical methods. Not only the range of acceptable sciences, and the choice of linguistic frameworks, but also—and this is related to the first two points—the type of formalization was thus very much an open issue at that time. Adhering to the ideal of a strictly logical attitude and to the necessity of using formal tools did not determine the kind of formalism one had to employ. Rather, logical formalisms remain a tool within a more comprehensively defined framework.

13.5 What Became of Oppenheimian Tolerance? Oppenheim on Reduction and on the Unity of the Sciences

Rescher characterizes Oppenheim's later work as being based upon a "shift" towards dealing with the problem of structuring the sciences in a "more general sense" as compared to Oppenheim's texts from the 1920s (Rescher 1997, 340). It is precisely this "shift" and the "more general sense," however, that are interesting. That is, is there a change in methods, in outlook, in problems, or are the later texts rather to be viewed as continuing his original interest? How could the scope of Oppenheim's 1920s publications possibly be surpassed towards something more general? The great question, then, remains: How much of Oppenheim's original approach can survive the intervention of analytical orthodoxy, as it is incorporated in the labels of "reduction," "unity of science" and the Hempel–Oppenheim doctrine of deductive-nomological explanations?

First, *Gestalt* properties. In the Rescher–Oppenheim paper on *Gestalt* qualities, which is explicitly introduced as taking the Oppenheim–Grelling paper "as point of departure" (Rescher and Oppenheim 1955, 89, n. 1), no critical consideration at all is offered with regard to the earlier papers by Oppenheim and Grelling. In particular, the key intention to incorporate a "broad range" of issues remains untainted.

More interesting are the Kemeny–Oppenheim and Oppenheim–Putnam[29] papers on reduction, and on the unity of science, respectively. Both papers rely heavily

[29]This paper is the only co-authored paper in which Oppenheim features as the first author; but then, that may just be a question of adhering to an alphabetical order that is broken only once, in the case of the Rescher–Oppenheim paper. It contains one of the more explicit discussions of Carnap's ideas, and clearly rejects Carnapian 'epistemological' reductionism (Oppenheim and Putnam 1957, 5). On the Oppenheim–Putnam conception of "unity of science," see Hacking (1996).

on the Hempel–Oppenheim account of scientific explanations; the notion of explanation becomes considerably clarified, and is distinguished in rather strong terms from translatibility (Kemeny and Oppenheim 1956, 16).[30] Both papers again show a remarkable degree of programmatic coherence, with the Oppenheim–Putnam paper explicitly claiming to continue the ideas of the paper on reduction. In their main outline, both of these papers still contain a remarkable degree of openness and tolerance that makes it understandable why Oppenheim in 1957 can rephrase his original project via the formal means he has acquired in these joint projects.

What is especially remarkable about both papers is that, in each case, one of the crucial notions remains undefined. Specifically, these are the notion of "systematization" in the Kemeny–Oppenheim paper on reduction, and the very idea of being "united" or "connected" in the paper on the unity of science. These notions can only be dealt with in "some intuitive sense" that the authors do not even attempt to spell out in detail (Oppenheim and Putnam 1957, 4; cf. also Kemeny and Oppenheim 1956, 11). Clearly, their intention is to grasp and analyze our intuitions rather than to correct them. Another joint feature of these papers is the emphasis that they place on issues in mereology that apparently has to serve as the most plausible candidate for a logical formalization of key ideas from metaphysics.

Even more clearly than many other papers by Oppenheim and his various co-authors, the paper on the unity of science wears continuity with his early work on its sleeve. In drawing up a concrete list of hierarchical levels (Oppenheim and Putnam 1957, 9), Oppenheim and Putnam directly mirror nineteenth-century hierarchies as they are to be found in, e.g., Auguste Comte or Wilhelm Ostwald. Without an explicit argument they state rather apodictically that "There must be several levels."[31] The reason, again, is pragmatic. For, without hierarchies, the idea of unification would be neither "*credible*" nor practicable (Oppenheim and Putnam 1957, 8). In the host of examples presented by Oppenheim and Putnam, they again draw on rather aged historical material and refer back to vintaged literature on the classification of the sciences.[32] In the summary of this work, the title phrase of Oppenheim's (1926) publication recurs, although this time with the indefinite article: "The idea of reductive levels employed in our discussion suggests what may plausibly be regarded as a *natural order of sciences*" (Oppenheim and Putnam 1957, 28). This topic remains central in Oppenheim's publications in which he again features as the sole author.

---

[30]Hilary Putnam considered this paper in 1969 as "still being the best paper on the subject" (Putnam 1969, 242).

[31]Jaegwon Kim refers to this paper as "the only explicit discussion of the levels picture I know of in contemporary analytical philosophy" (Kim 2002, 6), thereby extending the label "contemporaneous" to cover almost fifty years of philosophical development—and he also emphasises the close similarities with earlier hierarchical models.

[32]E.g. by Comte and Flint (Oppenheim and Putnam 1957, 34 note 45); on the context cf. Ziche (2008).

13.6 Oppenheim After Hempel–Oppenheim, Kemeny–Oppenheim and Oppenheim–Putnam

Two publications from the late 1950s mark the last of Oppenheim's independent attempts to realize his systematic program. By this point, Oppenheim's cooperative projects have afforded him many sophisticated logical tools. However, the topics and titles of these publications echo his ideas from the 1920s almost literally. How, then, are these later publications on the "dimensions of knowledge," and on giving a natural order of the sciences, to be characterized?

We focus on the earlier and longer paper, "Dimensions of Knowledge," from 1957.[33] "Dimensions of Knowledge" is a long and detailed article, comprising some 40 pages. It contains many parallels to Oppenheim's early autonomous work of the late 1920s, Specifically, the unity of science is stressed, the concept of order plays a central role, and different coordinate systems are introduced and their interpretations and transformations are discussed. Oppenheim thus seems to hold on to his earlier agenda. This said, there are also important differences. The most striking difference between Oppenheim's earlier work and "Dimensions of Knowledge" lies in the subject of investigation. In the 1920s Oppenheim is concerned with the definitive order of the sciences, starting with individual "*Wissenschaften*" as the units of consideration. In 1957, he focuses not on whole sciences but rather on specific products of scientific practice, namely scientific publications, which are identified as syntactic objects in a given, formalized language of the sciences. These differences are readily explainable by reference to the collaborative work that Oppenheim has done in between these works, and by the new formal methods that he has become acquainted with in the course of those collaborations. Thus, while Oppenheim often explicitly attributes the formal-logical parts of his joint papers to his collaborators (see, e.g., Hempel and Oppenheim 1948a, note 1), in "Dimensions of Knowledge" he employs those methods himself and shows a thorough acquaintance with the then prominent approach to philosophy of science as an outgrowth of logical empiricism. Indeed it seems fruitful to read "Dimensions of Knowledge" as Oppenheim's

[33]Oppenheims' later paper (Oppenheim 1959) is marked as a discussion note on the earlier one and is classified by Oppenheim as a "supplement." In it he does not refer back to his earlier autonomous work but only relates to the paper on the "Unity of Science" that he co-authored with Putnam.—Oppenheim's papers from the 1960s and 1970s focus on quantum mechanics. In Bedau and Oppenheim (1961), the discussion is based on the relationship between "phenomenal" and "interpretational" sentences, and attempts a precise definition of "compatibility." In cooperation with Brody (Brody and Oppenheim 1966, 1967, 1969), these ideas are extended into the field of psychology, together with Lindenberg (Lindenberg and Oppenheim 1974, 1978) he extends this notion even further to cover epistemological and economic issues, still remaining as untechnical as possible, and without making use of new developments in logic (such as modal logic).

attempt to link his own long-term project, the understanding of the order of the *Wissenschaften*,[34] with the emerging orthodoxy of philosophy of science that he himself has helped to build up.

According to that orthodoxy, slightly different versions of which have been called "the received view," a scientific theory is a linguistic object: a set of sentences in some language with a clearly identifiable structure.[35] Meaning postulates (analytical sentences) identify interrelations among the non-logical terms. Furthermore, some division between observational and theoretical sentences is presupposed, either via a division among the vocabulary, the domain of quantification, or simply as an irreducible division among the sentences. For his part, Oppenheim appears to adopt the received view. He subscribes to the requirement of a syntactic presentation (even suggesting that "syntactic" is a synonym for "precise") and the specification of meaning postulates (Oppenheim 1957, 154), and he takes a scientific publication—a set of sentences in some formal language of science—to be divided into three parts: theoretical, observational, and auxiliary. Furthermore, he suggests the broadest and most liberal solution for making this division by suggesting that the relevant criteria ultimately are pragmatic ones: "The only method by which we can discover which statements in the unformalized publication are to be placed in which category (in the process of formalization) is, in many cases, to ask the author" (Oppenheim 1957, 156).

Oppenheim is explicit about maintaining a link to the logical empiricist orthodoxy. Apart from the parallels just mentioned, this becomes very clear from the opening section of his paper. He states that "it is the *purpose of this paper* to give a rational reconstruction or 'explication' for concepts widely used in the literature of philosophy of science for many years and, indeed, centuries" (Oppenheim 1957, 152), having named concept pairs such as broad/deep, theoretical/observational, nomothetic/ideographic and referencing traditional authors as diverse as Pascal, Goethe, Windelband, Carnap and Kemeny. In this vein, he explicitly refers to Carnap in connection with the notion of explication and even names Carnap's general requirements for an explication (Oppenheim 1957, 152 note 3). Furthermore he stresses the logical and formal nature of his work and twice in the opening section, he defends his project by reference to the "well-established tradition" of logical empiricism (Oppenheim 1957, 152–153).

Oppenheim goes on to construct various coordinate systems for an ordering of scientific publications, and much like in his work in the 1920s he is concerned with the intuitive interpretation of the coordinates and their transformations. Thus, the formal work, which is limited to simple algebraic transformations, in this

[34]Oppenheim narrows down his field of study to publications in "empirical science" (Oppenheim 1957, 155), without however commenting on this any further. That he envisages a broad view of "science" becomes clear from his rejection of the distinction between natural sciences and humanities (Oppenheim 1957, 182 note 24); see also the programmatic reference to the unity of science at the end of the paper (Oppenheim 1957, 191).

[35]While some presentations of the received view narrow down this requirement to the requirement of an axiomatic presentation in first-order logic, this is historically inaccurate. Oppenheim is also explicit about admitting a broader range of logical languages; cf. Oppenheim (1957), 157.

respect remains at the same, rather elementary level as in his *Natürliche Ordnung*. The influence of the new logic is, however, clearly visible at a fundamental level. Oppenheim strives for precise quantitative measures that ascribe concrete numbers to well-defined properties of scientific publications, and bases all his coordinatizations on three formally perspicuous notions: a syntactically defined measure of *extensity*, a syntactically defined measure of *strength*, and the division of the language of science into a *theoretical* and an *observational* part. These notions form the interface between Oppenheim's background acquired in the collaborations of the 1940s and 1950s, and his own agenda as it had already presented itself in the 1920s. In this way he employs the syntactic notions of the logical empiricist orthodoxy of the time in an attempt to further his own ordering project.

Specifically, Oppenheim bases his new, explicit coordinatization on two syntactically defined measures that Kemeny had worked out within two of his own publications (Kemeny 1953, 1955), wherein he in turn acknowledges having been substantially influenced by Oppenheim (Kemeny 1953, 289 note 1, 1955, 722 note 1). The *strength* of a publication (a set of sentences) is a measure of its information content. Kemeny (1953) had defined a "logical measure function," building forth on a discussion in the Hempel–Oppenheim paper on scientific explanation. Roughly, this is the negative logarithm of the ratio of the number of models in which the sentences comprising the publication hold, and the total number of models fulfilling the meaning postulates. Oppenheim mentions the formal correspondence to the Shannon information measure (Oppenheim 1957, 188).[36] The *extensity* of a publication, or "breadth of subject matter" (Oppenheim 1957, 158), determines its place on a scale of simplicity vs. complexity. Oppenheim adopts Kemeny's (1955) explication of complexity as his official definition of extensity which, roughly, counts the number of different classificatory formulae that are consistent with the meaning postulates. He explains that this measure represents the possible fineness of classification of the vocabulary, and that, accordingly, spurious extensity is to be avoided by choosing as simple a vocabulary as possible (Oppenheim 1957, 162).

These two measures determine the place of any given publication on a two-dimensional plane, much as in the *Denkfläche*. Oppenheim derives a number of alternative coordinatizations as possible explicanda of, e.g., the thoroughness of a publication, or its degree of concentration (Oppenheim 1957, § 4). He then moves on—like in the move from *Denkfläche* to *Denkraum*—to introduce a third dimension. However, apart from the number of dimensions, there are hardly any parallels to the *Denkraum* construction, which was based on a distinction between diachronic vs. synchronic descriptions. In "Dimensions of Knowledge," Oppenheim introduces a third dimension by splitting the strength (i.e., information content) of a publication

[36]While he officially adopts Kemeny's explication of strength in terms of the "logical measure function," Oppenheim also considers the use of other, similar measures of strength (see Oppenheim 1957, 169–170).

into its theoretical and observational strength.[37] As the theoretical and observational information content are independent, and extensity is independent from strength, Oppenheim thus indeed succeeds in giving a formally perspicuous construction of a three-dimensional logical space in which every scientific publication is assigned a well-defined set of coordinates. From here, Oppenheim goes on to defend his fundamental dimensions (Oppenheim 1957, § 5–6) and then introduces a number of explications of intuitive classifications of publications such as universalist vs. specialist, or Windelband's nomothetic vs. ideographic. He remains faithful to his official aim of classifying publications, but by this point elements of psychological typology also come to play an important role in his work. Thus, in his defence of the explications, he explicitly states that he is moving from considering types of mentality to publications in order to be able to apply the new logical methods (Oppenheim 1957, 182 note 25), and he brings up the notion of a "*scientific-behavior space*," mentioning formal similarities between results from psychological factor analysis and his logical space and suggesting research into the "*psychology of scientific thought*" in order to, e.g., explicate the distinction between romanticists and classicists (Oppenheim 1957, 186–187).[38] Oppenheim even voices the hope that empirical research into typology may help to unmask apparent methodological differences among scientists as "pseudo-conflicts, i.e. as clashes of intellectual personality types" (Oppenheim 1957, 188). What we see here can be thought of as a kind of psychological transformation of logical empiricists' anti-metaphysics: An Oppenheim-based psychological typology can serve to unravel the pseudo-problems inherent in traditional science and philosophy.

13.7 Summary: Order, Tolerance, Orthodoxy

This last remark shows clearly that Oppenheim's original program has survived largely untainted, and perhaps even more interestingly, that Oppenheim manages to link his earlier interests in the psychology of personality types with some of the central ideas of orthodox logical empiricism, for instance with the empiricists' strong anti-metaphysical bias. The idea that one can indeed draw on fields such as psychological typology in order to clarify issues in the philosophy of science presupposes a very strong conviction as to the possibility of grounding these psychological topics on a secure foundation by means of what Oppenheim calls a "logical attitude." In fact, these discussions are allotted far more space in "Dimensions of Knowledge" than the explication of the more technical terms "confirmation" or "factual support." With respect to those latter notions, which are central to mainstream logical empiricism, Oppenheim only refers to a number

[37] See Oppenheim (1957), 165, for the official definition of the coordinates, which includes some normalization.

[38] On these typological categories cf. also Ostwald (1909).

of his co-authored papers and to some works by Carnap, without however going into any particular detail (Oppenheim 1957, § 12). We may, therefore, safely assume that it is here, in establishing the mutual usefulness of ideas stemming from logical empiricism and of the apparently rather different fields of psychology, *Gestalt* theories, or tolerant systematizations of the sciences, that Oppenheim's own interests surface. In the *Revue internationale de Philosophie*, Oppenheim's text from 1957 was accompanied by three companion pieces that underpin the impression that breadth and openness remain crucial to Oppenheim's program. Here, Charles Morris, colleague of Carnap, Neurath, Russell, Bohr and Dewey in editing the *Encyclopedia of Unified Science*, comments approvingly on the relevance of Oppenheim's ideas for psychological research. Frederick R. Kling from the "Educational Testing Service" in Princeton even devises empirical tests for the Oppenheimian dimensions. Finally, Sylvain Bromberger (to whom we owe one of the standard analyses of the so-called "asymmetry of explanation") compares Oppenheim's ideas to those of Duhem, and seriously considers Oppenheim's claims to rationally reconstruct certain ideas of Pascal.[39]

These continuities of Oppenheim's program can also be regarded as a survival of central ideas of the Berlin Group. Crucial for this program remains the motive of *tolerance*.[40] It plays a decisive role in a surprisingly large variety of contexts. We shall focus on two of them that are of particular interest for the historical and systematic analysis of logical empiricism. First, there are the various *pragmatic turns* that are ascribed to the protagonists of logical empiricism. Second, there is the more recent trend in the *historiography of logical empiricism* and analytical philosophy that increasingly tends to emphasize the surprising coalitions and historical continuities between seemingly very different positions.

In "Dimensions of Knowledge," Oppenheim's three basic dimensions are based on state-of-the-art formal methods, and he affirms many tenets of mainstream logical empiricism.[41] By comparison, it is striking how closely his discussion of alternative coordinatizations resembles his work from the 1920s, e.g. in the discussion of typification (Oppenheim 1957, § 7–11). His formal exposition of alternative coordinates remains at the same simple algebraic level as in the earlier publications, but while, in the 1920s publications, the numerical values of these coordinates were not based on formal analysis, he now offers a firm basis on which to link his equations to explicitly defined numbers. On this front, Oppenheim admits

[39]See the data in the bibliography. See also the review by Chisholm (1962), who briefly discusses all four texts.—French library catalogues (e.g. the catalogue of the Bibliothèque nationale) refer to Oppenheim's "Dimensions of Knowledge," together with the accompanying pieces by Morris, Kling and Bromberger, under the title "psychologie de la pensée," thus opening up yet another link to older issues in psychology, but also underlining the continuity in Oppenheim's work in which the psychology of thought had already featured prominently in the 1920s (see Sect. 13.3).

[40]On the related motive of disunity as a category in analyzing the sciences, see Galison and Stump (1996).

[41]Apart from what has been mentioned above, cf., e.g., also his denial of any realist taxonomy and his affirmation of pragmatic language choice in Oppenheim (1957), 173.

that his (or rather, Kemeny's) syntactically defined measures may be difficult to compute ("a tedious and thankless job," Oppenheim 1957, 186) and may therefore be inconvenient. Both pragmatic shortcuts and the employment of new empirical techniques are suggested to overcome this difficulty. It is fitting, then, that in closing Oppenheim sticks to the idea to extend his analysis to all sciences, and to provide "a natural order of the scientific disciplines" (Oppenheim 1957, 191). In fact, this very phrase is the title of his 1959 discussion note on "Dimensions of Knowledge," showing that the indefinite article is not an accident. Bringing the ordering project up to date thus means two things for Oppenheim: to employ the methods of formal logic, as urged by Hempel already as far back as his 1931 review, and to drop the uniqueness assumption with respect to such an ordering, in accordance with a pragmatic view on language choice.

Pragmatic tendencies have been discussed in connection with various classical authors of logical empiricist persuasion. An extreme version is presented by Richard Jeffrey who distinguishes changes in Hempel's views from the rather fixed ideas of Carnap:

> Hempel's view, however, changed. His defence [...] of Carnap's and Neurath's physicalism testifies, in a way, to the presence in logical empiricism of certain 'postmodern' themes: a textualist turn to sentences from the facts or reality they are said to report, a descriptive turn from logic to empirical sociology of science, and a pragmatic turn from truth to inclusion in the text as the basic scientific concern (Jeffrey 1995, 6).

A far more careful version of this idea is presented by Wolters who emphasizes that pragmatic motives have always been present in logical empiricism, and that as such the changes in Hempel's ideas (due, among other influences, to his contact with Kuhn) ought not be described as a radical "conversion." Rather, Hempelian "pragmatic empiricism" is, according to Wolters, best viewed as the "perfection" of logical empiricism (Wolters 2000, 224, 226).

These formulations suggest, with varying degrees of radicality, a change in the self-image of protagonists of logical empiricism that again and again, in various ways and with changing emphasis, recurs as a notion of pragmatic justification. Oppenheim's use of formal methods might be used to illustrate this point: The choice of a formalised presentation of scientific theories is a precondition for applying the tools of logical analysis to the sciences, but is not inherent in the concept of science itself. We are therefore at liberty—just as was the case with Carnap's own principle of tolerance—to choose from a rather broad range of possible formal modes of presentation. Oppenheim's switch from "the" order of the sciences in 1926 to "a" natural order in 1959 nicely illustrates this attitude.

Similar pragmatist ideas are voiced again and again in the discussions between the protagonists of this paper. To present just one further example: Kemeny gives a strangely ambivalent summary of the Hempel-Oppenheim-paper on scientific explanation that combines the idea of logic-based legislation with the great importance of concrete examples and the broad range of possible applications. The 1948 paper by Hempel and Oppenheim, according to Kemeny, achieves only a "semiformalized refinement of science," and in Kemeny's view it apparently does not have to aspire for more. For, the "proposed criterion went beyond ordinary usage and legislated as to what would be a reasonable, precise concept corresponding to our intuitive ideas. The fruitfulness of these attempts is amply demonstrated by the long list of extremely interesting and varied philosophical papers that have arisen as a result of these proposed definitions." (Kemeny 1964, 141) Thus, the theory of scientific explanations also bears witness to an increasing awareness of the relevance of the pragmatic aspects of explanations.

This leads to important historiographic questions concerning the extent and the background of the innovations brought about by logical empiricism. Traditionally, logical empiricism is viewed as a break with tradition. This holds for European as well as for American contexts; for the developments in the United States, Ronald Giere asks with some astonishment how one can understand that American traditions ("a naturalistic pragmatism incorporating an empirical theory of inquiry") "get replaced by a philosophy that regarded induction as a formal relationship between evidence and hypothesis?" The possible answer, "that pragmatism was mistaken and logical empiricism correct," is dismissed as being too simple. Rather, he asserts, one should ask "why philosophers then *believed* that pragmatism was so obviously wrong and logical empiricism so obviously right." He also asks the rather obvious question regarding how these developments in the United States might be related to earlier discussions in Europe (Giere 1996). Though clearly critical of the idea that logical empiricism broke with tradition on all levels, recent scholarship continues to be cast in terms of distinct lines that, however, can come together in a rather surprising "fraternal linking of hands" (Richardson 2000, 5) between seemingly divergent conceptions of philosophy, referring, in particular, to the links between Carnap and Morris. Richardson stresses that all these types of philosophy are united in aiming at transforming philosophy into a scientific endeavour; this goal, however, does not uniquely determine the form that a scientific philosophy must take.

The same basic structure can be seen in Oppenheim's works: Core notions such as "order" or formalization remain open in a very similar way. These ideals can be pursued without thereby restricting philosophy to just one line of argument, or one mode of presentation. The continuity within Oppenheim's program, however, sufficiently illustrates that in his case one cannot talk of a clear *turn* towards more pragmatic positions, or to an increase in openness and richness. These aspects have been typical of his work from the very beginning and have also been incorporated into his cooperation with his more classically logical-empiricist co-authors. It does, therefore, not really make sense to ask whether, and how, Oppenheim in "Dimensions of Knowledge" can be understood to have 'converted' to logical empiricism. Rather, his work with its ramifications for the philosophical community of the 1930s–1950s urges us to accept what Oppenheim called a *tolerant attitude* as one of the central ingredients of the relevant philosophical developments in this period. Looking at the work of Oppenheim, and comparing his oeuvre with that of the various Oppenheim+x-teams, it becomes clear that both trends—that toward pragmatic choices and the new openness characteristic for recent historiographic attitudes towards logical empiricism—are not forced upon logical empiricism by later insights into immanent problems of orthodox logical empiricism, or into the necessity to adopt more open-minded methodologies in the historiography of science. Rather, the openness and disunity that characterizes both trends is inscribed into the very genesis of logical empiricism and the corresponding ideas in the philosophy of science. Thus, the burden of argument shifts. Not a more liberal way of looking at the genesis and development of logical empiricism, but rather the narrowing down towards an orthodoxy require an argument.[42] What is surprising is not so much the tolerant attitude but rather the high degree of streamlining that results in the emergence of an orthodoxy with seemingly clear boundaries. Ironically, Oppenheim's works played a decisive role in these processes—but this just adds to the necessity to better understand the developments that lead to a narrowing down of the tolerant attitude that Oppenheim—and many others!—preserved from the 1920s up to at least the 1950s.

This emphasis on scientific tolerance fits very well with the basic commitments of the Berlin Group. Milkov emphasizes the continuity of a typically Berlin program of studying science "internally," i.e. from the perspective of the actual practice of scientists (what in modern sciences studies would rather be called an externalist attitude). It is then no longer surprising that Oppenheim feels free to refer to the projects of Derek de Solla Price—who was to become one of the great figures in early, sociologically oriented science studies—in order to empirically test fundamental assumptions in the philosophy of science by looking at the history of science (Oppenheim 1959). In line with this attitude, Oppenheim clearly supports Frederick Kling's idea to provide experimental support for the ideas voiced in "Dimensions of Knowledge".

Writing new types of histories about logical empiricism brings, as it were, some of the original ideas of the scientific philosophy of the Berlin and Vienna groups back home. Oppenheim's idea to produce, in the terms he employed in the 1920s, a "living whole" made up of all the sciences that there are, and at the same time to adhere to the strictest standards of logical rigour, may seem irritating if measured against the handbook versions of logical empiricism. However, what emerges as the far more irritating fact is that those versions could and did arise, and could develop a tight grip upon the understanding of twentieth-century philosophy. Oppenheim is the ideal witness for the fact that some of the central topics in twentieth-century philosophy—science and scientificity, formalization, order—do not uniquely determine the form of philosophy that can be built around these topics, and do not imply demarcationist tendencies or eliminative forms of reductionism.

[42]On the question as to how to study the emergence of an analytical orthodoxy, see again Giere (1996).

References

(a) Publications and Co-publications by Oppenheim in Chronological Order

Oppenheim, Paul. 1908. Der Abbau des Narceins. Diss., University of Gießen, Gießen.

Oppenheim, Paul. 1926. *Die natürliche Ordnung der Wissenschaften. Grundgesetze der vergleichenden Wissenschaftslehre*. Jena: Gustav Fischer.

Oppenheim, Paul. 1928. *Die Denkfläche. Statische und dynamische Grundgesetze der wissenschaftlichen Begriffsbildung*, Kant-Studien. Ergänzungshefte 62. Berlin: Kurt Metzner.

Oppenheim, Paul. 1930. Der Denkraum. *Kant-Studien* 35: 227–239.

Hempel, Carl Gustav, and Paul Oppenheim. 1936a. *Der Typusbegriff im Lichte der neuen Logik. Wissenschaftstheoretische Untersuchungen zur Konstitutionsforschung und Psychologie*. Leiden: A.W. Sijthoff.

Hempel, Carl Gustav, and Paul Oppenheim. 1936b. L'importance logique de la notion de type. In *Actes du congrès international de philosophie scientifique. Sorbonne. Paris 1935. II: Unité de la science*, 41–49. Paris: Hermann et Cie.

Oppenheim, Paul. 1937. Von Klassenbegriffen zu Ordnungsbegriffen. In *Travaux du IXe congrès international de philosophie. Congrés Descartes. VI: Logique et Mathématiques*, 69–76. Paris: Hermann et Cie.

Grelling, Kurt, and Paul Oppenheim. 1937a. Der Gestaltbegriff im Lichte der neuen Logik. *Erkenntnis* 7: 211–225.

Grelling, Kurt, and Paul Oppenheim. 1937b. Supplementary remarks on the concept of gestalt. *Erkenntnis* 7:357–359 (1937 a,b are reprinted, the first in translation, in Smith (ed.) 1988b, 191–209).

Grelling, Kurt, and Paul Oppenheim. 1939a. Logical Analysis of 'Gestalt' as 'Functional Whole'. *Paper, sent in for the fifth international congress for the unity of science,* Cambridge, MA, 1939. Repr., in Smith (ed.) 1988b, 210–226.

Grelling, Kurt, and Paul Oppenheim. 1939b. Concerning the structure of wholes. *Philosophy of Science* 6: 487–489.

Helmer, Olaf, and Paul Oppenheim. 1945. A syntactical definition of probability and of degree of confirmation. *The Journal of Symbolic Logic* 10: 25–60.

Hempel, Carl Gustav, and Paul Oppenheim. 1945. A definition of 'Degree of Confirmation'. *Philosophy of Science* 12: 98–115.

Hempel, Carl Gustav, and Paul Oppenheim. 1948a. Studies in the logic of explanation. *Philosophy of Science* 15: 135–175.

Hempel, Carl Gustav, and Paul Oppenheim. 1948b. Reply to David L. Miller's comments. *Philosophy of Science* 15: 350–352.

Kemeny, John G., and Paul Oppenheim. 1952. Degree of factual support. *Philosophy of Science* 19: 307–324.

Rescher, Nicholas, and Paul Oppenheim. 1955. Logical analysis of Gestalt concepts. *The British Journal for the Philosophy of Science* 6: 89–106.

Kemeny, John G., and Paul Oppenheim. 1955. Systematic power. *Philosophy of Science* 22: 27–33.

Kemeny, John G., and Paul Oppenheim. 1956. On reduction. *Philosophical Studies* 7: 6–19.

Oppenheim, Paul, and Hilary Putnam. 1957. Unity of science as a working hypothesis. In *Minnesota studies in the philosophy of science. Vol. II: Concepts, theories, and the mind–body problem*, ed. Herbert Feigl, Michael Scriven, and Grover Maxwell, 3–36. Minneapolis: University of Minnesota Press.

Oppenheim, Paul. 1957. Dimensions of knowledge. *Revue Internationale de Philosophie* 11: 151–191.

Oppenheim, Paul. 1959. A natural order of scientific disciplines. *Revue Internationale de Philosophie* 13: 354–360.
Bedau, Hugo, and Paul Oppenheim. 1961. Complementarity in quantum mechanics: A logical analysis. *Synthese* 13: 201–232.
Brody, Nathan, and Paul Oppenheim. 1966. Tensions in psychology between the methods of behaviorism and phenomenology. *Psychological Review* 73: 295–305.
Brody, Nathan, and Paul Oppenheim. 1967. Methodological differences between behaviorism and phenomenology in psychology. *Psychological Review* 74: 330–334.
Brody, Nathan, and Paul Oppenheim. 1969. Application of Bohr's principle of complementarity to the mind–body problem. *The Journal of Philosophy* 66: 97–113.
Oppenheim, Paul. 1969. Reminiscence to Peter. In *Essays in honor of Carl G. Hempel*, ed. Nicholas Rescher, 1–4. Dordrecht: Reidel.
Lindenberg, Siegwart, and Paul Oppenheim. 1974. Generalization of complementarity. *Synthese* 28: 117–139.
Lindenberg, Siegwart, and Paul Oppenheim. 1978. The bargain principle. *Synthese* 37: 387–412.

(b) Other Publications

Ash, Mitchell G. 1994. Gestalttheorie und Logischer Empirismus. In *Hans Reichenbach und die Berliner Gruppe*, ed. Lutz Danneberg, Andreas Kamlah, and Lothar Schäfer, 87–100. Braunschweig/Wiesbaden: Vieweg.
Ash, Mitchell G. 1995. *Gestalt psychology in german culture 1890–1967. Holism and the quest for objectivity*. Cambridge: CUP.
Beth, Evert W. 1959. Science and classification. *Synthese* 11: 231–244.
Bromberger, Sylvain. 1957. A note on 'Fort et Étroit' and 'Ample et Faible'. *Revue Internationale de Philosophie* 40: 205–210.
Cat, Jordi. 2007. Switching Gestalts on Gestalt psychology: On the relation between science and philosophy. *Perspectives on Science* 15: 131–177.
Chisholm, Roderick M. 1962. Review of Oppenheim 1957. *The Journal of Symbolic Logic* 27: 126.
Danneberg, Lutz. 1993. Logischer Empirismus in Deutschland. In *Wien–Berlin–Prag. Der Aufstieg der wissenschaftlichen Philosophie. Zentenarien Rudolf Carnap–Hans Reichenbach–Edgar Zilsel*, ed. Rudolf Haller and Friedrich Stadler, 320–361. Wien: Hölder–Pichler–Tempsky.
Danneberg, Lutz, and Wilhelm Schernus. 1994. Die *Gesellschaft für wissenschaftliche Philosophie*: Programm, Vorträge und Materialien. In *Hans Reichenbach und die Berliner Gruppe*, ed. Lutz Danneberg, Andreas Kamlah, and Lothar Schäfer, 391–481. Braunschweig/Wiesbaden: Vieweg.
Feigl, Herbert, and May Brodbeck (eds.). 1953. *Readings in the philosophy of science*. New York: Appleton–Century–Crofts.
Friedman, Michael. 2000. Hempel and the Vienna circle. In *Logical empiricism in North America*, ed. Gary L. Hardcastle and Alan W. Richardson, 94–114. Minneapolis/London: University of Minnesota Press.
Galison, Peter, and David J. Stump (eds.). 1996. *The disunity of science. Boundaries, contexts, and power*. Stanford: Stanford UP.
Giere, Ronald N. 1996. From *Wissenschaftliche Philosophie* to philosophy of science. In *Origins of logical empiricism*, ed. Ronald N. Giere and Alan W. Richardson, 335–354. Minneapolis: University of Minnesota Press.
Goodman, Nelson. 1946. A query on confirmation. *The Journal of Philosophy* 43: 383–385.
Goodman, Nelson. 1953. The new riddle of induction. In *Fact, fiction, and forecast*, 4th ed., ed. Nelson Goodman, 59–83. Cambridge, MA/London: Harvard UP. 1983, 59–83.
Gregg, John R., and F.T.C. Harris (eds.). 1964. *Form and strategy in science. Studies dedicated to Joseph Henry Woodger [. . .]*. Dordrecht: Reidel.

Hacking, Ian. 1996. The disunities of the sciences. In *The disunity of science. Boundaries, contexts, and power*, ed. Peter Galison and David J. Stump, 37–74. Stanford: Stanford UP.
Hahn, Hans, Otto Neurath, and Rudolf Carnap. 1929. Wissenschaftliche Weltauffassung. Der Wiener Kreis. In *Gesammelte philosophische und methodologische Schriften*, ed. Otto Neurath, Rudolf Haller, and Heiner Rutte, 299–336. Wien: Hölder–Pichler–Tempsky. 1981.
Hempel, Carl Gustav. 1931. Review of Oppenheim 1926. *Erkenntnis* 2: 473–474.
Hempel, Carl Gustav. 1991. Hans Reichenbach remembered. *Erkenntnis* 35: 5–10.
Jeffrey, Richard. 1995. A brief guide to the work of Carl Gustav Hempel. *Erkenntnis* 42: 3–7.
Kamlah, Andreas. 1993. Hans Reichenbach—Leben, Werk, Wirkung. In *Wien–Berlin–Prag. Der Aufstieg der wissenschaftlichen Philosophie. Zentenarien Rudolf Carnap–Hans Reichenbach–Edgar Zilsel*, ed. Rudolf Haller and Friedrich Stadler, 238–283. Wien: Hölder–Pichler–Tempsky.
Kemeny, John G. 1951. Carnap on probability. *The Review of Metaphysics* 5: 145–156.
Kemeny, John G. 1953. A logical measure function. *Journal of Symbolic Logic* 18: 289–308.
Kemeny, John G. 1955. Two measures of complexity. *Journal of Philosophy* 52: 722–733.
Kemeny, John G. 1964. Analyticity versus fuzziness. In *Form and strategy in science. Studies dedicated to Joseph Henry Woodger [. . .]*, ed. John R. Gregg and F.T.C. Harris, 122–145. Dordrecht: Reidel.
Kim, Jaegwon. 2002. The layered model: Metaphysical considerations. *Philosophical Explorations* 5: 2–20.
Kling, Frederick R. 1957. An empirical study related to 'Dimensions of Knowledge'. *Revue Internationale de Philosophie* 40: 194–205.
Kluge, Susann. 1999. *Empirisch begründete Typenbildung. Zur Konstruktion von Typen und Typologien in der qualitativen Sozialforschung*. Opladen: Leske + Budrich.
Köchy, Kristian. 2010. Vielfalt der Wissenschaften bei Carnap. Lewin und Fleck. Zur Entwicklung eines pluralen Wissenschaftskonzepts. *Berichte zur Wissenschaftsgeschichte* 33: 54–80.
Lewin, Kurt. 1925. Über Idee und Aufgabe der vergleichenden Wissenschaftslehre. In *Kurt-Lewin-Werkausgabe. Vol. 1: Wissenschaftstheorie I*, ed. Alexandre Métraux, 49–79. Bern/Stuttgart: Huber–Klett Cotta. 1981.
Lewin, Kurt. 1929. Review of Oppenheim. *Natürliche Ordnung, Kant-Studien* 34: 461–464.
Lewin, Kurt. 1936. *Principles of topological psychology*. New York/London: McGraw–Hill.
Luchins, Abraham S, and Edith H. Luchins. 2001. Kurt Grelling: Steadfast scholar in a time of madness. http://www.gestalttheory.net/archive/kgbio.html. Accessed 7 May 2012; expanded version of an article in *Gestalt Theory* 22:228–281.
Milkov, Nikolay. 2008. Die Berliner Gruppe und der Wiener Kreis: Gemeinsamkeiten und Unterschiede. In *Analysen, Argumente, Ansätze*, ed. Martina Fürst et al., 55–63. Frankfurt: Ontos Verlag.
Morris, Charles. 1957. A comment on Dr. Paul Oppenheim's 'Dimensions of Knowledge'. *Revue Internationale de Philosophie* 40: 192–193.
Ostwald, Wilhelm. 1909. *Große Männer*. Leipzig: Akademische Verlagsanstalt.
Peckhaus, Volker. 1993. Kurt Grelling und der logisch Empirismus. In *Wien–Berlin–Prag. Der Aufstieg der wissenschaftlichen Philosophie. Zentenarien Rudolf Carnap–Hans Reichenbach–Edgar Zilsel*, ed. Rudolf Haller and Friedrich Stadler, 362–385. Wien: Hölder–Pichler–Tempsky.
Putnam, Hilary. 1969. On properties. In *Essays in honor of Carl G. Hempel*, ed. Nicholas Rescher et al., 235–254. Dordrecht: Reidel.
Reichenbach, Hans. 1929. The aims and methods of physical knowledge. In *Hans Reichenbach. Selected writings 1909–1953*, vol. 2, ed. Maria Reichenbach and Robert S. Cohen, 120–225. Dordrecht: Reidel. 1978.
Reichenbach, Hans. 1951. *The rise of scientific philosophy*. Berkeley/Los Angeles: University of California Press.
Reichenbach, Hans. 1978. In *Selected writings 1909–1953*, vol. 1, ed. Maria Reichenbach and Robert S. Cohen. Dordrecht: Reidel.

Reichenbach, Maria. 1993. Erinnerungen und Reflexionen. In *Wien–Berlin–Prag. Der Aufstieg der wissenschaftlichen Philosophie. Zentenarien Rudolf Carnap–Hans Reichenbach–Edgar Zilsel*, ed. Rudolf Haller and Friedrich Stadler, 284–296. Wien: Hölder–Pichler–Tempsky.
Rescher, Nicholas. 1997. H_2O: Hempel–Helmer–Oppenheim, an episode in the history of scientific philosophy in the 20th century. *Philosophy of Science* 64: 334–360.
Rescher, Nicholas. 2006. The Berlin school of logical empiricism and its legacy. *Erkenntnis* 64: 281–304.
Richardson, Alan W. 1994. Carnap's principle of tolerance. *Proceedings of the Aristotelian Society* 68: 7–83.
Richardson, Alan W. 2000. Logical empiricism, American pragmatism, and the fate of scientific philosophy in North America. In *Logical empiricism in North America*, ed. Gary L. Hardcastle and Alan W. Richardson, 1–24. Minneapolis/London: University of Minnesota Press.
Salmon, Wesley C. (ed.). 1979. *Hans Reichenbach: Logical empiricist*. Dordrecht: Reidel.
Salmon, Wesley C. 1990. *Four decades of scientific explanation*. Minneapolis: University of Minnesota Press.
Schröder-Heister, Peter. 1984. Oppenheim, Paul. In *Enzyklopädie Philosophie und Wissenschaftstheorie*, vol. 2, ed. Jürgen Mittelstraß, 1083–1084. Mannheim/Wien/Zürich: Bibliographisches Institut.
Smith, Barry. 1988a. Gestalt theory: An essay in philosophy. In *Foundations of Gestalt theory*, ed. Barry Smith, 11–81. München/Wien: Philosophia.
Smith, Barry (ed.). 1988b. *Foundations of Gestalt Theory*. München/Wien: Philosophia.
Stadler, Friedrich. 1997. *Studien zum Wiener Kreis. Ursprung, Entwicklung und Wirkung des Logischen Empirismus im Kontext*. Frankfurt a.M: Suhrkamp.
Thiel, Christian. 1993. Carnap und die wissenschaftliche Philosophie auf der Erlanger Tagung 1923. In *Wien–Berlin–Prag. Der Aufstieg der wissenschaftlichen Philosophie. Zentenarien Rudolf Carnap–Hans Reichenbach–Edgar Zilsel*, ed. Rudolf Haller and Friedrich Stadler, 175–188. Wien: Hölder–Pichler–Tempsky.
Tucker, Albert. 1985. John Kemeny. http://www.princeton.edu/~mudd/finding_aids/mathoral/pmc22.htm. Accessed 7 May 2012.
van Kempski, J. 1952. Zur Logik der Ordnungsbegriffe, besonders in den Sozialwissenschaften. *Studium Generale* 5: 205–218.
Wolters, Gereon. 2000. Die pragmatische Vollendung des logischen Empirismus. In memoriam Carl Gustav Hempel (1905–1997). *Journal for General Philosophy of Science* 31: 205–242.
Woodger, Joseph H. 1939. The technique of theory construction. In *International encyclopedia of unified science*, vol. II, 5th ed. Chicago: University of Chicago Press.
Woodger, Joseph H. 1952. *Biology and language. An introduction to the methodology of the biological sciences including medicine*. Cambridge: CUP.
Ziche, Paul. 2008. *Wissenschaftslandschaften um 1900. Philosophie, die Wissenschaften und der nicht-reduktive Szientismus*. Zürich: Chronos.
Ziche, Paul. 2009. Wilhelm Ostwald als Begründer der modernen Logik. Logik und künstliche Sprachen bei Ostwald und Louis Couturat. In *Ein Netz der Wissenschaften? Wilhelm Ostwalds "Annalen der Naturphilosophie" und die Durchsetzung wissenschaftlicher Paradigmen*, ed. Pirmin Stekeler-Weithofer, Heiner Kaden, and Nikos Psarros, 46–66. Leipzig: Sächsische Akademie der Wissenschaften.
Ziche, Paul. 2011. Alternative discoveries of modern logic. Coincidences and diversification. In *Foundations of the formal sciences VII: Bringing together philosophy and sociology of science*, ed. Karen François, Benedikt Löwe, Thomas Müller, and Bart van Kerkhove, 243–267. London: College Publications.
Znaniecki, Florian. 1952. *Cultural sciences. Their origin and development*. Urbana: University of Illinois Press.

Paul Oppenheim in the early 1930s in Frankfurt

Chapter 14
Carl Hempel: Whose Philosopher?

Nikolay Milkov

> For most academics, even most philosophers, the individual who best personified logical empiricism in North America was neither Carnap nor Reichenbach, but Carl Hempel. ... Hempel's early papers, "Studies in the Logic of Confirmation" (1945) and "Studies in the Logic of Explanation" (1948, with Paul Oppenheim), effectively defined what by 1960 were arguably the two most active areas of research in North American philosophy of science.
>
> (Giere 1996, pp. 339–340)

14.1 Michael Friedman's Thesis

Recently, Michael Friedman has claimed that virtually all the seeds of Hempel's philosophical development trace back to his early encounter with the Vienna Circle (Friedman 2003, 94). Hempel, it is true, spent the fall term of 1929 as a student at the University of Vienna, and, thanks to a letter of recommendation from Hans Reichenbach, he even attended some sessions of the Vienna Circle. This gave the young Hempel the opportunity to witness firsthand what was called in the literature "stage one" of the debate that saw Moritz Schlick and Friedrich Waismann go head-to-head with Otto Neurath and Rudolf Carnap on the "protocol sentences."

As opposed, however, to Friedman's view of the principal early influences on Hempel, we shall see that those formative influences originated rather with the Berlin Group. The evidenced adduced here against Friedman on this score concentrates on his contention that Hempel's entire philosophical development, as well as the major themes that were his special concern, were colored by (i) the

N. Milkov (✉)
Department of Philosophy, University of Paderborn, Paderborn, Germany
e-mail: nikolay.milkov@upb.de

N. Milkov and V. Peckhaus (eds.), *The Berlin Group and the Philosophy of Logical Empiricism*, Boston Studies in the Philosophy and History of Science 273, DOI 10.1007/978-94-007-5485-0_14,

Neurath–Schlick debate over the nature of truth (1932–1934), in which Hempel sided with Neurath; and (ii) Neurath's disputation with Carnap of the mid-1930s. This latter contest saw Neurath defending a naturalist position on the issue of scientific investigation, whereas Carnap, who won Hempel's support, championed the philosophical logic pioneered by Frege and Wittgenstein. Carnap adopted the latter as a so-called "logic of science". Friedman himself alludes to Hempel's confession, in the 1980s, that he ultimately abandoned the practice of "Carnapian explications" (cf. § 5, below). In the end, Hempel turned to a variation of Neurath's naturalism, which took the form of Kuhnian historical and sociological studies that foreground the "pragmatic" factors of science.[1]

Hempel actually spent much less time in Vienna than in Berlin, where he studied under Reichenbach from 1926 till 1933 and wrote a dissertation on probability,[2] (Hempel 1935–36) Reichenbach's specialty. Hempel also attended seminars conducted by Walter Dubislav, another member of the Berlin Group. The seriousness of Hempel's involvement in the Dubislav's work is evidenced by the fact that together with Olaf Helmer, Hempel read the proofs of Dubislav's book *Contemporary Philosophy of Mathematics* (Dubislav 1932, p. v). As late as 1934, immediately prior to leaving for Brussels in April, Hempel wrote Reichenbach that he continued to find Dubislav's colloquium "very stimulating."[3]

Besides Hempel's presence at some Vienna Circle meetings, another factor that Friedman adduces in support of his thesis that Hempel's philosophy of science has its roots in the Vienna Circle is that Hempel first won his reputation as an author with his 1935–1936 *Analysis* papers on the Vienna-Circle theory of truth (Cf. Hempel 1935a, b, 1936). In reality, however, these publications moot topics that in the 1930s were being widely debated in the analytic literature and Hempel weighed in on them, as Friedman (2003, p. 99) himself informs us, only after Susan Stebbing invited him in January 1935 "to present a lecture in London on the latest developments within the Vienna Circle and in particular on the exchange between Neurath and Schlick that had just appeared in *Erkenntnis*." (The first of the three *Analysis* articles, "On the Logical Positivists' Theory of Truth" (1935a), is merely a revised version of the London lecture.) Consequently, that Hempel wrote the articles does not unequivocally support the view that the ideas of the Vienna Circle alone were the source of his interest in explicating the Vienna-Circle theory of truth in the three papers.

Tellingly, Hempel's crystal clear and comprehensive critical treatment of the Neurath–Schlick debate is characteristic of the Berlin Group. In form, Hempel's analysis closely approximates that of Kurt Grelling in the latter's reviews of Carnap's *Aufbau* (Grelling 1929) and Reichenbach's *The Philosophy of Space and Time* (Grelling 1930). Grelling himself, under Leonard Nelson, had mastered a discursive style distinguished for its high degree of clarity in thought and exposition,

---

[1]On the "pragmatism" of Hempel's later position see Wolters (2003).

[2]Since Reichenbach left Germany for Turkey in the summer of 1933, formally, Wolfgang Köhler, not Reichenbach, was the supervisor of Carl Hempel's dissertation.

[3]Carl Hempel's letter to Hans Reichenbach of 19.03.1934 [HR 013-46-30].

widely recognized as signature skills of Nelson's group of neo-Friesians. It is not accidental that it was Hempel as the *Berliner*—not as the member of the Vienna Circle—who produced the first and most perspicuous account of the dispute between Schlick and Neurath. This was thanks to training in Berlin that enabled Hempel to lay out the arguments and the issues at stake in the debate with great accuracy and lucidity.

Most importantly, from 1934 through 1936 Hempel also worked on the book *Der Typusbegriff im Lichte der neuen Logik* (Hempel and Oppenheim 1936b). As we shall see, at that point in time Hempel's real interest lay in these studies rather than in the Neurath–Schlick debate.

Friedman's finding is that in these years "the tension between a Carnapian and a Neurathian conception of philosophy of science, which had fundamentally shaped Hempel's earliest work but had long since lay dormant, was stimulated and came to life once again" in the last years of his philosophical development (Friedman, 110). On this account, Hempel, the "logician of science" and anti-naturalist, finally woke up from his "dogmatic slumber." The present chapter demonstrates that Friedman's reading of the facts is misleadingly one-sided.

14.2 Methodological Remark: Carl Hempel as a Historian of Philosophy

A major challenge to any effort at determining Carl Hempel's place on the map of the history of philosophy of science is that Hempel himself was a reluctant and unreliable historian of philosophy, his own philosophy in particular. Only in the early 1980s did friends manage to persuade him to grant Richard Nollan an interview and in that way leave us something of an autobiographical record. The interview shows Hempel to be an inexact chronicler. He reports, for example, that Herbert Feigl was "the first, or one of the first" of the Vienna Circle to leave Austria for the United States (Hempel 2000, p. 14). Historians of the philosophy of science, however, all know that Feigl immigrated to the USA a full 5 years before Carnap, the next member of the Circle to flee. Furthermore, in his recollections about the Vienna Circle and the Berlin Group, published in 1993, Hempel often mistakenly identifies the latter with the "Society for Empirical/Scientific Philosophy," which was an independent entity (Cf. Gerner 1997).[4] Occasionally Hempel says that the Society was a partner of the Vienna Circle, and sometimes he states that it was affiliated with the "Ernst Mach Association" (cf. Hempel 1993, pp. 3 and 4). The latter two, however, were different entities as well.

Besides a hazy historical memory, Hempel had no well-developed sense of "philosophical loyalty:" he never felt obliged to identify himself with a particular philosophical school—not the Berlin Group, not the Vienna Circle, nor any other philosophical coterie. In his last days, however, Hempel did concede that he was

[4]Cf. Chapter One, § 1.3.

"closely associated" with both Vienna and Berlin, and hence doubtless significantly influenced by them (cf. Wolters 2003, p. 111). One does also find only a few references to Neurath, the member of the Circle whom Hempel closely followed in the 1930s—and to whose position, according to Friedman, he returned in the 1980s: and these appear only in the first paper (published 1945) of a collection of 12 of Hempel's most distinguished essays, published under the title *Aspects of Scientific Explanation* (1965).

As the evidence suggests, then, we cannot take Hempel as a reliable source of the history of his own philosophical development. The objective here is thus to trace the verifiable lines of influence on Hempel in their wide variety.

14.3 Paul Oppenheim and Carl Hempel

When trying to determine Hempel's relation to the Berlin Group, one must bear in mind that the latter was neither limited to the city of Berlin, nor to the years 1926–1933 (cf. § 1.3). The Berlin Group has roots that extend to South Germany. Indeed, Reichenbach first formulated many of the ideas that appear in his mature thinking while serving as an Associate Professor (*Privatdozent*) in Stuttgart (1920–1926). In 1922 he began corresponding with Carnap, who was then living in Buchenbach, in the Black Forest, some 180 km (ca. 110 miles) southwest of Stuttgart. It was at Erlangen (Bavaria), at the cutting-edge conference on exact philosophy, that in March of 1923 Reichenbach and Carnap first met (cf. Thiel 1993).

The Berlin Group attracted other non-Berliners, as well, something that Hempel's late recollections confirm:[5]

> The discussion [of the Society] lasted for 4 h, the final two of them at a nearby café, where the excited participants—among them Reichenbach, Dubislav, Grelling, Heinrich Scholtz (who had come from Kiel, I believe), Kurt Lewin, and the very gentle Paul Bernays[6] [from Göttingen]—had become so agitated and noisy they almost caused a public nuisance and made young couples at neighboring tables break off their tender exchanges. (Hempel 1993, p. 4)

One of the external (or associate) members of the Berlin Group was Paul Oppenheim (1885–1977) of Frankfurt on Main. He was the product of the cross-fertilization of business, industry and scientific philosophy that was typical in Germany at the beginning of twentieth-century. (Another such "product" was Count Georg von Arco, one of the co-founders of the Society for Empiric Philosophy.) Around the turn of the twentieth century, Oppenheim was a student in Giessen, where his professor was Hermann Ernst Grassmann, a son and follower of Hermann

[5]This story refers to Hempel's letter to his friend, written in November 1929, and is thus reliable.

[6]David Hilbert's assistent Paul Bernays was sometime a member of the Leonard Nelson's "Jakob Friedrich Fries Society" in Göttingen (active between 1913 and 1921). In the mid-1930s Heinrich Scholtz set up what was later called the "Münster Group" of exact philosophy.

Heinrich Grassmann, the mathematician who authored important works in universal algebra (cf. Graßmann 1844). By the late 1920s Oppenheim also held an adjunct lectureship at the University of Frankfurt.[7]

Oppenheim started to work together with Reichenbach as early as 1921—a contact probably facilitated by Einstein, with whom Oppenheim was on good terms. This collaboration was intensive and lasted till the end of the 1920s. In the "Acknowledgements" of his book *Die natürliche Anordnung der Wissenschaft: Grundgesetze der vergleichenden Wissenschaftslehre* (1926), Oppenheim thanked Reichenbach for "his constant and most effective help by putting [his] ideas into a book form." The title of this volume itself reveals that in the 1920s, Oppenheim also worked with Kurt Lewin, who extensively explored the "comparative theory of sciences [*die vergleichende Wissenschaftslehre*]" (cf. Lewin 1920, 1925). Reflective of this relationship is the extended and highly laudatory review of Oppenheim's book which Lewin published in *Kant-Studien* (cf. Lewin 1929).[8]

During these years, Lewin worked on a program somewhat akin to that of Reichenbach.[9] Under the influence of Ernst Cassirer, Lewin strove to replace mainstream scientific concepts, such as causality, with other theoretical notions that served various complex functions. Lewin's concept of *genetic series*, for example, together with the related notion of *genidentity* as applied to biology and physics, elicited wide interest.[10] The concept of "genidentity" identifies the relation that secures the continuity of an object from one point in time to another; in other words, it explores the way in which objects preserve their identity over time.[11] Lewin's aim in formulating novel scientific concepts was to recast the epistemology of science, and with it scientific classification, along new lines.

[7]This point betrays Oppenheim's connection with another person close to the ideas of the Berlin Group—Franz Oppenheimer (1864–1943). Oppenheimer was the first professor of sociology in Germany and a close friend of Leonard Nelson: in the mid-twenties Oppenheimer invited Nelson's former doctoral student Julius Kraft to become his assistant. (Kraft was also close friend of Karl Popper with whom he launched in 1957 the journal *Ratio* o.s. Cf. Popper 1962) Among Oppenheimer's students were Theodor Adorno and Ludwig Eckhart (the "father" of the West-German *Wirtschaftswunder* after World War Two). Interestingly enough, Oppenheimer spoke about "united science [*Einheitswissenschaft*]" much before either the Berlin Group or the Vienna Circle did so. (Cf. Oppenheimer 1922, pp. xiv f., 10 f) This point was noted in Neurath 1932, p. 271, with reference to Kurt Lewin as a source of information.

[8]Reviews of Oppenheim's book were also published by Hempel (cf. Hempel 1931) and the mathematician of the Hilbert's group in Göttingen, Richard Courant (cf. Courant 1927), who was sometime also a member of the Jakob Friedrich Fries Society around Leonard Nelson.

[9]In the already mentioned paper of Kurt Grelling, "Philosophy of the Exact Sciences: Its Present Status in Germany," he presented Reichenbach and Lewin as two alternative philosophers of exact science. Cf. Grelling (1928), p. 98.

[10]That concept was used in Reichenbach (1928, 1956), Carnap (1928a) and Hermes (1938). See Chap. 5.

[11]Cf. with the theory of rigid designators of Hilary Putnam (one of Reichenbach's students at the University of California at Berkeley) and Saul Kripke.

What drew Reichenbach to Lewin was the interdisciplinary character of Lewin's thinking, something reflected in his conviction that every branch of science produces knowledge that can be of philosophical value. There was an important difference between Lewin and Reichenbach, though. Whereas the former was most interested in the ordering of the new scientific theories and their concepts, Reichenbach concentrated his efforts on bringing to light the ever-changing principles that mark the evolution of scientific theory. Moreover, whereas Lewin conceived of his comparative science of the sciences as a discrete discipline, Reichenbach's philosophy merged with the sciences.[12] More precisely, Reichenbach was convinced that philosophy and science focus on different facets of one and the same subject: nature.

Particularly noteworthy for our concern here is that in the late 1930s Carl Hempel more closely followed Lewin's Cassirer-inspired project than he did that of Hans Reichenbach. Still, Hempel never lost sight of what he learned from Reichenbach, Dubislav and Grelling in Berlin. All three continued to influence Hempel's thinking over the course of a long academic career.

Oppenheim first met Hempel about 1930 through Reichenbach.[13] After playing instrumental role in the Berlin Group's takeover of the Society for Empirical Philosophy in the summer of 1929, Reichenbach (who began to lose interest in Lewin's program) ceased working with Oppenheim and referred him to his promising student and follower Carl Hempel.

From 1934 through 1939 Hempel worked with Oppenheim in Brussels as his "scientific secretary." The issues they explored were clearly closer to Lewin's program than to that of Reichenbach, ranging as they did from the logic of classification and the systematic ordering of science, to taxonomy and the theory of ordering concepts that reflects conceptual isomorphism among different sciences.

In 1938, Kurt Grelling joined Oppenheim and Hempel in Brussels. Together Oppenheim and Grelling explored Lewin's theme of gestalt-theory. In effect, the trio of Oppenheim, Hempel, and Grelling thus constituted an independent satellite unit of the Berlin Group. As already said, along with its varying alignments of affiliated thinkers, the Group was clearly not limited geographically or to a particular time-frame.

In the fall of 1939 Oppenheim and Hempel immigrated to the USA. (Hempel had previously traveled to America, working in Chicago for 9 months in 1937–1938 as an assistant of Carnap's.) It was in America that a new cohort of the Berlin Group came to life in the years1942 through 1944, this time at Princeton, where Hempel joined Oppenheim and Helmer. Hempel's most influential papers, "Studies in the Logic of Confirmation" (1945) and "Studies in the Logic of Explanation" (1948,

[12]This difference is underlined in Grelling (1928), p. 98.

[13]Hempel himself remembers that he first met Oppenheim immediately after the former returned from Vienna, i.e. in Spring 1930, while Oppenheim dated this event in 1933 (Oppenheim 1969, p. 1).

with Paul Oppenheim), reflect this collaboration. After a period of intensive joint study, Hempel stopped working together with Oppenheim and instead served as his "philosophical advisor, talent scout, and professional agent" (Rescher 1997, p. 157).

Nicholas Rescher provides a firsthand description of the way that Oppenheim worked with his Berlin colleagues. Rescher joined Oppenheim in 1952 to investigate the logic of Gestalt theory, a topic that Oppenheim had earlier explored with Grelling in Brussels.[14]

Typically, Oppenheim raised

> (1) the topic of the investigation, and (2) [evinced] a guiding concern for structural issues that reflected a conceptual isomorphism among different scientific disciplines, specifically the view that there is a concept (e.g. that of *gestalt*) which, despite its origin in one particular science, was in principle a versatile instrumentality with useful applications in other branches of science. (Rescher 1997, p. 159)

In view of this, it seems evident that the topic of "scientific explanation" for which both Oppenheim and Hempel became famous in the philosophy of science originated with Oppenheim.

We should say a word at this juncture about Oppenheim's academic project from the 1930s through 1950s. In contrast to the "encyclopedic" program of the Vienna Circle, Oppenheim was not reductionist but

> looked to a more stylistic and structural unity of science, ... [thus he] proposed to search for shared elements of epistemic process among substantively diverse sciences ... for commonalities among the sciences that abstracted from substantive differences and looked at structural uniformities. ...
>
> Oppenheim, in sum, was convinced that various guiding concepts of scientific thought (classification, confirmation, explanation) reflected a fundamental structural community—an isomorphism of concepts of order—that runs across different branches of science. (ibid., pp. 161–162)

Hempel's idea of a generalized "logic of confirmation," which formalized evidential processes of thought common to all forms of scientific reasoning, appears to have been closely connected to this programmatic vision. It was not simply a further development of the Vienna Circle idea of epistemic significance. The same goes for the generalized "logic of explanation."

To be sure, in the early 1940, Hempel and Oppenheim concentrated their efforts on providing a definition of "degree of confirmation" measure for simple formalized languages as the quotient of two range measures. According to Rescher the issue of confirmatory strength of a theory soon transmuted into one of assessing the explanatory adequacy of a theory. The disadvantage of the old approach was the enormous "gap between the inevitably fragmentary observational evidence we actually have and the vast (literally unending) claims that are implicit in any general theory" (ibid., p. 168). The new approach claimed that "the best standard of theory assessment is one that proceeds not in terms of evidential support, but rather in terms

[14]This work resulted in Oppenheim and Rescher (1955).

of the extent to which the theory correctly directs and canalizes our observational expectations" (ibid., p. 169).

In fact, however, theoretical interest in the concept of scientific explanation was long an element of Oppenheim's thinking. It already played an important role in his two books, *Die natürliche Anordnung der Wissenschaft* (1926), and *Die Denkfläche* (1928). Oppenheim later recollected having "worked for years on the possibility of a systematic ordering of the sciences" in which the concept of explanation also played a role (cf. Oppenheim 1969, pp. 1, 3). Discussions of explanation also appear in Dubislav's *Naturphilosophie* (Dubislav 1933, pp. 93 ff.) and, more notably, in Hempel and Oppenheim's book *Der Typusbegriff im Lichte der neuen Logik* (1936b), where the antithesis between description and explanation plays an important role. Moreover, by 1936, these authors connected explanations with the covering laws that group together the empirical data that are to be explained (cf., esp., pp. 102 ff.). Roughly in the same period, Reichenbach, too, spoke about explanation as "summarizing the data under one law" (Reichenbach 1930, p. 55).

Years before Oppenheim, Dubislav and Reichenbach, however, the Southwestern neo-Kantian Heinrich Rickert (1863–1936) treated at length the factors distinguishing descriptions from explanations. Rickert held that there is an important difference (albeit not one in principle) between explanation and description: When we "explain," declared Rickert, we refer to the "generality of the necessity" that has no empirical sources (Rickert 1896, p. 81)—this in contrast with descriptions, which, according to Rickert, are limited to the empirical domain.

The foregoing history establishes how Hempel and Oppenheim's theory of explanation developed not only along the route from "confirmation" to "explanation" but also the other way round: from "explanation" to "confirmation." In other words, when Hempel and Oppenheim first explored the "logic of confirmation," they already had in mind the option to assess scientific theories through their explanatory power.

14.4 Hempel and Carnap

We turn now to the relationship between Hempel and Carnap. Michael Friedman rightly notes that "of all the leading members of the logical empiricist movement, Hempel had always been on closest terms, from a personal point of view, with Carnap" (Friedman 2003, p. 109). The concern here, however, is whether this supports Friedman's claim that the Vienna Circle exercised a continuing influence on Hempel. That it does not becomes apparent when one takes into account Carnap's philosophical character.

Among Carnap's distinguishing traits as a thinker was his predisposition readily to assimilate alternative theoretical doctrines. This is not to suggest that as a scientific philosopher Carnap was an indecisive thinker. Rather, as the memoir left by one of his students at the University of Chicago makes clear, the changeability of his theoretical position was a function of

> his almost selfless drive for truth. He really took seriously the idea that there is progress in human knowledge, that science is a cooperative enterprise whose protagonists share a common goal. He absolutely submerged his ego in that enterprise, more than anyone ..., and he would generally give others the benefit of the doubt—assuming that they too were joining in a selfless and disinterested search for truth.[15] (Sharpless 2009)

The profile that Seth Sharpless sketches above finds substantiation virtually throughout Carnap's career. In the *Aufbau* (1928a), for instance, besides the profound impact of Russell's program for the logical construction of physical objects, neo-Kantian and Husserlian currents also inform Carnap's discussion. Additionally, the work discloses anti-Kantian influences originating with Greifswald Realists, Hans Driesch, and others (cf. Milkov 2004). The reassessments of the logical positivism, which appeared in the 1990s (Friedman 1999; Richardson 1998), drew attention to only one aspect of Carnap's *Aufbau*, namely its connections with the Marburg neo-Kantians.

Further substantiating Sharpless's sketch are events that occurred in the early 1930s, when Carnap readily followed Neurath's lead in subscribing to physicalism (cf. Carnap 1932, p. 338) and promoting the project for an encyclopedia of the sciences. Indeed, the pair worked in close collaboration for about 2 years, with Neurath clearly setting the agenda. A final example of evidence that confirms the accuracy of Sharpless's portrait of his distinguished teacher is the formative impact that Russell's logicism had on Carnap at the very outset of his professional career. Following his move first to Vienna and later to Prague, however, Carnap increasingly viewed that logicism from the perspective of Ludwig Wittgenstein's pan-linguisticism—Wittgenstein insisted that all problems of philosophy are problems of language. Carnap's scholarship at the time featured this pan-linguisticism with such literal fidelity that in 1932 Wittgenstein accused him of downright plagiarism (Nedo and Ranchetti 1983, p. 381). This orientation finds its culminating expression in Carnap's *Logical Syntax of Language* (1934).

In addition to being susceptible to modes of thought originating in widely divergent philosophical currents, and in some respects *in spite* of it, Carnap exhibited an unmistakable "Berlin side." As the last living member of the Berlin Group, Olaf Helmer recalled, "The most prominent members of that group, aside from Hans Reichenbach himself, were Hempel, Dubislav, and (when he came to Berlin on a lecture visit) Rudolf Carnap."[16] Until 1926, Carnap's ideas owed a great deal to the thinking of Reichenbach. Even when, in Vienna, he came under the influence of Wittgenstein, Schlick, Waismann, and Neurath, particularly between 1926 and 1929, Carnap continued to pursue projects associated with Reichenbach, Dubislav and Lewin. Among other research initiatives that evidence this continuing affinity with the Berlin Group are Carnap's explorations in axiomatic (Carnap 1928b)

---

[15]In this kind of selfless pursuit of truth, Carnap is reminiscent of Bertrand Russell and strongly opposed Husserl and Wittgenstein who insisted that the truth they discovered are "eternal" and thus cannot be corrected or supplemented by their critics. Cf. Milkov (2012).

[16]Email communication of Olaf Helmer to the author from July 27, 2009

(Dubislav's and Reichenbach's subject), even as he called axiomatic "applied logistic" (Carnap 1929). Further, he investigated the topic of "definitions" (Carnap 1927) (Dubislav's theme), and he wrote about genidentity (Lewin's brainchild) in *Aufbau*.

This Berlin side of Carnap's scientific philosophy is what led Friedman to propose that the *Aufbau* is above all a book in philosophy of science, its aim being

> to redefine its [philosophy's] own task in the light of the recent revolutionary scientific advances that have made all previous philosophies untenable; ... to use recent advances in the science of logic together with advances in the empirical sciences (Gestalt psychology, in particular) to fashion a scientific replacement for traditional epistemology.[17] (Friedman 1991, pp. 508–509)

Be this as it may, between 1929 and 1936 Carnap and the Berlin Group were at odds over many points in philosophy of science. This surfaced dramatically in Carnap's debate with Grelling and Reichenbach on probability, which took place in September 1929 (cf. § 1.6 (b)). The venue was the "Conference of Epistemology of exact Sciences" in Prague, where the Berlin philosophers declared war on the "principle of verification."[18] Another sign of the conflict between Carnap and the Berliners was Reichenbach's attack on Carnap's "logical positivism" in the early 1930s. This challenge found its most incisive expression in the former's first book in English, *Experience and Prediction* (1938). In yet another symptom of fundamental disagreement with the Berliners, Carnap sharply questioned Reichenbach's naturalistic stance. As Carnap saw it, the philosopher of science is tasked with logically analyzing the language of science. Reichenbach, by contrast, understood his purpose as analyzing the facts of science. Carnap regarded this as overstepping the bounds proper to philosophy and unwarrantably proposing to engage in the practice of science (Carnap 1936a, b).

From the time shortly before he left Prague at the end of 1935, however, Carnap drifted closer to the Berlin Group's philosophical program. Two constellations of developments bear witness to this philosophical realignment:

(i) First and foremost, Carnap gradually shifted his attention away from the theory of verification as applied to the *sentences* of science and more toward exploring the confirmation of scientific *theories*. Two papers that he wrote at this time document this change: "Wahrheit und Bewährung" (1936) and "Testability and Meaning" (1936–1937). Reichenbach saw this development as marking Carnap's transition from "dogmatic" positivist of the Vienna Circle to scientific philosopher, one who critically analyzes the latest advances in the sciences (Reichenbach 1938, pp. 76 f.).

---

[17]In support of this claim we would like to note that between 1926 and 1935 Carnap taught philosophy at the University of Vienna and then at the University of Prague. When he started to teach at the University of Chicago, however, he invited (in 1937) Reichenbach's students Hempel and Helmer, and not some of his own students, to become his assistants. This also explains why Hempel and Helmer so easily started to work together with Carnap.

[18]Cf. Diskussion über Wahrscheinlichkeit, *Erkenntnis* 1 (1930): 260–287.

(ii) From 1942 through 1944, Carnap's thinking came under the spell of a new Berliner current through his contacts with the "H_2O philosophers:" Hempel, Helmer and Oppenheim, whom as we've seen for all intents and purposes constituted a later variant of the Berlin Group. Carnap's interest in induction and the later focus on "comparative concepts"[19] (cf. Carnap 1950, §§ 4 f.) clearly betray the Group's impact on him. Furthermore, during this period Carnap's studies in the logical foundation of probability reflect the thinking of Reichenbach. This is not to suggest that Carnap followed Reichenbach in the treatment of probability—he did not. Rather, Carnap simply began serious work on a topic—probability—that happened to have preoccupied Reichenbach from the beginning of his philosophical career. (Carnap himself reported that his interest in probability originated with lectures he audited by Richard von Mises.)

Among other things, what obscures the Berlin Group's impact on Carnap is that Hempel, that unreliable historian of philosophy and reluctant autobiographer, stressed Carnap's continuing influence on *himself*. Evidently Hempel could not entertain the thought that in many instances it was he and Oppenheim who significantly influenced *Carnap*. Lamentably, it is Hempel's account that has become the accepted view in the literature.

14.5 The Method of Explication

A good deal of evidence unquestionably points to Carnap's influence on Hempel. This influence, however, did not invariably reflect ideas that preoccupied the Vienna Circle. One example is Carnap's so-called "method of explication" (cf. Carnap 1945) which he propounded while he was writing *Meaning and Necessity* (1942–1944). He described it as the process of "transforming a given more or less inexact concept into an exact one" (Carnap 1950, p. 3). Hempel adopted this method, first thematizing "explication" in his *Fundamentals of Concept Formation in Empirical Science* (1952).

Actually, both Carnap and Hempel employed the method prior to taking it up as a formal theme of philosophical analysis. It is notable in this connection that Carnap's conception of explications underwent considerable transformation. While he formulates it in *Meaning and Necessity* (1947) along roughly Fregean/Russellian lines, in *Introduction to Semantic Theory* (1950) Carnap presents it more from a late-Wittgenstein standpoint. To be more exact, while in 1947 Carnap sought to develop a formal means of replacing vague concepts with precise ones by applying the exactitude of the scientific method, by 1950 he concluded that explication cannot

---

[19]Comparative concepts were already discussed in Hempel and Oppenheim (1936a, b). Cf. also Tegtmeier (1981).

be decided in an exact way. Carnap thus ultimately arrived at the position that the theory of explications, rather than being able to invoke the standard of scientific exactitude, is perforce limited to the simply satisfactory (cf. ibid., p. 4).

Historians of the philosophy of science tend to assume that it was the thinking of British ordinary-language philosophers that was largely behind the Wittgensteinian shift in Carnap's approach to the method of explication. But this assumption rests solely upon references Carnap made in *Meaning and Necessity* (pp. 8, 42, and 63n. 7) to C.H. Langford's paper on G.E. Moore's "paradox of analysis," which appeared in Schilpp's *G.E. Moore* volume. The influence of Wittgenstein is much more long standing and direct, however, tracing back to the early 1930s and Carnap's wholehearted subscription to Wittgenstein's "enchantment with words." Under this influence, Carnap held that language is philosophy's subject matter and that the subject-matter of scientific philosophy is the language of science. Wittgenstein himself developed the technique of conceptual analysis, arguably his main preoccupation in those of his later works that investigate the necessary and sufficient conditions of language applications. And exactly Wittgenstein's method of conceptual analysis is what grounds Carnap's (and thus Hempel's) method of explication.

Moreover, in the 1930s, this method figured as the principal topic of debate among British philosophers such as Susan Stebbing, John Wisdom and Max Black. Hempel readily joined this discussion. In fact the first paper he published after relocating to the USA focuses upon Max Black's treatment of vagueness, not philosophy of science (cf. Hempel 1939). This redirection of his thinking set the stage for Hempel readily to adopt, some 10 years later, Carnap's method of explication.

14.6 Carl Hempel Between External and Internal Philosophy of Science

Oppenheim's Lewin-inspired approach to structural issues of science at once postulated a conceptual isomorphism among different scientific disciplines and so helped to introduce the method of explication. This doctrine had two contrasting consequences by the mid 1960s. One the one hand, it led Hempel to conceive a discrete new discipline, *philosophy of science*, with clear-cut themes and a compelling program of theoretical research. This innovation of Hempel's found its classic expression in his *Philosophy of Natural Sciences* (1966). It signaled the birth of the philosophy of science as a discipline and was a significant development of modern intellectual history, effectively narrowing the gap between modern science and philosophy.

On the other hand, however, by the mid 1960s, it became clear that the method of explications which Carnap and Hempel employed[20] was not interdisciplinary but

[20]As already seen, Carnap and Hempel practiced it from the beginning of the 1940s onward.

rather "non-disciplinary" (Rescher 1997, p. 162). What's more, it had no "connection to any scientific theory. The concepts to be analyzed [by it] were general, methodological concepts supposedly common to all the sciences." (Giere 1996, p. 340) In other words, they patently belonged to what Ernan McMullin termed "external philosophy of science."[21] The result for pedagogy and scholarly praxis was "an increasing separation between philosophy of science and the content of the sciences. People trained in philosophy, but with little knowledge of any science, could write article after article with titles like 'The Paradoxes of Confirmation' or 'The Symmetry between Explanation and Prediction'." (ibid., p. 341)

The prevalent view today is that the situation started to change only with the appearance and the assimilation of the ideas of Thomas Kuhn's *Structure of Scientific Revolutions* (1962). Kuhn's work impelled authors of philosophical studies to acquire detailed knowledge in the special sciences and exhibit it in their writings. Arguably, this was a turn to "internal philosophy of science."

In truth, however, Reichenbach had already acquired and worked with such detailed knowledge back in the 1920s. On this score, he was the first contemporary philosopher of science—or, more precisely, the first philosopher of physics, a discipline that presupposes "detailed investigations into the particular aspects or interpretations of physical theories" (Ryckman 2007, p. 193). The claim in these pages is that the philosophy of science, a vital and thriving sub-discipline today, was born in Berlin in the 1920s and early 1930s.

14.7 Epilogue

By way of conclusion, we should note that when Hempel abandoned Carnapian explications and embraced naturalism in the 1980s, he was not, *pace* Friedman (cf. § 14.1, above), simply following Neurath. Substantiating this fact is unimpeachable evidence that in his last years Hempel, returning to his philosophical roots, propounded a modified form of Reichenbach's Berlin naturalism. In fact, Friedman's reading of the history hinges on four matters with respect to which Neurath, curiously enough, happened to be close to Reichenbach and his colleagues in Berlin:

- Like Reichenbach, Neurath had called for philosophers strongly interested in scientific theories as instruments for prediction. That said, however, while Neurath discussed sciences in quite general terms (vis-à-vis the unity-of-science project), Reichenbach undertook detailed investigations into particular themes of physical theories.

[21] On Ernan McMullin's terms "external" and "internal" philosophy of science see Chapter One, § 1.9.

- Like Reichenbach, Neurath (after 1931) fought logical *positivism*, criticizing it for straying from scientific praxis. To differentiate their positions from the positivists, both Reichenbach and Neurath began characterizing their work and defending it as *logical empiricism*. (Reichenbach also used the term *logistic empiricism*.)
- Like Reichenbach, Neurath embraced a program for the unity of science. But whereas Neurath championed an "encyclopedia of sciences" that employed unific concepts, Reichenbach advocated parallel investigations of the "relativised a priori" principles in different sciences.
- Lastly, like Reichenbach, Neurath embraced, after 1931, the program for physicalism (Carnap following suit in 1932).

It should be clear, then, that while Friedman rightly sees Carl Hempel as at last siding, in the 1980s, with Neurath, the evidence suggests that what motivated Hempel's move was not Neurath himself but rather Reichenbach and the spirit of the Berlin Group in general, the milieu in which Hempel served his philosophical apprenticeship.

References

Carnap, Rudolf. 1927. Über eigentliche und uneigentliche Begriffe. *Symposion* 1: 355–374.

Carnap, Rudolf. 1928a. *Der logische Aufbau der Welt*. Berlin: Weltkreis-Verlag.

Carnap, Rudolf. 1928b. *Untersuchungen zur allgemeinen Axiomatik*. Hrsg. von Thomas Bonk. Darmstadt: Wissenschaftliche Buchgesellschaft, 2000.

Carnap, Rudolf. 1929. *Abriss der Logistik*. Wien: Springer.

Carnap, Rudolf. 1932. Die physikalische Sprache als Universalsprache der Wissenschaft. *Erkenntnis* 2: 432–465.

Carnap, Rudolf. 1936a. Von der Erkenntnistheorie zur Wissenschaftslogik. In: *Actes du Congrès international de philosophie scientifique, Paris 1935*, Fasc. 1, *Philosophie scientifique er l'empirisme logique*, 36–41. Paris: Hermann.

Carnap, Rudolf. 1936b. Wahrheit und Bewährung, in: *Actes du Congrès international de philosophie scientifique, Paris 1935, Fasc. 4, Induction et probabilité*, Paris: Hermann, pp. 18–23.

Carnap, Rudolf. 1936–1937. Testability and meaning. *Philosophy of Science* 3 (1936): 419 ff.; 4 (1937): 1 ff.

Carnap, Rudolf. 1945. The two concepts of probability. *Philosophy and Phenomenological Research* 5(1945): 513–532.

Carnap, Rudolf. 1947. *Meaning and necessity: A study in semantics and modal logic*. Chicago: University of Chicago Press.

Carnap, Rudolf. 1950. *Logical Foundations of Probability*. Chicago: University of Chicago Press.

Courant, Richard. 1927. Paul Oppenheim. *Die Naturwissenschaften Die natürliche Ordnung der Wissenschaft* 15: 655.

Dubislav, Walter. 1932. *Die Philosophie der Mathematik in der Gegenwart*. Berlin: Junker & Dünnhaupt.

Dubislav, Walter. 1933. *Naturphilosophie*. Berlin: Junker und Dünnhaupt.

Friedman, Michael. 1991. The re-evaluation of logical positivism. *The Journal of Philosophy* 88: 505–519.

Friedman, Michael. 1999. *Reconsidering logical positivism*. Cambridge: Cambridge University Press.

Friedman, Michael. 2003. Hempel and the Vienna circle. *Minnesota Studies in the Philosophy of Science* 18: 94–114.
Gerner, Karin. 1997. *Hans Reichenbach: sein Leben und Wirken*. Osnabrück: Phoebe.
Giere, Ronald. 1996. From *wissenschaftliche Philosophie* to philosophy of science. *Minnesota Studies in the Philosophy of Science* 16: 335–354.
Graßmann, Hermann Günther. 1844. *Die Wissenschaft der extensiven Größe oder die Ausdehnungslehre*. Leipzig: Wigand.
Grelling, Kurt. 1928. Philosophy of the exact sciences: Its present status in Germany. *The Monist* 38: 97–119.
Grelling, Kurt. 1929. Realism and logic: An investigation of Russell's metaphysics. *The Monist* 39: 501–520.
Grelling, Kurt. 1930. Die Philosophie der Raum-Zeit-Lehre. *Philosophischer Anzeiger* 4: 101–128.
Hempel, Carl. 1931. Review of Oppenheim 1926. *Erkenntnis* 2: 473–474.
Hempel, Carl. 1935a. On the logical positivists' theory of truth. *Analysis* 2: 49–59.
Hempel, Carl. 1935–1936. Über den Gehalt von Wahrscheinlichkeitsaussagen. *Erkenntnis* 5: 228–260.
Hempel, Carl. 1935b. Some remarks on 'Facts' and propositions. *Analysis* 2: 93–96.
Hempel, Carl. 1936. Some remarks on empiricism. *Analysis* 3: 33–40.
Hempel, Carl. 1939. Vagueness and logic. *Philosophy of Science* 6: 163–180.
Hempel, Carl. 1945. Studies in the logic of confirmation. *Mind* 54: 1–26 and 97–121.
Hempel, Carl. 1952. *Fundamentals of concept formation*. Chicago: University of Chicago Press.
Hempel, Carl. 1993. Empiricism in the Vienna circle and in the Berlin society for scientific philosophy. Recollections and reflections. *Institute of the Vienna Circle Studies* 1: 1–9.
Hempel, Carl. 2000. Intellectual autobiography—The interview with Richard Nollan. In *Science, explanation, and rationality*, ed. J.H. Fetzer, 3–35. Oxford: Oxford University Press.
Hempel, C., and Oppenheim, P. 1936a. *L'importance logique de la notion de type. Actes du Congreés International de Philosophie Scientifique*, vol. 2, 41–49. Paris: Hermann.
Hempel, C., and Oppenheim, P. 1936b. *Der Typusbegriff im Lichte der neuen Logik*. Leiden: Sijthoff.
Hempel, C., and Oppenheim, P. 1948. Studies in the logic of explanation. *Philosophy of Science* 15: 135–175.
Hermes, Hans. 1938. *Eine Axiomatisierung der allgemeinen Mechanik*. Leipzig: Hirzel.
Lewin, Kurt. 1920. *Die Verwandtschaftsbegriffe in Biologie und Physik und die Darstellung vollständiger Stammbäume*. Berlin: Bornträger.
Lewin, Kurt. 1925. Über Idee und Aufgabe der vergleichenden Wissenschaftslehre. *Symposion* 1: 61–94.
Lewin, Kurt. 1929. Review of Oppenheim 1926. *Kant-Studien* 34: 461–464.
Milkov, Nikolay. 2004. G. E. Moore and the Greifswald objectivists on the given, and the beginning of analytic philosophy. *Axiomathes* 14: 361–379.
Milkov, Nikolay. 2012. The construction of the logical world: Frege and Wittgenstein on fixing boundaries of human thought. In: *Crossing borders*, eds. Alfred Dunshirn et al., 151–161. Vienna: University of Vienna.
Nedo, Michael, and Michele Ranchetti. 1983. *Ludwig Wittgenstein: Sein Leben in Bildern und Texten*. Frankfurt: Suhrkamp.
Neurath, Otto. 1932. In *Sociology in physicalims*, ed. M. Stölzner and T. Uebel, 269–314. Wiener Kreis/Hamburg: Meiner, 2006.
Oppenheim, Paul. 1926. *Die natürliche Anordnung der Wissenschaft: Grundgesetze der vergleichenden Wissenschaftslehre*. Jena: Fischer.
Oppenheim, Paul. 1928. *Die Denkfläche: Statische und dynamische Grundgesetze der wissenschaftlichen Begriffsbildung*. Berlin/Charlottenburg: Pan.
Oppenheim, Paul. 1969. Reminiscences of Peter. In *Essays in Honour of Carl G. Hempel*, ed. N. Rescher, 1–4. Dordrecht: Reidel.
Oppenheim, P., and N. Rescher. 1955. Logical analysis of Gestalt concepts. *The British Journal for the Philosophy of Science* 6: 89–106.

Oppenheimer, Franz. 1922. *Allgemeine Soziologie*. Stuttgart: Fischer, 1964.
Popper, Karl. 1962. Julius Kraft. *Ratio* 4: 2–10.
Reichenbach, Hans. 1928. *Philosophie der Raum-Zeit-Lehre*. Berlin: de Gruyter.
Reichenbach, Hans. 1930. Die philosophische Bedeutung der modernen Physik. *Erkenntnis* 1: 49–71.
Reichenbach, Hans. 1938. *Experience and prediction*. Chicago: Chicago University Press.
Reichenbach, Hans. 1956. In *The direction of time*, ed. M. Reichenbach. Berkeley: University of California Press.
Rescher, Nicholas. 1997. H_2O: Hempel–Helmer–Oppenheim: An episode in the history of scientific philosophy in the 20th century. *Philosophy of Science* 64: 779–805.
Richardson, Alan. 1998. *Carnap's reconstruction of the world: The Aufbau and the emergence of logical positivism*. Cambridge: Cambridge University Press.
Rickert, Heinrich. 1896. *Die Grenzen der naturwissenschaftlichen Begriffsbildung*. Tübingen: J.C.B. Mohr, 1921.
Ryckman, Thomas. 2007. Logical empiricism and the philosophy of physics. In *The Cambridge companion to logical empiricism*, ed. A. Richardson and T. Uebel, 193–227. Cambridge: Cambridge University Press.
Sharpless, Seth. 2009. Reminiscences about Carnap at Chicago (1945–1951). URL = http://www.sethsharpless.com/papers/Reminiscences.htm
Tegtmeier, Erwin. 1981. *Komparative Begriffe: Eine Kritik der Lehre von Carnap und Hempel*. Berlin: Dunker & Humblot.
Thiel, Christian. 1993. Carnap und die wissenschaftliche Philosophie auf der Erlanger Tagung. In *Wien, Berlin, Prag. Der Aufstieg der wissenschaftlichen Philosophie*, ed. R. Haller and F. Stadler, 175–188. Vienna: Hölder–Pichler–Tempsky.
Wolters, Gereon. 2003. Carl Gustav Hempel—Pragmatic empiricist. In *Logical empiricism: Historical and contemporary perspectives*, ed. P. Parrini et al., 109–122. Pittsburgh: University of Pittsburgh Press.

Carl Hempel, Spring 1958, at the University of Yale (by Veli Valpola)

The editors and publishes would like to thank the following for permission to use photographs: Fred Stein Archive, Stanfordville, NY (for 2); Fotoagentur Ullstein Bild, Berlin (for 3); Nicholas Rescher (for 1 and 5); Karin Gimple-Grelling, Zürich (for 4); The Special Collections Department, University Library System of the University of Pittsburgh (for 6).

Chapter 15
Hempel, Carnap, and the Covering Law Model

Erich H. Reck

Carl G. Hempel (1905–1997) is usually not taken to be a philosopher of the same stature as Hans Reichenbach, the central figure in the Berlin Group and his doctoral advisor, or Rudolf Carnap, the leading member of the Vienna Circle and another important influence on him. Yet Hempel's impact on philosophy was almost as widespread and lasting as theirs, particularly in the United States where he emigrated and where his career flourished. Hempel was educated at the Universities of Göttingen, Heidelberg, Vienna, and Berlin (Ph.D. in 1934). He first visited the US in 1937–1938 to work as Carnap's research assistant at the University of Chicago. He came back in 1939, as a refugee, so as to stay permanently. His first teaching positions were in New York, at City College (1939–1940) and Queens College (1940–1948). Later he taught at Yale (1948–1955), Princeton (1955–1975), and the University of Pittsburgh (1976–1985). Over the course of his long career Hempel had many students. He was also active in the profession in other ways, e.g., as Vice-President of the Association for Symbolic Logic and as President of the American Philosophical Association. In retrospect, he has been called "one of the principal figures of scientific philosophizing in the twentieth century" (Rescher 2005, 127).[1]

Hempel's main contributions concern the philosophy of science.[2] He is most well known for his writings on the notions of confirmation, explanation, rationality, cognitive significance, and scientific theory. In the present essay I will focus on

---

An early version of this paper was presented at the conference, "Die Berliner Gruppe", Paderborn, September 5, 2009. I would like to thank Nikolay Milkov and Volker Peckhaus for inviting me to it. I am also grateful to various audience members for criticisms, comments, and encouragements.

[1]For biographical information, cf. Fetzer (2000b, 2010), Rescher (2005), also Friedman (2000).

[2]For overviews of Hempel's main works, cf. Salmon (2000), Kitcher (2001), Fetzer (2010), and Curd (2012).

E.H. Reck (✉)
Department of Philosophy, University of California, Riverside, CA, USA
e-mail: erich.reck@ucr.edu

N. Milkov and V. Peckhaus (eds.), *The Berlin Group and the Philosophy of Logical Empiricism*, Boston Studies in the Philosophy and History of Science 273, DOI 10.1007/978-94-007-5485-0_15,

his work on scientific explanation and its impact on philosophy in the English-speaking world. Central in this connection is Hempel's article (co-written with Paul Oppenheim), "Studies in the Logic of Explanation" (1948), which Wesley Salmon, another main contributor to the corresponding debates, characterized as "epoch making" (Salmon 2000, 311). Concerning Hempel's subsequent collection of essays, *Aspects of Scientific Explanation* (Hempel 1965b), James Fetzer has remarked that it "became a scholar's bible for generations of graduate students" (Fetzer 2010, 1). Similarly, Hempel's textbook, *Philosophy of Natural Science* (Hempel 1966), was read by generations of undergraduate students and it is still sometimes assigned today. My main goal in this essay will be to get clearer about why exactly these texts were so influential and, more basically, what their philosophical significance is. The quick answer, to be elaborated in what follows, is that this is where Hempel's Covering Law Model for scientific explanation was presented and elaborated.

***

"Studies in the Logic of Explanation" (1948) is not the first work in which the Covering Law Model (CL model, for short) appeared. Its core idea had been suggested by other philosophers, e.g., by Karl Popper, Richard Braithwaite, and John Stuart Mill. In fact, it can be traced back all the way to Aristotle (Fetzer 2000a). And as far as Hempel's own publications are concerned, the idea was already presented in "The Function of General Law in History" (1942). Nevertheless, it is the 1948 article that primarily set the stage for later discussions.[3] It starts as follows:

> The present essay [provides] an elementary survey of the basic patters of scientific explanation and a subsequent more rigorous analysis of the concept of law and the logical structure of explanatory arguments (Hempel and Oppenheim 1948, 567).

In Part I of their essay Hempel and Oppenheim then introduce several motivating examples of scientific explanations and, most importantly, the following schema:

$$\begin{array}{ll} C_1, C_2, \ldots, C_k & \text{Statement of antecedent conditions} \\ \underline{L_1, L_2, \ldots, L_r} & \text{General laws} \\ E & \text{Description of the phenomenon to be explained} \end{array}$$

The "basic pattern of scientific explanations" is thus: E (the "explanandum") is deduced logically from $C_1, C_2, \ldots, C_k$ and $L_1, L_2, \ldots, L_r$ (the "explanans"). The two authors go on to spell out several additional requirements for explanation, divided into two groups. The "logical conditions of adequacy" are: (i) The corresponding argument has to be valid, i.e., E has to be in fact derivable from $C_1, C_2, \ldots, C_k$ and $L_1, L_2, \ldots, L_r$; (ii) at least one general law "must be required for the derivation"; (iii)

[3]In Wesley Salmon's words: "The 1948 Hempel–Oppenheim article marks the division between the pre-history and the history of modern discussions of scientific explanation." (Salmon 1990, 10)

the explanans must have "empirical content", i.e., "be capable, at least in principle, of test by experiment or observation". The one "empirical condition of adequacy" is: (iv) "The sentences constituting the explanans must be true", so that the argument is sound (*ibid.*, 569–570). Hempel and Oppenheim also argue that, because of their underlying logical forms, there exists a "symmetry" between explanation and prediction in science. In later parts of the essay they develop, among others, a "more rigorous analysis of the concept of law" by applying the concepts and tools of modern logic (syntax and formal semantics).

Implicit in the schema from "Studies in the Logic of Explanation" is that the explanandum is derived from deterministic laws together with relevant initial conditions. But Hempel acknowledged quickly that in science there are explanations based on statistical or probabilistic laws as well. In many of them the explanandum is not a deductive consequence of the explanans, but it follows only with a certain probability. Strictly speaking, the schema above applies thus only to "deductive-nomological" explanations, while "inductive-statistical" explanations have to be treated separately. Moreover, there are scientific explanations in which statistical claims are derived deductively from more general statistical laws, in which case we are dealing with "deductive-statistical" explanations. Then again, in all three kinds of cases the explanandum is subsumed under, or "covered" by, general laws; and hence, what is crucial for scientific explanations generally is "nomic expectability". In Hempel's later publications this view is articulated in terms of an all-encompassing "Covering Law Model", based on a schema that generalizes the one from the 1948 essay. The most systematic, mature treatment of his position occurs in "Aspects of Scientific Explanation" (Hempel 1965a), the centerpiece of Hempel (1965b), while a simpler and more accessible discussion lies at the heart of Hempel (1966).

***

It took some years after the publication of Hempel and Oppenheim (1948) for the CL model to attract much attention. However, from the 1960s on it became a central and entrenched part of "scientific philosophy"—it became the "received view" on explanation, the position against which all alternatives were measured. Why did it have such an impact? Hempel's steadily increasing personal influence was important, no doubt, i.e. his recognition as a main player in the field. Yet there were more philosophical reasons as well, including the following: First, Hempel and Oppenheim's careful, formally precise treatment rehabilitated the notion of explanation among scientifically oriented philosophers. While this may be surprising from today's point of view, in the early twentieth century that notion was widely seen as problematic, e.g., as too subjective (too much anchored in a "feeling" of insight). One benefit of the CL model, in the eyes of many, was to secure its objectivity and rationality.[4] Second, Hempel's account of scientific laws

[4]As Salmon later put it: "[T]he Hempel-Oppenheim 1948 article forced scientific explanation onto the attention of a wide class of logicians and philosophers of science. There was an explicit proposal regarding the nature of scientific explanation on the table, and it challenged philosophers

was carefully crafted to get around Humean scruples concerning the notion of causation, as shared by many empiricists.[5] Consequently the CL model could be taken to provide an indirect but respectable way of talking about causation in terms of law-based explanations. In both respects, the approach was perceived as leading to substantive philosophical progress.

It was not just among philosophers that the CL model was noted and admired. The model also exerted a significant influence on other disciplines, such as history and some of the social sciences. In those contexts it was taken to be normative, i.e., as telling researchers to produce explanations of CL form.[6] But soon its alleged universal applicability was called into question. (Eventually it came to be seen as a central part of the "positivist" legacy, where ideas and methods from one field, namely mathematical physics, were imposed on others in counterproductive ways; but that took a while). Within the philosophy of science doubts about the CL model also started to emerge. The initial ones concerned the specifics of the formal account of laws in Hempel and Oppenheim (1948), which were shown to lead to paradoxical consequences.[7] While it may have appeared for a while that some minor tinkering would get around these problems, gradually further criticisms of the CL model arose, often in the form of "counterexamples" to it. These examples—many of which became classics in themselves (the flagpole, the moon and tides, syphilis and paresis, etc.)—called the CL model into question in a number of ways. Some challenged Hempel and Oppenheim's "symmetry thesis" for explanation and prediction; others were meant to establish, very fundamentally, that for a scientific account to be explanatory it was neither necessary nor sufficient to have CL form; etc.[8]

While the CL model kept having defenders, including Hempel himself (who worked on improving his treatment of inductive-statistical explanations[9]), it began to be seen, more and more, as the foil against which to pit alternative accounts. The two primary alternatives became the "causal model", with Wesley Salmon as the main initial proponent, and the "unification model", represented by Michael

---

to respond either positively or negatively. It elicited alternative analyses. The temptation to say that there is no such a thing as scientific explanation seems to have vanished." (Salmon 2000, 315)

[5]I take the logically based account of scientific laws in the later parts of Hempel and Oppenheim (1948) to be due mostly to Hempel. I will come back to Oppenheim's role briefly later in this essay.

[6]Cf. the discussion of archeology in Salmon (1990), 25–26. Hempel had applications to history in mind from early on; cf. Hempel (1942). Some of its earliest and longest lasting criticisms concern that intended application.

[7]Cf. the discussion of the early, internal criticisms of the CL model (by R. Eberle, D. Kaplan, R. Montague, and others) in Salmon (1990), chapter 2.

[8]The corresponding counterexamples focused attention on, among others, the necessity of laws, the deductive structure of explanations, and certain explanatorily relevant causal asymmetries not captured by the CL schema. For overviews, cf. Salmon (1990), chapters 2–3, and Fetzer (2000a).

[9]Many of the "second-wave" challenges to the CL model concerned the inductive-statistical case. One attempt to improve on Hempel's position was Salmon's "statistical-relevance" model, which was subsequently also found wanting. Cf. Salmon (1990), chapter 3 and Fetzer (2000a).

Friedman and Philip Kitcher. In some respects these were not outright rejections of the CL model but modifications of it (especially the unification model). However, more radical alternatives also appeared, e.g., Bas van Fraassen's "pragmatic" model (based on a formal analysis of explanation-seeking why questions) and, already earlier, a more informal, contextual approach to explanation championed by Michael Scriven (guided by a radically different methodology).[10] It seems fair to say that, as a result of the proliferation of alternative approaches, there is no "received view" about scientific explanation any more today, even though causal models tend to be more prominent than others. Some would even argue that it is misguided to look for a universal model and that what is needed, instead, is a plurality of models, since explanations come in a variety of different forms.

***

It is not my goal here to provide a comprehensive overview of the debate about scientific explanation, much less a resolution for it.[11] After having sketched at least some relevant developments, I want to return to Hempel, the CL model, and its significance. Often the attitude with respect to that model, especially by critics, appears to be the following: What Hempel and Oppenheim did, in their classic essay and elsewhere, was to start with some representative examples of scientific accounts (by Kepler, Galilei, Newton, Einstein, etc.) and then distill out their essential form, i.e., the aspect that makes them "explanatory". If successful, this procedure would have provided us with an analysis of the notion of explication in a very strong sense: an articulation of jointly necessary and sufficient conditions for explanations in general. And as these conditions were formulated in terms of modern (deductive and inductive) logic, it would have amounted to the reductive analysis of a notion central to science. This is, then, what the significance of the CL model is typically taken to amount to. It is just that the analysis it embodies does not work, as the counterexamples are supposed to have shown.

Two different reactions to the resulting situation are possible. First, one can hold on to the goal of providing a reductive analysis, and in particular, of articulating necessary and sufficient conditions for explanation. That is to say, while it may be true that the Hempel and Oppenheim's model does not work as such, one can take modify it or replace it by a better analysis (along causal or unification lines, say). As a second and more radically reaction, one can take the "counterexamples" to the CL to have shown, not only that this model is inadequate, but that the whole approach underlying it, in terms of a formal and reductive analysis, needs to be abandoned. That would not necessarily mean that we have to give up analyzing the notion of explanation; but we should, so the suggestion here, proceed in a non-reductive, contextual way. Now, these two kinds of reactions are not only quite different, they

[10]Pitt (1988) contains representative texts; cf. also Kitcher and Salmon (1989) and, again, Salmon (1990). For the basic difference between Scriven's and the other approaches, cf. Reck (2012).

[11]An authoritative recent discussion of the topic, as presented by the proponent of a causal account, can be found in Woodward (2003). For a more general overview, see also, e.g., Psillos (2007).

are opposed to each other.[12] At the same time, they rely on a shared assumption about the CL model, namely: that it has been refuted, in some fairly direct way, by the "counterexamples". Or more generally, it is assumed that the model has been refuted by the careful description of scientific practice.

However, do the standard criticisms of the CL model really refute it so directly? First doubts arise when one takes seriously Hempelian remarks such as the following:

> [T]hese models are not meant to describe how working scientists actually formulate their explanatory accounts. Their purpose is rather to indicate in reasonably precise terms the logical structure and the rationale of various ways in which empirical science answers explanation-seeking why-questions. The construction of our models therefore involves some measure of abstraction and of logical schematization (Hempel 1965a, b, 412).

Moreover, it is not just that the CL model (the deductive-nomological, inductive-statistical, and deductive-statistical models taken together) involves "abstraction" and "schematization", as Hempel readily admits. If the model is taken to provide a reductive analysis of explanation, one misrepresents its nature and purpose more fundamentally—or so the argument I want to consider next. But if the CL model is not meant to constitute a reductive analysis, how else could we think about it? An answer to that question is provided by Carnap's notion of explication. (I will consider a second, different answer later in the essay as well.)

***

Rudolf Carnap introduced the notion of explication for the first time in his book, *Meaning and Necessity* (1947); he then discussed it in more detail in his next book, *Logical Foundations of Probability* (1950). As he writes in the former:

> The task of making more exact a vague or not quite exact concept used in everyday life or in an earlier stage of scientific or logical development, or rather of replacing it by a newly constructed, more exact concept, belongs among the most important tasks of logical analysis and logical construction. We call this the task of explicating, or of giving an *explication* for, the earlier concept; this earlier concept, or sometimes the term used for it, is called the *explicandum*; and the new concept, or its term, is called an *explicatum* of the old one (Carnap 1947, 7–8; original emphasis).

If one adopts Carnapian explication as one's methodology, this does lead to abstraction and schematization, along Hempelian lines. But beyond that, descriptive accuracy is rejected, or downplayed, in an even stronger sense. The sense at issue is flagged by Carnap's talk of "replacing" an earlier, vague concept by a new, more exact one. Here Carnap points to the fact that the main thrust in giving an explication, in his sense, is revisionary and normative rather than descriptive. And this makes it significantly different from reductive analysis.

---

[12]The first reaction, or the first kind of alternative, is much more common in the literature on scientific explanation. Even van Fraassen's pragmatic model can be seen as falling into this first camp. I take Michael Scriven's approach to be an example of the second kind of response, in the sense that he provided what Peter Strawson would later call a "connective analysis" of the notion of explanation. For further discussion of the latter point, cf. Reck (2012); for another, more recent representative of Scriven's camp, cf. Wright (2011).

Two closely related aspects of the relevant difference are the following: First, in an explication we start with a vague notion and replace it by a more exact one; and because of the vagueness of the former, it is misguided to judge the latter in terms of whether it "fully captures" what was there before. Second and more positively, what the new notion should be judged by instead is its usefulness. As Carnap writes in *Logical Foundations of Probability*:

> Strictly speaking, the question whether the solution [the explicatum, thus the explication overall] is right or wrong makes no good sense because there is no clear-cut answer. The question should rather be whether the proposed solution is satisfactory, whether it is more satisfactory than another one, and the like (Carnap 1950, 4).

Shortly after this passage Carnap list four main criteria for evaluating an explicatum: "(1) similarity to the explicandum; (2) exactness; (3) fruitfulness; (4) simplicity". Note here that, while "similarity" is the first of the desiderata listed, there are three others; and those criteria typically bear more weight in Carnap's and later applications of explication. Note also that, by only requiring "similarity" in a sense left fairly unspecific, descriptive adequacy with respect to earlier practice appears to be required only in a very weak sense.

Returning to Hempel, there are a number of reasons for regarding the CL model, as well as his approach more generally, as an instance of Carnapian explication. To begin with, many of the features distinctive of explication are present, e.g., the insistence on exactness and the use of formal tools (syntax and formal semantics). There were also personal connections between Hempel and Carnap, including during the period when both Carnap's notion of explication and Hempel's CL model took shape (the late 1930s and the 1940s). More concretely, Carnap is one of the people Hempel and Oppenheim thank explicitly for "stimulating discussions and constructive criticisms" in the first footnote of Hempel and Oppenheim (1948). In addition, Hempel mentions Carnap and the notion of explication positively in some of his later reflections on his work (Hempel 1973, 1988). Finally, other central participants in the ensuing debate about the CL model describe the underlying approach in Carnapian terms; thus Salmon writes: "The Hempel-Oppenheim article is an outstanding example of the use of an artificial language for the purposes of explicating a fundamental scientific concept." (Salmon 1990, 35)

***

Suppose therefore that we interpret the CL model as a case of explication in Carnap's sense. What exactly follows about that model, especially concerning how to evaluate it? As already noted, for Carnap "similarity" between the explicatum and the explicandum is a desideratum, but only one that plays a minor and subordinate role. Beyond that, the only guidance with which he provides us in this connection is the following:

> An indication of the meaning with the help of some examples for its intended use and other examples for uses not now intended can help the understanding. An informal explanation in general terms may be added (Carnap 1950, 1).

Notice the emphasis on "intended use" in this passage, which signals what is really crucial. Namely, in the end the evaluation of an explicatum is thoroughly pragmatic;

if it serves its purpose its adoption is justified, even if this means discarding much of the old, vague "meaning" in the process. Now, if that is the underlying assumption, another question arises: What exactly is the purpose, or what are the purposes, in play here? Neither Hempel nor Carnap are very explicit in that connection (nor are many of their followers). This is partly because a thorough discussion of goals, thus of teleology and normativity, would not fit well into their empiricist framework, partly also, presumably, because an open-ended variety of goals is at issue. Yet specifying the relevant goals is crucial for present purposes.

Let us assume, for example, that the primary goal in employing the CL model is the characterization of scientific practice, after all. In that case we are clearly back to descriptive accuracy as the main yardstick; and all the putative "counterexamples" are directly relevant. In contrast, the force of the usual criticisms appears to be considerably weaker if what we are aiming at is one of the following: (a) to contribute to the advancement of science, e.g., by clarifying its basic concepts or by improving its methodology; (b) to contribute to the advancement of philosophy, by answering some distinctively philosophical questions. Yet even along such lines, one may wonder whether Carnap marginalizes descriptive accuracy, or what he calls "similarity", too much. After all, might the right kind of similarity not play an important role for the effectiveness of the explicatum, as it takes over the role of the explicandum, in science? And might it not be crucial in philosophy too, depending on which particular questions we ask there? In either case, it would seem that at some point in the process there has to be a careful evaluation of whether, and to what degree, the "abstraction and logical schematization" involved in an explication do serve our purposes, whatever those are (cf. Reck 2012).

Let us suppose that, at least for some explications, questions about their descriptive accuracy, about the appropriateness of idealizations, etc. do remain. Arguably it is still the case that a Carnapian explication cannot be refuted by examples in any strict sense, because it is not meant to be right or wrong, only more or less useful, as we saw. This applies to the CL model, at least in contexts where the description of scientific practice is not our main goal. Thinking about it in such terms helps to clarify the model's significance. It also allows us to make sense of what has happened since various alternatives to the CL model took center stage, thereby depriving it of its status as "the received view". Assume here, as is usual nowadays, that one or several of the counter-models are superior, in one way or another. This leaves us with the question: Why are we still talking about the CL model at all, i.e., why hasn't it simply been discarded?

The answer is, as I would suggest, that the CL model has remained useful in various ways even after its "refutation". For one thing, it is still frequently taken to be a suitable starting point for introducing students to the explanation debate (as in Pitt 1988); similarly for giving retrospective accounts of the debate's development (cf. Salmon 1990; Psillos 2007, etc.). Along less historical and more systematic lines, Hempel's model has continued to play the role of a useful object of comparison too. As Philip Kitcher puts it:

> The many-sided character of Hempel's lucid discussions, especially in the title essay of *Aspects of Scientific Explanation*, provides a model for philosophical exploration of an important metascientific concept (Kitcher 2001, 156).

And with the current situation in the explanation debate in mind, he adds:

> If there is a consensus, its central tendency is that, while Hempel's covering-law model is inadequate, it is exemplary in demonstrating the range, rigor, and clarity that any satisfactory theory of explanation should strive for (*ibid.*, 158).

In passages such as these, the CL model is put forward as exemplary for how philosophy of science, or analytic philosophy more generally, is to be done. Likewise, but with an opposite valence, one can use the CL model for illustrating the limitations of analytic philosophy, of formally oriented approaches more generally, or of Carnapian explication in particular, at least if they are understood too narrowly (Reck 2012). Finally, might it even be possible to argue that, by locating generality at the core of explanation, there is something right about the CL model, something to be rescued, even if Hempel articulated it in a misleading way?[13]

***

In the last two sections I considered reasons for interpreting the CL model as an explication in Carnap's sense. This leads to insights concerning the model's significance, as I argued, and contributes to a more adequate evaluation of it. Now I want to turn the tables. That is to say, I want to challenge a Carnapian interpretation of the CL model. I also want to reconsider Hempel's attitude towards Carnap's philosophical methodology more generally. In the end the situation is more complex and more interesting, both with respect to Hempel and the CL model.

Let us start with Hempel. A first observation in that connection is that, while Hempel was indeed close to Carnap at certain points in his career, including in the late 1930s and the 1940s, there were other influences on him too. Hempel met Carnap in 1929, while spending a semester in Vienna as a student. But not only Carnap had an impact on him then, other members of the Vienna Circle did so too, especially Otto Neurath (Friedman 2000; Wolters 2003). And while in Berlin, Reichenbach influenced Hempel's research strongly, as evidenced by his acknowledgment of the role probabilistic laws play in science, a point often highlighted by Reichenbach. Even more importantly for present purposes, there was Hempel's collaboration with Paul Oppenheim. Commentators uniformly mention the latter as a co-author of "Studies in the Logic of Explanation"; but the general tendency is to ascribe most of the ideas in this essay to Hempel. Might there not have been more to Oppenheim's input? Note, for example, that the notion of explanation is much less central in Carnap's, Neurath's, or Reichenbach's writings than in Hempel's. The specific focus on

[13]Note, moreover, that outside of philosophy something close to the CL model is still often taken for granted when people talk about scientific explanation, especially in the natural sciences.

that notion, also the strong emphasis on "covering laws", as well as their application to history and the social science could well be due to Oppenheim, at least in part.[14]

Such additional influences on Hempel await further exploration.[15] But we can already note now that he did not remain a strict Carnapian later on in his career. As a first piece of evidence, consider Hempel's answers to some related questions in an interview from 1982 to 1983. In that context he states the following about the goals and the methodology of the philosophy of science: "[We must] come very close to what we find as a matter of fact in the actual research activities of scientists" (Hempel 2000a). Similarly, in one of his later published articles, entitled "On the Cognitive Status and the Rationale of Scientific Metholodogy" (1988), he declares:

> [An explicatory theory] should not just prescribe norms for rational research procedures but should also have the potential for providing at least an approximate descriptive and explanatory account of some aspects of actual scientific practice (Hempel 1988, 209).

Such declarations are far from Carnap's relaxed attitude towards descriptive accuracy, as part of his more normative and revisionist methodology. It is tempting to read the last quotation even as a direct rebuttal, or disavowal, of Carnap's relatively cavalier stance towards "similarity" in explication. But maybe that is reading too much into the passage.

Beyond such evidence, it is well known that Hempel was influenced by Thomas Kuhn's work in the history and sociology of science later in his career (from the mid-1960s on), partly also by Quine's philosophical naturalism.[16] It may be that encountering their approaches reawakened the influence of Neurath in him, who had emphasized sociological aspects in the study of science and promoted his own form of naturalism earlier.[17] Hempel's parallel interactions with more descriptively oriented philosophers of science, such as Michael Scriven and N. R. Hanson, might also have played a role in his increasing emphasis on staying close to "the actual research activities of scientists". In those respects, the development of Hempel's views illustrates broader trends in the philosophy of science, from the 1960s to the 1980s. But actually, even in his earlier, classic work on explanation Hempel displays a significant amount of attention to examples and to scientific practice already. Insofar as that is the case, seeing Hempel and the CL model purely in the light (or in the shadow?) of Carnap is too quick and somewhat misleading.

---

[14]It is worth adding here that Oppenheim didn't just collaborate with Hempel but with other philosophers as well (including Kurt Grelling, Olaf Helmer, Nicholas Rescher, and Hilary Putnam). And often these collaborations involved working out Oppenheim's ideas (cf. Rescher 2005).

[15]For further forays in that direction, compare the two essays by Nikolay Milkov (Chaps. 1 and 14) in this volume.

[16]Hempel started interacting with Kuhn in 1963–1964, when both spent time at the Center for Advanced Studies in the Behavioral Sciences in Palo Alto. Subsequently, they became colleagues at Princeton. Quine's views were very prominent in the US during the 1960s and later, of course.

[17]As Michael Friedman reports, Hempel himself later talked about "his conversion from the point of view of Carnapian 'explication' or 'rational reconstruction' to the point of view of Kuhnian historical and sociological naturalism as a return to Neurath's original conception" (Friedman 2000, 45).

Finally, an aspect of "Studies in the Logic of Explanation" (1948) that tends to be overlooked might be even more relevent here. In that essay it is the last part, in which Hempel and Oppenheim develop their "more rigorous analysis of the concept of law", that makes it look most Carnapian. Yet what is usually discussed under the label "CL model" in the literature—essentially the Hempel-Oppenheim explanation schema, divorced from their formal account of laws—occurs much earlier, right after the survey of motivating examples. What should one say, then, about that schema, especially from a Carnapian perspective: Is it part of the "clarification of the explicandum", like the initial discussion of examples; or is it part of the formal explicatum instead? The answer is not clear, it seems to me, since the CL schema hovers somewhere in-between these two sides. And insofar as that is the case, it constitutes yet another non-Carnapian side of Hempel and "the CL model".

***

To round off this essay, I want to reconsider the CL model one more time, from a slightly different angle, and so as to give my interpretation of Hempel yet another twist. My cue now is the fact that this account of scientific explanation is almost uniformly called a "model". Usually not much is made of that fact; but might it not deserve separate attention? In my discussion so far, I contrasted two general perspectives on the Hempel-Oppenheim account: seeing it as a reductive analysis, thus as aiming at necessary and sufficient conditions for being an explanation; and seeing it as a Carnapian explication, to be evaluated pragmatically and not, or not primarily, in terms of descriptive adequacy. The CL model is very vulnerable to counterexamples if we adopt the first perspective, while these examples may be discounted to a considerable degree if we take up the second perspective. However, one might respond that neither perspective is entirely adequate, since both lead to significant distortions. Is there no third alternative? I now want to indicate, that there is indeed room for such an alternative, or for an in-between position that might be truer to the CL account and is interesting in its own right.

So far we encountered two reasons for why the Hempel-Oppenheim account should not be seen as straightforwardly descriptive: it involves "abstraction and logical schematization"; it might be seen more as a useful tool, along Carnapian lines, than as a faithful description of scientific practice. Recall also that, even after its demise as the "received view", the CL account has continued to be used fruitfully as an "object of comparison". All three points suggest to me a comparison to the use of models in scientific research. What I have in mind here is not so much "models" in the sense of mathematical logic (set-theoretic structures), but, say, the Bohr model of the atom, Maxwell's vortex model for the electromagnetic field, and similar models in biology and the social sciences. Just like the CL account, models in that sense involve idealization; they too are primarily tools; and here again, an old model may profitably be compared to a newer one even after its demise. In recent philosophy of science there has been a significant amount of discussion concerning scientific models; the corresponding literature can thus be taken as a reference point (see, e.g., Morgan and Morrison 1999; Bailer-Jones 2009).

Within the philosophy of science, serious interest in the topic of models arose as part of the move from a "syntactic" to a "semantic" view of scientific theories. This move was meant to shed light on certain aspects of scientific research, especially current research, which would have remained obscure otherwise. What I am suggesting is a parallel shift with respect to the CL model. And in that case, the shift involves getting clearer about certain aspects of philosophical research. This is not to say that conceiving of the CL model as an explication, rather than as a reductive analysis, is not illuminating at all. Still, bringing in the notion of model can be used to correct distortions introduced by that conception. In particular, it provides a way in which the descriptive dimension of the CL model can be taken more seriously after all. My new suggestion is this, then: Hempel and Oppenheim's model is descriptive of scientific practice in roughly the same (indirect, complex) way in which scientific models are representative of corresponding phenomena; and saying that is compatible with accepting, indeed emphasizing, its role as a tool, also with idealization and abstraction, etc.[18]

If this suggestion is on the right track—if it is appropriate to conceive of the CL account as a model in something like the scientific sense—the insights gained may apply more broadly. I do not mean to suggest that every treatment of a philosophical problem, or every case of philosophical "analysis", can and should be re-described as the use of a model; yet perhaps some can (including appeals to the unification "model", the causal "model", etc.). And if so, the CL model may serve as a paradigmatic example here too, thus adding another dimension to its continuing usefulness. Actually, I suspect that significant differences in the uses of models—between science and philosophy, comparing different cases within each discipline, etc.—will emerge along such lines. For example, the CL model seems to be more a meta-theoretic tool than, say, the Bohr model; it also appears to be normative in a different sense.[19] These are all initial, rough-and-ready suggestions, of course. Much more will have to be done, in terms of thinking through their implications, to make my suggestion really convincing. My hope is that I have said enough to make doing so look like a potentially profitable project.

***

If anything has become evident in this essay, it should be that the Hempel-Oppenheim model for scientific explanation is not as straightforward to categorize or evaluate as one might have thought initially. Often it is taken to constitute a strong kind of analysis and, as such, it is the target of various "counterexamples" intended to refute it. Yet that seems not entirely fair to the approach. A more appropriate way to conceive of it is arguably as an explication in Carnap's sense. Indeed, there

---

[18]For a survey of ways in which scientific models are representative, cf. Bailer-Jones (2009), chapter 8.

[19]For a few more comments on the meta-theoretic and the normative role of the CL model, cf. Reck (2012). For comparisons of different kinds of models within science (physical, mechanical, set-theoretic, etc.), see again Morgan and Morrison (1999), Bailer-Jones (2009), and the literature referred to in them.

are several good reasons for doing so. But in the end that conception seems also distorting in certain ways. In particular, it downplays the model's descriptive side too much. As a third, and in some ways intermediate, alternative I suggested viewing the CL model as functioning like a model in science, similar to the Bohr model of the atom, say. Admittedly, I did not spell out this alternative in any detail here. Nor has anyone else done so in the literature until now, as far as I am aware, in spite of the fact that it is almost universally called the CL "model". My suggestion was that it may be worth doing so, also beyond the case of Hempel.

In conclusion, let me return to Hempel's stature as a philosopher. I started out this essay by noting that Hempel is typically not regarded as a thinker of the same caliber as, say, Reichenbach and Carnap. Nevertheless, he too exerted a strong and lasting influence in philosophy, especially with his work on scientific explanation. As Nicholas Rescher wrote aptly:

> [Hempel & Oppenheim's "Studies in the Logic of Explanation" was] one of those unusual publications that set the agenda for a whole generation of investigators. It set in train an enormous body of discussions and publications which shaped the course of deliberations about scientific explanation over the next decades [. . .] (Rescher 1997, 334)

Similarly, James Fetzer has talked about Hempel's "enormous influence", especially in the English-speaking world; as Fetzer puts it, "during his two decades at Princeton [1955–1975], Hempel's approach dominated the philosophy of science" (Fetzer 2010, 12). It seems to me that such claims about the significance of Hempel's contributions, while somewhat partisan, are basically correct. Indeed, one goal of the present paper was to establish that fact. Then again, it remains true that Hempel was not as original and radical as Reichenbach or Carnap, including methodologically, which justifies granting them an even higher status in the pantheon of twentieth-century "scientific philosophers". Perhaps for that very reason, Hempel's approach was easier to emulate by others, and that may have contributed to his widespread influence.

References

Bailer-Jones, Daniela. 2009. *Scientific models in philosophy of science*. Pittsburgh: Pittsburgh University Press.

Carnap, Rudolf. 1947. *Meaning and necessity*. Chicago: University of Chicago Press.

Carnap, Rudolf. 1950. *Logical foundations of probability*. Chicago: University of Chicago Press.

Curd, Martin. 2012. Carl G. Hempel: Logical empiricist. In *Key thinkers in the philosophy of science*, ed. James R. Brown, 83–111. London: Continuum Press.

Fetzer, James. 2000a. The paradoxes of Hempelian explanation. In Fetzer 2000b, 111–137. Oxford: Oxford UniversityPress.

Fetzer, James H. (ed.). 2000b. *Science, explanation, and rationality: The philosophy of Carl G. Hempel*. Oxford: Oxford University Press.

Fetzer, James H. 2010. Carl Hempel. *The Stanford Encyclopedia of Philosophy (Summer 2012 Edition)*, ed. Edward N. Zalta, http://plato.stanford.edu/archives/sum2012/entries/hempel/.

Friedman, Michael. 2000. Hempel and the Vienna Circle. In Fetzer 2000b, 39–64. Oxford: Oxford University Press.

Hempel, Carl G. 1942. The function of general law in history. *The Journal of Philosophy* 39: 35–48; repr. in Hempel 1965b. 231–243.
Hempel, Carl G. 1965a. Aspects of Scientific Explanation. In Hempel 1965b, 331–496. New York: Free Press.
Hempel, Carl G. 1965b. *Aspects of scientific explanation and other essays in the philosophy of science*. New York: The Free Press.
Hempel, Carl G. 1966. *Philosophy of natural science*. Englewood Cliffs: Prentice-Hall.
Hempel, Carl G. 1973. Rudolf Carnap: Logical empiricist. *Synthese* 25: 256–268; repr. in Hempel 2000b. 253–267.
Hempel, Carl G. 1988. On the cognitive status and the rationale of scientific methodology. *Poetics Today* 9: 5–27; repr. in Hempel 2000b. 199–228.
Hempel, Carl G. 2000a. An intellectual autobiography. In Hempel 2000b, 3–35. Oxford: Oxford University Press.
Hempel, Carl G. 2000b. *Carl G. Hempel: Selected philosophical essays*, ed. Richard Jeffrey. Cambridge: Cambridge University Press.
Hempel, Carl G. 2001. *The philosophy of Carl G. Hempel: Studies in science, explanation, and rationality*, ed. James H. Fetzer. Oxford: Oxford University Press.
Hempel, Carl G., and Paul Oppenheim. 1948. Studies in the logic of explanation. *Philosophy of Science* 15: 135–175; repr. In Hempel 1965b. 245–290; also in Pitt 1988. 9–51.
Kitcher, Philip. 2001. Carl G. Hempel (1905–1997). In *A companion to analytic philosophy*, ed. Aloysius P. Martinich and David Sosa, 148–159. Oxford: Blackwell.
Kitcher, Philip, and Wesley C. Salmon (eds.). 1989. *Scientific explanation. Minnesota studies in the philosophy of science XIII*. Minnesota: University of Minnesota Press.
Morgan, Mary S., and Margaret Morrison (eds.). 1999. *Models as mediators: Perspectives on natural and social science*. Cambridge: Cambridge University Press.
Pitt, Joseph (ed.). 1988. *Theories of explanation*. Oxford: Oxford University Press.
Psillos, Stathis. 2007. *Past and contemporary perspectives on explanation*. In *General philosophy of science: Focal issues*, ed. T. Kuipers. 97–174. Amsterdam: North Holland.
Reck, Erich. 2012. Carnapian explication: A case study and critique. In *Carnap's ideal of explication and naturalism*, ed. P. Wagner, 96–116. London: Palgrave Macmillan.
Rescher, Nicholas. 1997. H2O: Hempel-Helmer-Oppenheim. An episode in the history of scientific philosophy in the 20th century. *Philosophy of Science* 64: 334–360.
Rescher, Nicholas. 2005. The Berlin school of logical empiricism and its legacy. In *Studies in 20thcentury philosophy. Collected papers*, N. Rescher, 119–147. Frankfurt: Ontos Verlag.
Salmon, Wesley. 1990. *Four decades of scientific explanation*. Pittsburgh: University of Pittsburgh Press.
Salmon, Wesley. 2000. The spirit of logical empiricism: Carl G. Hempel's role in twentieth-century philosophy of science. In Fetzer 2000b, 309–324. Oxford: Oxford University Press.
Wolters, Gereon. 2003. Carl Gustav Hempel: Pragmatic empiricist. In *Logical empiricism. Historical and contemporary perspectives*, ed. P. Parrini, W. Salmon, and M. Salmon, 109–122. Pittsburgh: Pittsburgh University Press.
Woodward, James. 2003. *Making things happen: A theory of causal explanation*. Oxford: Oxford University Press.
Wright, Larry. 2011. Explanation, contrast, and the primacy of practice. *European Journal of Philosophy*. doi:10.1111/j.1468-0378.2011.00472.x.

Index

N. Milkov and V. Peckhaus (eds.), *The Berlin Group and the Philosophy of Logical Empiricism*, Boston Studies in the Philosophy and History of Science 273, DOI 10.1007/978-94-007-5485-0,

W

Printed by Printforce, the Netherlands